Vol. 32. **Determination of Organic Compounds: Methods and Procedures.** By Frederick T. Weiss

Vol. 33. **Masking and Demasking of Chemical Reactions.** By D. D. Perrin

Vol. 34. **Neutron Activation Analysis.** By D. De Soete, R. Gijbels, and J. Hoste

Vol. 35. **Laser Raman Spectroscopy.** By Marvin C. Tobin

Vol. 36. **Emission Spectrochemical Analysis.** By Morris Slavin

Vol. 37. **Analytical Chemistry of Phosphorus Compounds.** Edited by M. Halmann

Vol. 38. **Luminescence Spectrometry in Analytical Chemistry.** By J. D. Winefordner, S. G. Schulman, and T. C. O'Haver

Vol. 39. **Activation Analysis with Neutron Generators.** By Sam S. Nargolwalla and Edwin P. Przybylowicz

Vol. 40. **Determination of Gaseous Elements in Metals.** Edited by Lynn L. Lewis, Laben M. Melnick, and Ben D. Holt

Vol. 41. **Analysis of Silicones.** Edited by A. Lee Smith

Vol. 42. **Foundations of Ultracentrifugal Analysis.** By H. Fujita

Vol. 43. **Chemical Infrared Fourier Transform Spectroscopy.** By Peter R. Griffiths

Vol. 44. **Microscale Manipulations in Chemistry.** By T. S. Ma and V. Horak

Vol. 45. **Thermometric Titrations.** By J. Barthel

Vol. 46. **Trace Analysis: Spectroscopic Methods for Elements.** Edited by J. D. Winefordner

Vol. 47. **Contamination Control in Trace Element Analysis.** By Morris Zief and James W. Mitchell

Vol. 48. **Analytical Applications of NMR.** By D. E. Leyden and R. H. Cox

Vol. 49. **Measurement of Dissolved Oxygen.** By Michael L. Hitchman

Vol. 50. **Analytical Laser Spectroscopy.** Edited by Nicolo Omenetto

Vol. 51. **Trace Element Analysis of Geological Materials.** By Roger D. Reeves and Robert R. Brooks

Vol. 52. **Chemical Analysis by Microwave Rotational Spectroscopy.** By Ravi Varma and Lawrence W. Hrubesh

Vol. 53. **Information Theory As Applied to Chemical Analysis.** By Karel Eckschlager and Vladimir Stepánek

Vol. 54. **Applied Infrared Spectroscopy: Fundamentals, Techniques, and Analytical Problem-solving.** By A. Lee Smith

Vol. 55. **Archaeological Chemistry.** By Zvi Goffer

Vol. 56. **Immobilized Enzymes in Analytical and Clinical Chemistry.** By P. W. Carr and L. D. Bowers

Vol. 57. **Photoacoustics and Photoacoustic Spectroscopy.** By Allan Rosencwaig

Vol. 58. **Analysis of Pesticide Residues.** Edited by H. Anson Moye

Vol. 59. **Affinity Chromatography.** By William H. Scouten

Vol. 60. **Quality Control in Analytical Chemistry.** By G. Kateman and F. W. Pijpers

Vol. 61. **Direct Characterization of Fineparticles.** By Brian H. Kaye

Vol. 62. **Flow Injection Analysis.** By J. Ruzicka and E. H. Hansen

Vol. 63. **Applied Electron Spectroscopy for Chemical Analysis.** Edited by Hassan Windawi and Floyd Ho

Vol. 64. **Analytical Aspects of Environmental Chemistry.** Edited by David F. S. Natusch and Philip K. Hopke

Vol. 65. **The Interpretation of Analytical Chemical Data by the Use of Cluster Analysis.** By D. Luc Massart and Leonard Kaufman

Vol. 66. **Solid Phase Biochemistry: Analytical and Synthetic Aspects.** Edited by William H. Scouten

Inorganic Chromatographic Analysis

CHEMICAL ANALYSIS

A SERIES OF MONOGRAPHS ON ANALYTICAL CHEMISTRY AND ITS APPLICATIONS

VOLUME 78

A WILEY-INTERSCIENCE PUBLICATION

JOHN WILEY & SONS

New York / Chichester / Brisbane / Toronto / Singapore

Inorganic Chromatographic Analysis

Edited by

JOHN C. MACDONALD

Department of Chemistry
Fairfield University
Fairfield, Connecticut

A WILEY-INTERSCIENCE PUBLICATION

JOHN WILEY & SONS

New York / Chichester / Brisbane / Toronto / Singapore

Library of Congress Cataloging in Publication Data:

Main entry under title:

Inorganic chromatographic analysis.
(Chemical analysis ; v. 78)
"A Wiley-Interscience publication."
Includes index.
1. Chromatographic analysis. I. MacDonald, John C.
II. Series.

QD79.C4156 1985 543'.089 84-22911
ISBN 0-471-86263-0

Printed in the United States of America

10 9 8 7 6 5 4 3 2 1

CONTRIBUTORS

Richard J. Laub, San Diego State University, San Diego, California

John C. MacDonald, Fairfield University, Fairfield, Connecticut

Harold M. McNair, Virginia Polytechnic Institute and State University, Blacksburg, Virginia

James W. Mitchell, AT&T Bell Laboratories, Murray Hill, New Jersey

Joan T. Overfield, Fairfield University, Fairfield, Connecticut

Christopher A. Pohl, Dionex Corporation, Sunneyvale, California

John M. Riviello, Dionex Corporation, Sunneyvale, California

Joseph Sherma, Lafayette College, Easton, Pennsylvania

Evelyn R. Savitzky, Perkin-Elmer Corporation, Norwalk, Connecticut

Peter C. Uden, University of Massachusetts, Amherst, Massachusetts

Roy A. Wetzel, Dionex Corporation, Sunneyvale, California

PREFACE

The use of chromatography for organic analyses is routine and most chromatographic applications are for organic substances. Although inorganic applications of chromatography date back to the work of Schwab in 1937, and later to the use of ion exchange during the Manhattan Project of World War II, inorganic chromatographic applications were not as routine as organic applications until recently.

The change results from developments in high-performance liquid chromatography (HPLC), along with the invention of ion chromatography and the subsequent publication of the first ion chromatographic paper in 1975 (H. Small, T. S. Stevens, and W. C. Bauman, *Anal. Chem.*, **47,** 1801 (1975)). Prior to that time, most chemists did not associate continuous monitoring of the eluting mobile liquid phase with practical inorganic chromatographic analyses. Fraction collection of a liquid effluent was common, but not as efficient as continuous monitoring of the effluent. Continuous detection was difficult because of the lack of a convenient probe to directly monitor the eluting liquid.

As a result of that first ion chromatographic paper, which demonstrated the practical use of conductivity for continuous and universal detection, chemists became more aware of the possibilities of applying continuous monitoring of column effluents in inorganic chromatography. Applications of continuous monitoring of column effluents of inorganic chromatography did exist prior to these innovations. However, these procedures, such as the gas chromatography of metal chelates, were not routine ordinarily. Hundreds of research papers applying gas chromatography and thin-layer chromatography to inorganics also existed prior to 1975. These fields of inorganic chromatography continue to be important and are presented in Chapters 5 and 7. Liquid column inorganic chromatographies are presented in Chapters 6, 8 and 9.

Following the above innovations, chemists began to adapt a wide variety of standard analytical techniques to continuous monitoring of inorganic effluents. Many of these detectors, adapted for chromatography, are now available from instrument companies. These detectors utilize a variety of analytical principles: visible and ultraviolet light absorption, amperometry, fluorescence, flame emission, atomic absorption, and inductively coupled plasma (ICP) spectroscopy. Indeed, ICP detection revitalized inorganic gas chromatography. Further, chemists now realize more fully the possibilities of chemical reactions to convert inorganic effluents to detectable species. The availability (also from instrument companies) of chemical reactors that couple to chromatographs aided this growth of inorganic chromatographic analysis. Inorganic analyses that were previously difficult if not impossible are now routine. Speciation—for example, the individual determination of Cr(III) and Cr(VI) in the same sample—is an area where modern inorganic chromatographic analysis offers great value.

A need existed for an introductory text summarizing inorganic chromatographic analysis. Most of the development and subsequent applications of chromatography had been to solve problems of organic chemistry. As detailed in Chapter 1, the possibilities of inorganic chromatography had been largely overlooked. Many protocols for inorganic chromatographic analyses now exist and should be used more frequently. This volume in this series is intended to yield such an increase.

The subject matter of inorganic chromatography is wide in scope and an edited multiauthor volume was most feasible. As editor, I deeply appreciate the kindness, patience, and universal cooperation of the authors. A brief statement as to each chapter following my introductory chapter is appropriate.

Professor R. J. Laub (Chapter 2, "Theory of Chromatography") teaches this course at San Diego State University from fundamental physiocochemical principles. This complete approach to chromatographic theory is not available anywhere else in the literature, and will likely be incorporated by many teachers into their chromatography courses. Furthermore, that author has pioneered in the use of window diagrams for chromatographic optimization, and includes his computer programs for use by others. Professor H. M. McNair (Chapters 3 and 4, "Instrumentation for Gas Chromatography" and "Instrumentation for High-Performance Liquid Chromatography") teaches these subjects not only at Virginia Polytechnic Institute and State University but also in the courses of the American Chemical Society and overseas under international auspices. His audiovisual materials on these subjects are in common use worldwide. Professor P. C. Uden (Chapter 5, "Gas Chromatography of Inorganic Compounds, Organometallics, and Metal Complexes") of the University of Massachusetts is the acknowledged expert in this area. The chapter updates his previous review with D. E. Henderson (*Analyst,* **102,** 889 (1977)). In addition to ICP detection, the author lists as factors contributing to increased interest in gas chromatography of inorganics: inert fused silica capillary columns, and now supercritical fluid chromatography. Chapter 6, "High-Performance Liquid Chromatography of Inorganics and Organometallics," is by the editor and demonstrates that the high resolution of modern HPLC, although developed for organic analysis, is now equally applicable to inorganic species. HPLC is now the method of choice for the determination of uncharged inorganics by liquid chromatography. I especially appreciate the service to chromatography provided by Professor J. Sherma (Chapter 7, "Thin Layer Chromatography of Inorganic Ions and Compounds") of Lafayette College, a liberal arts institution, in preparing his chapter. His accomplishments in this area are world renowned and bring much satisfaction to believers (like myself) in a more prominent role for science in undergraduate liberal arts education. The chapter is a definitive summary of inorganic application of thin-layer chromatography.

Although classical ion exchange is fully developed, there are areas of application in inorganic chromatographic analysis where this technique remains irreplaceable. One such area is radiochemistry, and Dr. J. W. Mitchell of AT&T Bell Laboratories demonstrates this from personal experience (Chapter 8, "Ion Exchange in Radiochemistry"). An important extension of classical ion-exchange chromatography is ion chromatography as pioneered by Dow Chemical and licensed exclusively to the Dionex Corporation. This technique is presented mainly by scientists C. A. Pohl, J. M. Riviello, and R. A. Wetzel from Dionex Corporation (Chapter 9, "Ion Chromatography"). The chapter presents not only the original ion chromatography but also the now preferred use of the new fiber suppressors, as well as the newer techniques of ion chromatography: ion chromatography exclusion (ICE) and mobile phase ion chromatography (MPIC).

The literature of inorganic chromatographic analysis covers many areas, and the typical user is not likely to be completely familiar with all of this literature. This volume is intended as an introduction to modern inorganic chromatographic analyses and to direct the prospective user to the most useful procedures. Modern literature surveys will likely be necessary and the computer should be used. Given such computer access and the help of an information specialist, huge data banks can be rapidly surveyed. These procedures are introduced here (Chapter 10, "Computer Online Database Literature Searching and Examples") by Research Librarians J. T. Overfield of Fairfield University and E. R. Savitzky of Perkin-Elmer Corporation, who include computer printouts of literature searches in inorganic chromatography.

Since 1975, inorganic chromatographic analysis using both gas and liquid mobile phases has expanded dramatically. New instruments, detectors, stationary phases, and reactors now exist, and protocols using these devices are newly available for hundreds of inorganic species. This volume documents these advances in a single introductory source and will result in a more extensive use of inorganic chromatographic analyses.

John C. MacDonald

Fairfield, Connecticut
April, 1984

CONTENTS

CHAPTER

1

INTRODUCTION AND BRIEF HISTORY

JOHN C. MacDONALD

Department of Chemistry
Fairfield University
Fairfield, Connecticut

1.1 CHEMICAL SEPARATIONS

Chemical separations commonly occur as a result of the distribution of chemical components between two phases of matter. In distillation, these phases are liquid and gas. In precipitation, these phases are liquid and solid. In solvent extraction, these phases are liquid and liquid. In ordinary solvent extraction and precipitation the separations are discontinuous; the components distribute between the two phases and when chemical equilibrium exists, the separation procedure ends. In distillation, the separation is continuous; the distillate becomes more concentrated in the lower boiling component as this distillate rises in the distillation column. This concentration occurs as a result of the continuous vaporization and condensation of the components in the distillation column. In distillation, the success of the chemical separation is dependent on the difference in the boiling points of the separating components. With each vaporization and condensation, the lower boiling component becomes more concentrated in the vapor phase and less concentrated in the liquid phase. The greater the difference in boiling points and the greater the number of times the components move between the two phases, the more successful will be the separation. The fundamentals of chemical separations are presented in several recommended texts (1–3).

1.2 CHROMATOGRAPHY

Chromatography is also a method of chemical separation that is continuous. The separating components distribute between two phases. One phase is fixed in place and is called the stationary phase. The other flows over this stationary phase and is called the mobile phase. The separating components continuously distribute between these two phases. The separation occurs because each component distributes differently between the two phases. The components enter the stationary phase at the same place and at the same time but are carried along in the mobile phase at different rates. Separation by distillation occurs because of the continuous evaporation and condensation in the distillation column. Separation by chromatography also occurs because of the continuous movement of the separating components between two phases of matter, the stationary and mobile phases. In column chromatography, for example, the component that has the property of spending more time in the mobile phase will exit the column first. The longer the time spent in the stationary phase by a component, the later will that component exit the column. The factors influencing this distribution between the mobile and stationary phases are presented in Chapter 2, "The Theory of Chromatography."

1.3 CLASSIFICATION OF CHROMATOGRAPHIES

The different chromatographies are characterized by the nature of the stationary and mobile phases. Figure 1.1 shows these chromatographies. The phenomena upon which these chromatographic separation techniques are based are adsorption, partition, size exclusion, and ion exchange. The nature of the phenomenon in a particular separation is dependent upon the nature of that stationary phase. The theory for understanding these phasic differences is presented in Chapter 2, and some of the required instrumentations for applying these principles are presented in Chapters 3 and 4.

1.3.1 Adsorption Chromatography

In adsorption chromatography, the stationary phase is a surface active granular solid. Silica is now the most common stationary phase in adsorption chromatography and a variety of silicas and other products, porous and pellicular, are available from manufacturers. Some of these adsorbents are listed in the applications in Table 5.1. Choice of silica and other adsorbents is based upon particle size, surface area, and surface activity. Other commonly used adsorbents are alumina, charcoal, magnesia, and organic polymers such as styrene/divinylbenzene and derivatives. Adsorption chromatography is the oldest of chromatographies and excellent references are available (4–6).

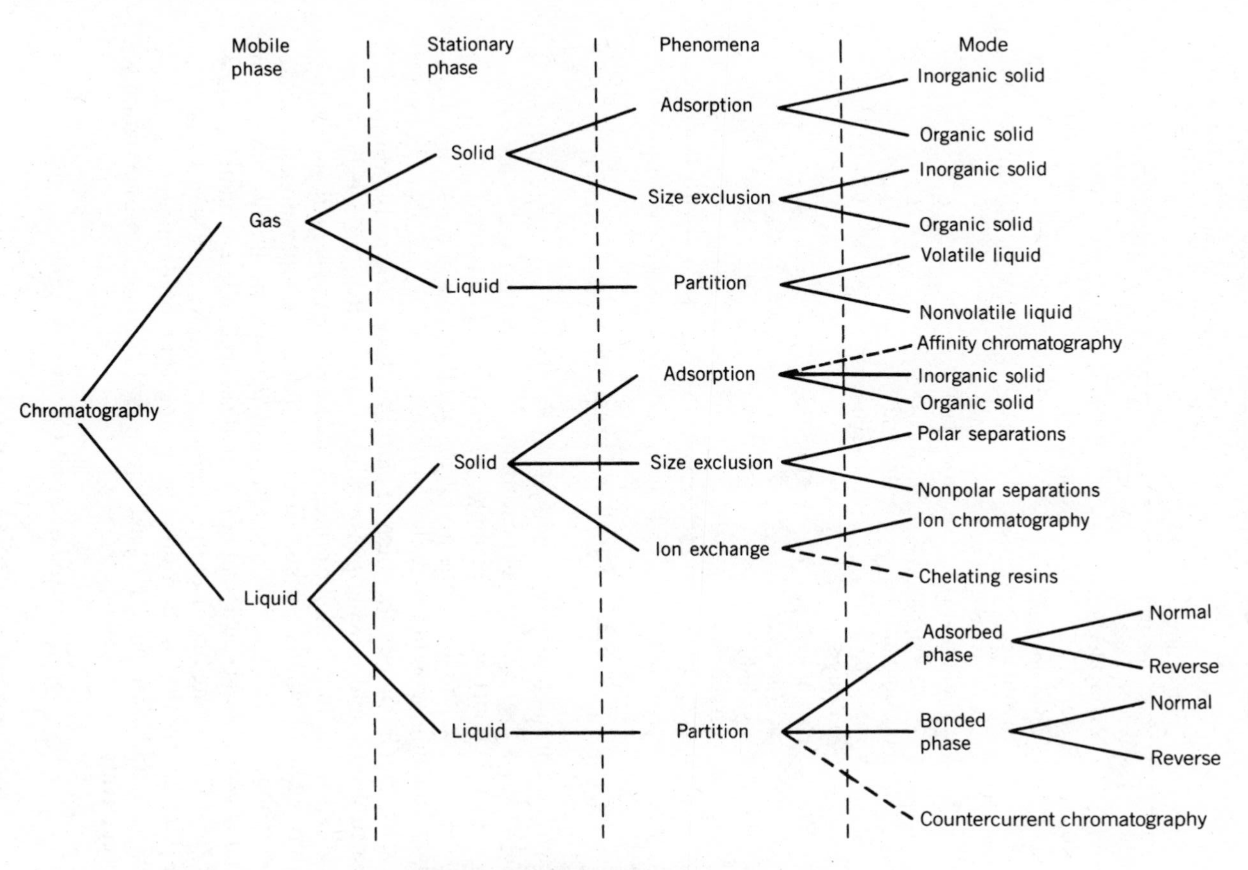

Figure 1.1. Outline of chromatographies.

Table 1.1 Johns-Manville Chromosorb Porous Polymers

Series	Structure	Surface Area (m^2/g)	Average Pore Diameter (Å)
101	Styrene-divinylbenzene	<50	30–40
102	Styrene-divinylbenzene	300–400	0.85
103	Cross-linked polystyrene	15–25	30–40
104	Acrylonitrile-divinylbenzene	100–200	6–8
105	Polyaromatic	600–700	4–6
106	Cross-linked polystyrene	700–800	50
107	Cross-linked acrylic ester	400–500	90
108	Cross-linked acrylic ester	100–200	235

1.3.2 Partition Chromatography

Partition chromatography was invented by Martin and Synge in 1941 (7) and extends the capability of adsorption. A solvent is fixed to the surface of a solid support as, for example, when silica is converted by water to silica gel. The equilibrium is no longer a surface phenomenon but a solubility phenomenon in which the solute partitions between the stationary and the mobile phases. In going to partition chromatography, Martin and Synge used liquid moving phases and improved their efficiency of separations many fold. Similar principles are used in gas–liquid chromatography wherein liquids are coated on a support for separations in which the mobile phase is a gas. Hundreds of partitioning liquids have been characterized and typical examples are discussed in later chapters. Some of the reference texts for partition chromatography are listed at the end of this chapter (8–15). Partition chromatography predominates in Chapters 5, 6, and 7 on inorganic gas, liquid, and thin-layer chromatography, respectively.

1.3.3 Size Exclusion Chromatography

In exclusion chromatography, separation is by molecular size. The stationary phase consists of small particles having pores. If certain molecules

Table 1.2 Gel Filtration Media

	Fractionation Range (daltons)	Bed Volume (ml/g dry gel)
Bio-Rad Polyacrylamide Gels		
Bio-Gel P-2	100–1,800	3.5
Bio-Gel P-6	1,000–6,000	7
Bio-Gel P-10	1,500–20,000	9
Pharmacia Dextran Gels		
Sephadex G-10	–700	2–3
Sephadex G-15	–1,500	2.5–3.5
Sephadex G-25	100–5,000	4–6
Sephadex G-50	500–10,000	9–11
Sephadex G-75	1,000–50,000	12–15

are sufficiently small to move into the pores, these molecules will exit the stationary phase later than those molecules that are too large to enter the pores readily. The larger molecules are carried more rapidly through the stationary phase by the mobile phase. The term "molecular sieve" is applied to zeolites, which are commonly used for the separation of the permanent gases in gas–solid chromatography. Figure 5.1 shows separations of permanent gases by molecular sieves. Polar gases, such as CO_2, are adsorbed and do not elute under these conditions. Temperature elevation desorbs these adsorbable gases. The use of temperature gradients in gas chromatography is akin to the use of liquid gradients in liquid chromatography.

In gas–solid chromatography, larger volatile molecules are separated by the use of porous polymers such as those summarized in Table 1.1. Size exclusion chromatography is of greater used in liquid chromatography for the separation of molecules of vapor pressure too small for separation by gas chromatography. As mentioned later in Chapter 6, size-exclusion liquid chromatography is viable also for molecules of very low molecular weight; for example, 100 daltons. Properties of typical exclusion materials in liquid chromatography are summarized in Table 1.2. The recent book by Yau, Kirkland, and Bly (16) is now the standard reference in this field.

1.3.4 Ion-Exchange Chromatography

In ion-exchange chromatography, the stationary phase is most usually styrene/divinylbenzene carrying ionic groupings and the mobile phase is

Table 1.3 Ion-Exchange Resins

Trade Name	Exchange Groups	Classification
Dowex 1	$\phi-CH_2N^+(CH_3)_3$ $\boxed{Cl^-}$	Strongly basic
Dowex 2	$\phi-CH_2N^+(CH_3)_2(C_2H_4OH)$ $\boxed{Cl^-}$	Strongly basic
Dowex WGR	$\phi-CH_2N^+(R)_2$ $\boxed{Cl^-}$	Weakly basic
Dowex 50W	$\phi-SO_3^+$ $\boxed{H^+}$	Strongly acidic
Dowex CCR-2	$\phi-R-COO^-$ $\boxed{Na^+}$	Weakly acidic
Dowex A-1, Chelex 100	$\phi-CH_2(CH_2COO^-\ \boxed{H^+})(CH_2COO^-\ \boxed{H^+})$	Chelating resin

an ionic solution. Competition between the ions in solution and the ionic sites on the resin governs the chromatographic separation.

Some of the more common ionic sites on ion-exchange resins are summarized in Table 1.3. In addition, there are specialized phases that are occasionally used; chelating resins are water insoluble compounds in which the attached group is a chelate. Typical examples, also in Table 1.3, are Dow Chemical Dowex A-1 and Bio-Rad Chelex 100, in which the iminoacetic acid chelating group is attached to the base polymer. These chelating resins are useful for inorganic separations (17–20).

The invention of ion chromatography has revitalized ion-exchange separations. Earlier work in ion exchange is in standard references (21–23). Recent utilization of classical ion exchange in radiochemistry is presented in Chapter 8 and the new field of ion chromatography is discussed in Chapter 9.

1.4 INORGANIC CHROMATOGRAPHY

Although applications of chromatography preceded Tswett, his paper given in 1903 at the meeting of the Biological Section of the Warsaw Society of Natural Sciences is generally accepted as the beginning of chromatography. Apart from Palmer's use of chromatography (24), little of chromatographic significance occurred after the work of Tswett until Kuhn, Winterstein, and E. Lederer used the technique in 1931 (25). Since then the growth of chromatography has been exponential. The adsorption chromatography of Tswett as revived by Kuhn et al. was soon followed by the development of the many chromatographies of Figure 1.1. Histories of this growth exist elsewhere (26–29) and need not be reproduced here. Most of this development in chromatography was for the separation of organic compounds.

The first application of column chromatography in inorganic chemistry is attributed to Schwab in Munich; resulting papers appeared in 1937 (30,31). Although Runge used paper chromatography in 1850 and Beyerinck used thin layers of gelatin in 1889 for inorganic separations, Schwab's use of alumina columns to separate inorganic ions is rightfully accepted as the beginning of modern inorganic chromatography. That technique with alumina was soon replaced by the use of organic ion exchangers (32), which were used extensively in the Manhattan Project during World War II. The publication of that research (33) demonstrated to a wide audience the practical value of inorganic chromatography.

By then the value of partition paper chromatography for the separation of amino acids was readily available (34) and M. Lederer used this chromatography to separate the inorganic ions of Sb (35) and Au, Pt, Pd, Ag and Cu (36). Shortly thereafter, paper electrophoresis was available (37) and Lederer also applied this form of chromatography to inorganic ions (38). Lederer's many contributions are seen in his books (39,40). The early contributions of Pollard to popularizing inorganic chromatography are summarized in his book (41).

By 1953, gas chromatography was available and volatile inorganic compounds could be determined. Although the first gas chromatogram (42) was the inorganic separation of air and CO_2 on charcoal, most of the applications of gas chromatography are to organic compounds. The limitation of gas chromatography is the need for volatility and stability of the inorganic compound at the temperature of the separation. Early work in this area of gas chromatography is summarized in the monograph of Guiochon and Pommier (43) and in the review paper of Uden and Henderson (44). Chapter 5 brings the review of inorganic gas chromatography up to date.

Inorganic liquid chromatography is best presented here in terms of the stationary phase. Chapter 6 emphasizes the increasing use of reversed-phase HPLC columns for the separation of inorganics and organometallics. Chapter 7 emphasizes the use of thin layers of stationary phases for inorganic separations. Of particular note is the presentation of the use of high-performance thin-layer chromatography (HPTLC) for inorganic separations wherein the small particle size yields higher separating efficiency. In both organic and inorganic separations the rapid development of HPLC column protocols for separation may possibly be done on comparable HPTLC layers. Note also that thin-layer chromatography has the unique advantage of allowing analyses in parallel. All the other chromatographies offer only serial analyses. A thin-layer chromatographic plate may contain strips of stationary phase along the length of the plate. In screening large numbers of organic samples routinely, thin-layer chromatography is frequently the method of choice in clinical laboratories. This use for screening purposes is less frequent in inorganic chromatography.

A variety of detectors are available for continuous monitoring of inorganics in the eluting mobile phase in gas and liquid chromatography. Some of these detectors are presented in Chapters 3 and 4. In the separation of ions no such continual monitoring was generally available until 1975. As a result, the use of ion exchange for direct inorganic analysis was not usually a routine procedure. There are special applications in which ion exchange is of great value and Chapter 8 presents its use in radiochemistry. The application of ion exchange to stable isotopes requires another means of detection.

Prior to 1975 there was no convenient general procedure for anion analysis and particularly so for analysis at low concentrations. Cations could be determined by use of emission spectral lines, but most anions could not be so conveniently determined. Researchers at Dow Chemical Company realized that if the background electrolytes eluting from an ion-exchange column could be removed, the remaining conductivity was due to the ions separated by the use of ion exchange. This removal was done (45) and conductivity-based detection was then used. This new field of chromatography, in which the effluent is continuously monitored, is called ion chromatography and is the topic of Chapter 9.

A few decades ago, a scientist could easily remain current in an area of expertise such as inorganic chromatographic analysis. This is no longer possible because of the exponential growth in research and publications. A scientist must now use data banks and a computer to be maximally effective. Fortunately, this can be done at minimal cost over telephone lines. Chapter 10 summarizes this aspect of using the literature to search a database and demonstrates this searching applied to inorganic chromatography with specific examples.

The successful literature search requires the use of appropriate language. The terminology of chromatography has been standardized through ASTM, the American Society for Testing Materials. These Standard Terms and Relationships are published in the *Annual Book of ASTM Standards* as E355 for gas chromatography and E682 for liquid chromatography, and are equally applicable to organic and inorganic chromatography.

1.5 CLOSING REMARKS

The tendency to associate chromotography with organics and not with inorganics is lessening. The environmental movement has contributed to this. Because of the increasing need for analyses of inorganics, particularly at low concentrations, the literature of inorganic chromatographic analysis has been reviewed and condensed.

Fishbein in Volume 2 (46) of his four-volume series, *Chromatography of Environmental Hazards*, reviews the inorganic chromatography of Be, Cr, Mn, Co, Ni, Cd, Sn, Pb, As, P, Se, Te, Hg, Tl, V, Sr, Y, Cs, actinides, nitrogen-containing gases, sulfur-containing gases, halogen-containing gases, CO, CO_2, O_3, peroxyacyl nitrates, phosgene, cyanogen, and hydrogen cyanide. That presentation covers over 350 pages and presents 1,393 references. The value and scope of the volume appears to be overlooked, perhaps because the word "inorganic" does not appear in the title. A more recent book on environmental analysis also contains inorganic chromatography, but again "inorganic" is not in the title (47) and this valuable source may also be overlooked.

In *75 Years of Chromatography—A Historical Dialogue* (29), C. S. G. Phillips, the well-known inorganic chemist (48) and author of the first book on gas chromatography (49), states ". . . I have always felt that the potential of gas chromatography and now HPLC has not been fully realized by those who work on the preparation of new inorganic compounds, particularly in the organometallic field" (50). In editing this volume I have come to agree with Professor Philips, and to believe this lack of utilization reflects the unavailability heretofore of a practical introductory summary of modern inorganic chromatographic analysis. The availability of these plenary chapters in this Series on Chemical Analysis should aid in the realization of this potential of inorganic chromatography. Inorganic and organometallic chemists will find the literature survey of Schwedt in 1981 (51) to be a very useful adjunct to this volume and the environmental books (46,47).

REFERENCES

1. J.A. Dean, *Chemical Separation Methods*, Van Nostrand Reinhold, New York, 1969.
2. B.L. Karger, L.R. Snyder, and C. Horvath, *An Introduction to Separation Science*, Wiley, New York, 1973.
3. J.M. Miller, *Separation Methods in Chemical Analysis*, Wiley, New York, 1975.

4. H.H. Strain, *Chromatographic Adsorption Analysis*, Wiley, New York, 1942.
5. L.R. Snyder, *Principles of Adsorption Chromatography*, Dekker, New York, 1968.
6. L.R. Snyder and J.J. Kirkland, *Introduction to Modern Liquid Chromatography*, 2nd Ed. Wiley-Interscience, New York, 1979.
7. A.J.P. Martin and R.L.M. Synge, *Biochem. J.*, **35,** 1358 (1941).
8. A.J.M. Keulmanns, *Gas Chromatography*, Reinhold, New York, 1957.
9. S. Dal Nogre and R.S. Juvets, *Gas-Liquid Chromatography—Theory and Practice*, Interscience, New York, 1962.
10. J.H. Purnell, *Gas Chromatography*, Wiley, London, 1962.
11. J.C. Giddings, *Dynamics of Chromatography*, Dekker, New York, 1965.
12. A.B. Littlewood, *Gas Chromatography: Principles, Techniques and Applications*, Academic Press, New York, 1970.
13. E. Heftmann, *Chromatography*, 3rd Ed., Van Nostrand Reinhold New York, 1974.
14. C.F. Simpson, ed., *Practical High-Performance Liquid Chromatography*, Heyden, London, 1976.
15. R.L. Grob, ed., *Modern Practice of Gas Chromatography*, Wiley-Interscience, New York, 1977.
16. W.W. Yau, J.J. Kirkland and D.D. Bly, *Modern Size Exclusion Liquid Chromatography*, Wiley-Interscience, New York, 1979.
17. Dow Chemical Company, *Dowex A-1 Chelating Resin*, 1964.
18. Bio-Rad Laboratories, *Separating Metals Using Chelex 100 Chelating Resin*, March, 1981.
19. H.S. Mahanti and R.M. Barnes, *Anal. Chem.*, **55,** 403 (1983).
20. R.R. Greenberg and H.M., Kingston, *Anal. Chem.*, **55,** 1160 (1983).
21. W.H. Reiman, III and H.F. Walton, *Ion Exchange in Analytical Chemistry*, Pergamon, New York, 1970.
22. F. Helffereich, *Ion Exchange*, McGraw-Hill, New York, 1972.
23. H.F. Walton (Ed.), *Ion Exchange Chromatography*, Academic Press, New York, 1976.
24. L.S. Palmer, *Carotenoids and Related Pigments: The Chromolipids*, American Chemical Society Monogram Series, Chemical Catalog Co., New York, 1922.
25. R. Kuhn, A. Winterstein and E. Lederer, *Hoppe-Seyler's Z. Physiol. Chem.*, **197,** 141 (1931).
26. V. Heines, *Chem. Technol.*, **1,** 280 (1971).
27. L.S. Ettre, *Anal. Chem.*, **43** (14), 20A (1971).
28. G. Zweig and J. Sherma, *J. Chromatog.*, **11,** 279 (1973).
29. L.S. Ettre and A. Zlatkis, *75 Years of Chromatography—A Historical Dialog*, Elsevier, New York, 1979.

30. G.-M. Schwab and K. Jockers, *Angew. Chem.*, **50,** 546 (1937).
31. G.-M. Schwab and G. Dattler, *Angew. Chem.*, **50,** 691 (1937).
32. B.A. Adams and E.L. Holmes, *J. Soc. Chem. Ind.*, **54,** 1–6T (1935).
33. W.C. Johnston, L.L. Quill and F. Daniels, *Chem. Eng. News*, **25,** 2494 (1947).
34. A.J.P. Martin, *Endeavour*, **6,** 21 (1947).
35. M. Lederer, *Anal. Chim. Acta*, **2,** 261 (1948).
36. M. Lederer, *Nature*, **162,** 776 (1948).
37. E.L. Durrum, *J. Amer. Chem. Soc.*, **72,** 2943 (1950).
38. M. Lederer and F.L. Ward, *Aust. J. Science*, **13,** 114 (1951).
39. E. Lederer and M. Lederer, *Chromatography—A Review of Principles and Application*, Elsevier, Amsterdam, 1955.
40. M. Lederer, *Introduction to Paper Electrophoresis and Related Methods*, Elsevier, Amsterdam, 1955.
41. F.H. Pollard and J.F.W. McOmie, *Chromatographic Methods of Inorganic Analysis*, Academic Press, New York, 1953.
42. F. Prior, Ph.D. Thesis, Innsbruck, 1947.
43. G. Guiochon and G. Pommier, *Gas Chromatography in Inorganics and Organometallics*, Ann Arbor Scientific Publishers, Ann Arbor, 1973.
44. P.C. Uden and D.E. Henderson, *Analyst*, **102,** 889 (1977).
45. H. Small, T.S. Stevens and W.C. Bauman, *Anal. Chem.* **47,** 1801 (1975).
46. L. Fishbein, *Chromatography of Environmental Hazards, Volume 2: Metals, Gaseous and Industrial Pollutants*, Elsevier, New York, 1973.
47. R.L. Grob (Ed.), *Chromatographic Analysis of the Environment*, 2nd Ed. Dekker, New York, 1983.
48. C.S.G. Phillips and R.J.P. Williams, *Inorganic Chemistry*, Volumes 1 and 2, Oxford, London, 1966.
49. C.S.G. Phillips, *Gas Chromatography*, Butterworths, London, 1956.
50. Reference 29, p. 319.
51. G. Schwedt, *Chromatographic Methods in Inorganic Analysis*, Huthig, New York, 1981.

CHAPTER

2

THEORY OF CHROMATOGRAPHY

R. J. LAUB

Department of Chemistry
San Diego State University
San Diego, California

2.1 INTRODUCTION

There are today a wide variety of static as well as dynamic (flow) techniques utilized for separations. In many of these, the desired analysis is effected on the basis of physical and/or chemical interactions of the analytes (solutes) with the system during the course of their passage through it. If during the elution process the rate of migration differs from one solute to the next, each can be distinguished by the time of its emergence from the system.

Chromatography encompasses a broad class of methodologies based upon this simple principle, as indicated in the definition provided by Laub and Pecsok (1):

> All chromatographic methods have the following features in common: two mutually immiscible phases are brought into contact (possess a common interface) wherein one (the mobile phase) is made to flow over the other (the stationary or static phase). The surface area of the latter that is exposed to the former generally is large. When a third component (the solute) is introduced (injected) into the system, it partitions (equilibrates) between the two phases. Because the partitioning process recurs many times during the course of elution of solutes through the system, the chromatographic technique can be made to be considerably more efficient than tandem arrangement of separatory funnels or gas bottles.
>
> The mobile and stationary phases can in principle be gases, liquids, or solids; in practice, the mobile phase either is a gas or a liquid while the stationary phase either is a liquid or a solid. Several variants thereby arise: gas/liquid (GLC), gas/solid (GSC), liquid/liquid (LLC), and liquid/solid (LSC) chromatography.

Thus, inherent in chromatography are equilibrium thermodynamic factors that govern solute/system interactions and also, kinetic factors that enter into the transport of solutes between the mobile and stationary phases as well as contribute to the overall rate of solute longitudinal migration. Moreover, although solutes are made to flow through the system, their rate of movement is taken to be slow compared to the time required to reach equilibrium with each phase at any point within the apparatus.

Consideration of each of these fundamental aspects of the chromatographic technique provides considerable insight into means whereby separations can be achieved and where, in situations where resolutions are already partially satisfactory, they can be substantially improved.

2.2 PHYSICOCHEMICAL PROPERTIES OF PURE AND BLENDED SUBSTANCES

Discussion of the chromatographic process is most conveniently begun with a brief overview of the various properties of pure substances that come into play, these then being extrapolated to mixtures. Broadly speaking, the several bulk properties of concern here can be categorized according to changes of physical state (vaporization, condensation, and fusion), and diffusion. All are apparently of importance in GC, while only fusion and diffusion appear at this time to be of consequence in LC.

2.2.1 The Vapor State

The Ideal Gas

The feature that distinguishes the vapor state from other states of matter is the property of compressibility at moderate (10,000 psig) pressure (supercritical fluids can therefore be considered to be vapors even though they deviate dramatically from various forms of the ideal gas law). This observation is stated formally in terms of Boyle's law: the volume of one mole of an ideal vapor $\overline{V}$ varies inversely as the pressure at constant temperature

$$p\overline{V} = k_1 \tag{1}$$

where k_1 is a constant that is temperature-dependent.

Variation of the volume of ideal vapor with temperature t (°C) at constant pressure is described in terms of the law of Gay-Lussac:

$$\overline{V} = k_2 + k_2\alpha t \tag{2}$$

where k_2 is a constant that is temperature-independent, and where the variable α is called the coefficient of thermal expansion of the gas. Extrapolation of this linear relation for many gases provides a value of α of 3.66×10^{-3} deg^{-1}, and eq. 2 may therefore be cast as

$$\overline{V} = k_2 \left(1 + \frac{t}{273}\right) \tag{3}$$

Setting $t = 0°C$, the value of k_2 for 1 mole of vapor is found at 1 atm to be 22,414 cm^3, and the quotient $k_2/273$ to be 82.1 cm^3 atm/deg/mol.

Combination of eqs. 1 and 3 thus provides the well-known relation between the pressure, molar volume and temperature (°K) of one mole of an ideal vapor:

$$p\overline{V} = RT \tag{4}$$

where the terms k_1/T and $k_2/273$ have been identified as R, the universal gas constant.

Real Gases

The gas law is of course only an idealized relation. A convenient initial measure of deviations of real gases from eq. 4 is the compressibility factor:

$$Z = \frac{p\overline{V}}{RT} \tag{5}$$

where Z (a function both of temperature and pressure) reduces to unity in the limit of ideality. Cartesian plots of Z against p or of surfaces of Z against p and T readily reveal that real gases can be made to approach ideality only under conditions of low pressure and high temperature. Figure 2.1 provides two plots that illustrate qualitatively this behavior. Over a range of low pressure at moderate to low temperatures T_1, pronounced negative curvature is found. As the temperature is increased the curve flattens, as indicated by that at T_2. The tangent to this curve is the ideal case, $Z = 1$.

The fact that all gases exhibit a minimum in such plots (provided that the temperature is reduced sufficiently) is taken to imply that below some T, vapors are more compressible than an ideal gas, that is, the volume at p,T is less than expected ($Z \propto \overline{V}/\overline{V}_{\text{ideal}}$). (Compressibility factors have been tabulated for very many gases, and it is in fact found that Z deviates most sharply from unity for those gases that are the easiest to liquefy.) Implicit in this finding is a major defect of eq. 4 ($\overline{V} = RT/p$), that is, its failure in predicting that the volume of a mole of gas $\overline{V}$ remains finite at

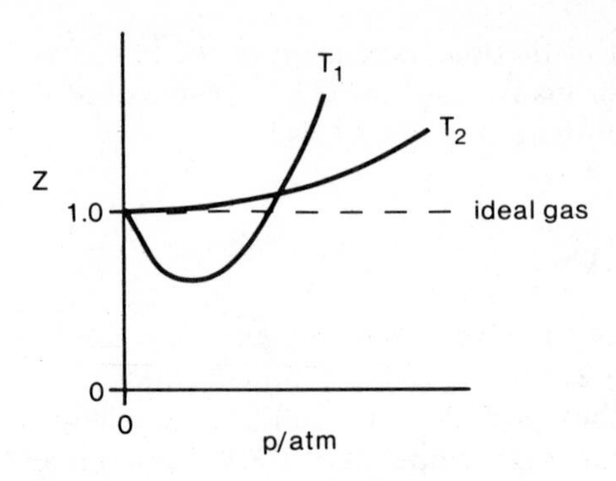

Figure 2.1. Plots of Z against p for a gas at temperatures T_1 and T_2, where $T_1 \ll T_2$.

$T = 0$. In order to take into account liquefaction (followed eventually by solidification), eq. 4 is modified as follows:

$$p\overline{V} = RT + pb \tag{6}$$

where b is a constant (with units of molar volume), and where $p\overline{V}$ reduces to the product pb at zero temperature. This version predicts that molecules retain a given size even under enormous pressure or conversely upon liquefaction with eventual freezing. Unfortunately, however, the generality of the ideal gas law is lost since there must apparently be a separate value of b for each gaseous species.

Even when molecular size is taken into account, it is found experimentally that the observed pressure of a gas is less than that predicted by eq. 6 by an amount that is proportional to the inverse square of the volume. Equation 6 is therefore modified further to

$$p\overline{V} = RT + pb - a\left(\frac{1}{\overline{V}} - \frac{b}{\overline{V}^2}\right) \tag{7}$$

which reduces to $p\overline{V} = RT + pb$ when a is set to zero. Equation 7 is one form of the well-known van der Waals relation that takes into account both molecular size (the b term) and intermolecular attraction (the a term). It is a remarkably simple equation that describes the pressure–volume product of gaseous species very nearly to their critical points.

Of prime concern in gas chromatography is the virial alternative to eq. 6 and 7, that is, expression of the $p\overline{V}$ product of a gas as a power series:

$$p\overline{V} = RT + pB + p^2C + \ldots \tag{8}$$

where the constants B, C, . . . , are termed the second, third, . . . , virial coefficients. When C, . . . = 0, eq. 8 reduces to eq. 6, that is, the van der Waals equation where only the volume occupied by molecules is taken into account. Moreover, reference to eq. 7 indicates that unless a is very large, the right-hand term of the relation will likely be negligible since $\overline{V}$ is generally much larger than unity or b. This holds most especially at high temperatures, and eqs. 6 and 8 can therefore be expected to converge under such conditions.

Equation 7 can also be rearranged to the approximate form

$$p\overline{V} = RT + \left(b - \frac{a}{RT}\right)p \tag{9}$$

where the quantity ($b - a/RT$) must evidently be equal to the second virial coefficient B of eq. 8. Values of B can therefore be estimated from tabulated values of a and b. For example, the $p\overline{V}$ product for n-pentane at 1 atm and 25°C is calculated from eq. 8 to be 23.29 L atm/mol (B = 1.194 L/mol experimentally), whereas eq. 9 yields a value of 23.82 L atm/mol (a = 19.01 L^2 atm/mol^2; b = 0.1460 L/mol). This level of agreement is very good considering the simplicity of the two equations and the extent of nonideality of n-pentane. Table 2.1 lists van der Waals constants and calculated and experimental second virial coefficients for a variety of vapors and permanent gases.

Mixtures of Real Gases

The virial description of blends of gases is given by the expression (5):

$$B_{1+2} = (x_1)^2 B_{11} + 2x_1(1 - x_1)B_{12} + (1 - x_1)^2 B_{22} \tag{10}$$

where B_{1+2} refers to the mixed gas; x_1 is the mole fraction of component 1; B_{11} and B_{22} are the virial coefficients of pure gases 1 and 2; and B_{12} is the so-called mixed virial coefficient that takes into account interactions between species 1 and 2. The latter three terms B_{11}, B_{22}, and B_{12} are of some importance both in analytical and physicochemical applications of gas chromatography. The first two of these can be calculated from the McGlashan-Potter relation (Table 2.1) if unavailable in the literature, or estimated from the respective van der Waals constants. B_{12} either can be measured directly by static or chromatographic means (cf. ref. 1, Ch. 4) or can be approximated from the relation (2)

Table 2.1 Van der Waals Constants *a* and *b*, and Calculated and Experimental Second Virial Coefficients *B* at 25°C

Gas	a (L^2 atm/ mol^2)	b (L/mol)	$-B$ (L/mol) Calc.[a]	Calc.[b]	Exptl.[c]
H_2	0.2442	0.02661	−0.01664	−0.01432	−0.01471
He	0.03412	0.02371	−0.02232	−0.01452	−0.01155
N_2	1.390	0.03912	0.01766	0.006171	0.004820
CO_2	3.592	0.04267	0.1040	0.1265	0.1246
CH_4	2.253	0.04278	0.04925	0.04158	0.04340
C_2H_6	5.489	0.06380	0.1604	0.1845	0.1870
C_3H_8	8.664	0.08445	0.2694	0.3719	0.3914
n-C_4H_{10}	14.47	0.1226	0.4684	0.7137	0.7100
n-C_5H_{12}	19.01	0.1460	0.6305	1.197	1.194

[a] $B = (b - a/RT)$.
[b] McGlashan-Potter (2)/Guggenheim-McGlashan (3) equation:

$$\frac{B}{V^c} = \left[0.430 - 0.886\left(\frac{T^c}{T}\right) - 0.694\left(\frac{T^c}{T}\right)^2 - 0.0375\,(n - 1)\left(\frac{T^c}{T}\right)^{4.5}\right]$$

where V^c and T^c are the critical volume and temperature (4), and where n is a constant (1 for most fixed gases; the carbon number for n-alkanes, and 4 for CO_2).
[c] Averaged values from J. H. Dymond and E. B. Smith, *The Virial Coefficients of Gases*. Copyright 1969 Clarendon Press, Oxford.

$$\frac{B_{12}}{V_{12}^c} = \left[0.430 - 0.886\left(\frac{T_{12}^c}{T}\right) - 0.694\left(\frac{T_{12}^c}{T}\right)^2 - (0.0375)\,(n_{12} - 1)\left(\frac{T_{12}^c}{T}\right)^{4.5}\right] \quad (11)$$

The "effective" critical volume V_{12}^c is given by

$$V_{12}^c = \frac{1}{8}\,[(V_1^c)^{1/3} + (V_2^c)^{1/3}]^3 \quad (12)$$

However, calculation of the "effective" critical temperature T_{12}^c is not nearly so straightforward. The geometric mean rule gives

$$T_1^c = (T_1^c T_2^c)^{1/2} \quad (13)$$

Table 2.2 Comparison (7) of Experimental B_{11} Data with Values Calculated from Eqs. 11–13

Solute	T (°C)	$-B_{11}$ (L/mol)		Δ(%)[a]
		Calc.	Exptl.	
BF_3	−12	0.137	0.174	21.3
$AsCl_3$	381	0.867	0.628	−38.1
$SbCl_3$	521	0.886	1.745	49.2
$SiCl_4$	234	0.899	2.675	66.4
$GeCl_4$	279	0.759	1.433	47.0
$SnCl_4$	319	1.164	1.192	2.35
$TiCl_4$	365	1.018	3.193	68.1
$ZrCl_4$	504	0.584	0.575	−1.57
$HfCl_4$	452	0.495	0.824	39.9
$HgCl_2$	699	0.532	0.701	24.1
			Average:	27.9%

[a] 100 (Exptl − Calc)/Exptl.

whereas the Hudson-McCoubrey combining rule (6) provides

$$T_{12}^c = 2(T_1^c\ T_2^c)^{1/2} \left[\frac{(I_1^d I_2^d)^{1/2}}{I_1^d + I_2^d}\right] \left\{\frac{64 V_1^c V_2^c}{[(V_1^c)^{1/3} + (V_2^c)^{1/3}]^6}\right\} \tag{14}$$

where I^d is an ionization potential. Equation 14 is superior to eq. 13 for systems comprising permanent gas + hydrocarbons when results calculated from the two relations are substituted into eq. 11. Far less is known about the accuracy of each with inorganic and organometallic solutes. The data of Zado and Juvet (7) provided in Table 2.2 illustrate the situation for inorganic chlorides, where experimental values of B_{11} are compared with those calculated with eqs. 11–13. The errors (last column of Table 2.2) are on the order of those found when eqs. 13 and 14 are compared for organic solutes with several carriers (8). Moreover, the matter is by no means a trivial one since reversals in GC elution order can arise if the magnitude of B_{12} is sufficiently large (9), as demonstrated most recently by Pretorious (10). However, because of the paucity of information regarding I^d, T^c, and V^c for most inorganic and organometallic vapors, it appears at this time that B_{12} must either be measured directly or estimated from eqs. 11–13, where an average positive error of ~ 30% should apparently be factored into the calculated results (11).

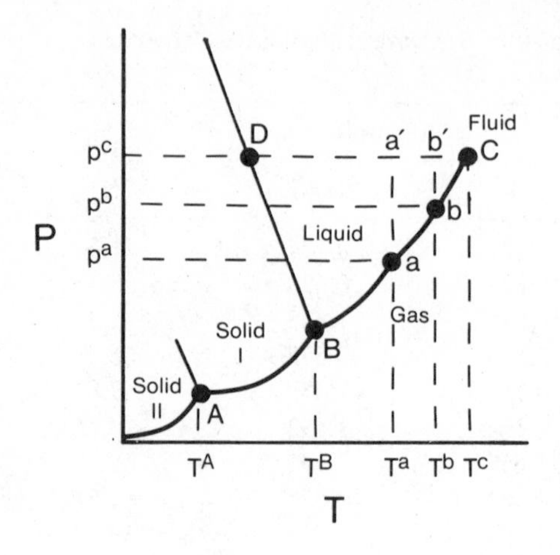

Figure 2.2. *p-T* diagram for a hypothetical one-component system.

Supercritical Fluids

The phase diagram (*p-T* plane) for a hypothetical substance in a closed container is presented in Fig. 2.2, Three phases coexist at point *A*, namely, solid II, solid I, and vapor. Any decrease along the isotherm T^A causes the substance to sublime completely. At point *B*, on the other hand, liquid, gas, and solid I coexist. A decrease in *p* from this point causes conversion of all material present within the container to a gas, while an increase in pressure induces liquefaction. The curve *BC* represents the set of *p-T* points at which both liquid and gas coexist. Picking that at *a*, any increase in *p* above p_a results in complete liquefaction. If the temperature is increased to T^b, we see that a greater pressure p_b is required to liquefy the substance. Eventually at temperatures above T^c, it is found that the material does not liquefy no matter how high the pressure. Instead of a liquid or a gas, the system contains a fluid. The fluid retains the physical appearance of a gas but, unlike a gas, it cannot be induced to liquefy upon the application of pressure of any amount. The temperature and pressure at which the boundary between the liquid and gaseous states disappears, *C*, are called the critical temperature T^c and critical pressure p^c. Along the curve *BD*, however, liquid and solid coexist even beyond point *D*, that is, a distinction can always be observed between liquid and solid phases.

The phases of a one-component system in a closed container of adjustable volume can also be represented by the isotherms of a *p-V* plane, as shown in Fig. 2.3. The curve T^c corresponds to the critical temperature from which can be obtained the critical volume and pressure as shown.

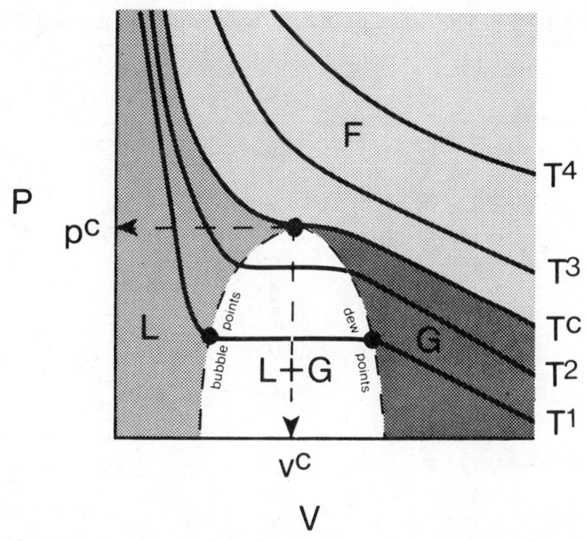

Figure 2.3. Isotherms of a p-V plane for a hypothetical substance.

Liquid and gas coexist in the region L + G but only liquid or gas are found in the respective shaded areas labeled L and G. Everywhere above the T^c isotherm the material exists as a fluid. The loci of bubble and dew points describe the L + G region and are interpreted as follows. Following the isotherm T^1 from left to right, some gas begins to form from the liquid when the bubble point is encountered at point 1. Upon enlargement of the container volume, an isobaric isothermal expansion of the system takes place such that more and more gas is formed until, at point 2, the last remains of liquid disappear (the dew point).

Plots of these types are often cast in terms of reduced variables, where $X^R = X/X^c$, X being the value of the parameter of interest and X^c the respective critical value of the variable (i.e., temperature, volume, pressure, compressibility, or density). Very many gases exhibit qualitatively the same shapes of plots of reduced variables, which is a consequence of the law of corresponding states, that is, $Z^c = p^c V^c/RT^c$ is very nearly a constant for all gases, as indicated in Table 2.3. The critical properties of a few compounds of interest here are provided also in Table 2.3, where T^c spans the range -268 (He) to 374°C (H_2O).

Planes of pressure against temperature offer an alternative view of the bubble and dew lines, and also reveal the possibility of unusual vapor–liquid equilibrium situations. Three such plots are shown in Fig. 2.4a–c. In each, the bubble and dew points coincide at the critical point, C. Two

Table 2.3 Critical Properties of Selected Compounds in Order of T^c

Compound	T^c (°C)	p^c (atm)	V^c (L/mol)	Z^c
He	−267.94	2.26	0.058	0.304
H_2	−239.9	12.8	0.065	0.304
Ne	−228.7	26.9	0.042	0.307
N_2	−147.0	33.5	0.090	0.291
CO	−140.2	34.5	0.093	0.294
Ar	−122.0	48.0	0.075	0.290
O_2	−118.8	49.7	0.074	0.290
NO	−94.0	65.0	0.058	0.256
CO_2	31.3	72.9	0.094	0.275
C_2H_6	32.4	48.3	0.148	0.285
C_3H_8	96.8	42.0	0.200	0.277
NH_3	132.3	111.3	0.073	0.243
n-C_4H_{10}	152.0	37.5	0.254	0.273
SO_2	157.5	77.8	0.12	0.269
CS_2	278.8	78.0	0.17	0.293
H_2O	374.1	218.2	0.056	0.230

separate points are also identified in Fig. 2.4*a*, one representing the maximum pressure C_1 and the other the maximum temperature C_2 at which liquid and vapor can coexist. The critical point need not occur between points C_1 and C_2, however, as shown in (*b*) and (*c*).

Consider the isothermal compressions shown in Fig. 2.5*a* for the first type of *p-T* curve of Fig. 2.4: for change 1, the gas is compressed to point *a* (on the dew line) where liquid begins to form. Further compression to a point a' results in disappearance of the last trace of gas at the bubble line. Change 2 produces a different result: isothermal compression to point *b* causes liquid to form which continues to increase in volume to point *m*. A tie line through this point gives the relative amounts of substances in each phase. Thereafter, the volume of liquid decreases until point b' is reached. This is the second point at which the dew line is intersected, and all liquid is again converted into vapor. The interesting situation thereby arises where an *increase* in pressure brings about volatization. For the curve in Fig. 2.5*b*, both isothermal compressions 1 and 2 intersect the dew line twice at a,a' and at b,b'. However, consider the isobaric expansions 1 and 2 in Fig. 2.5*c*. At point *a* the bubble line is encountered and vapor begins to form. The material is converted completely into gas at point a'. In contrast, expansion 2 yields a maximum of vapor at point

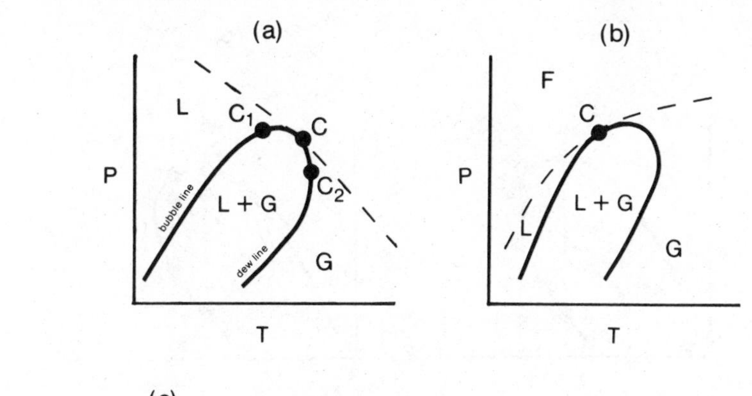

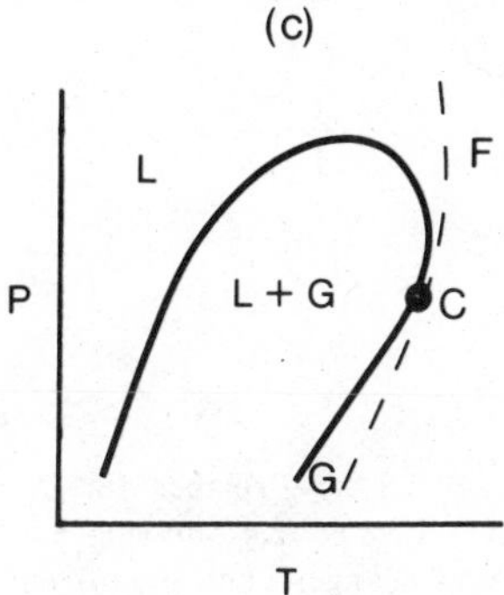

Figure 2.4. *p-T* curves for hypothetical substances where the critical point C lies between the point of maximum pressure C_1 and the point of maximum temperature C_2 (*a*); to the left (*b*); and to the right (*c*).

m, which is converted back completely to liquid at b'. Thus, an increase in temperature along the path mb' results in a *decrease* of volatization.

Evaporation upon an increase of pressure that results from two intersections of a dew line has been called retrograde behavior of the first kind (12), isotherms 2 and 1 and 2 respectively of Figs. 2.5*a* and *b*. Isothermal intersection twice of a bubble line (for example, isotherm 3, path Cb') is termed retrograde behavior of the second kind and is possible only with curves of the type shown in Fig. 2.5*c*: an increase in pressure brings about the formation of some vapor with eventual return completely to the liquid state at point b'. The isobaric change 2 in the figure also results in retrograde behavior of the second kind, and comprises in this instance increasing condensation with increasing temperature.

Mixtures of Supercritical Fluids

In the simplest representation of critical phenomena for binary-component blends of gases, the critical pressure varies smoothly as a function of temperature as shown in Fig. 2.6. However, this hardly represents the

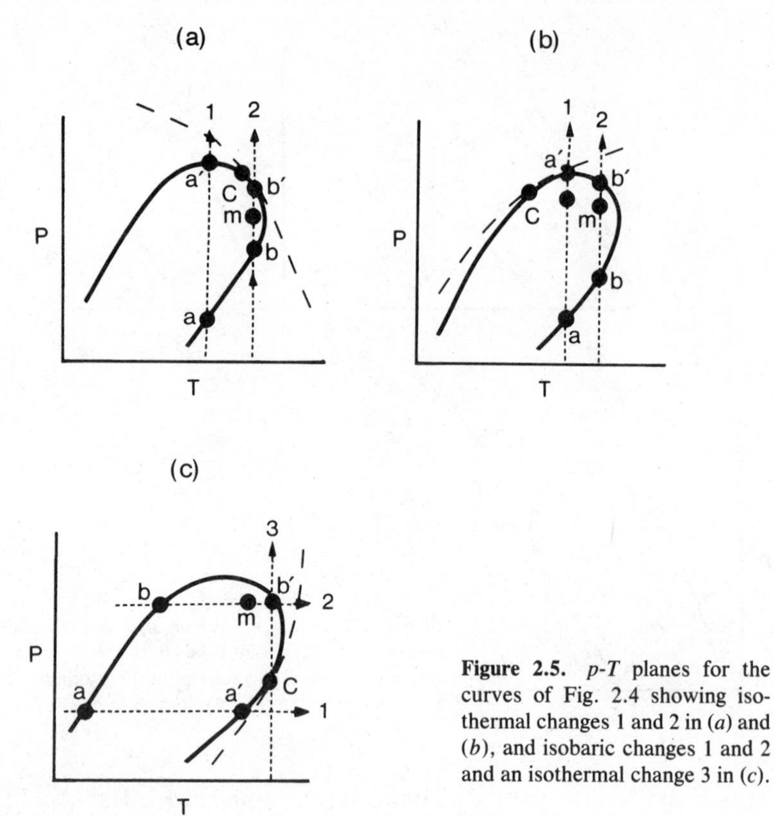

Figure 2.5. *p-T* planes for the curves of Fig. 2.4 showing isothermal changes 1 and 2 in (*a*) and (*b*), and isobaric changes 1 and 2 and an isothermal change 3 in (*c*).

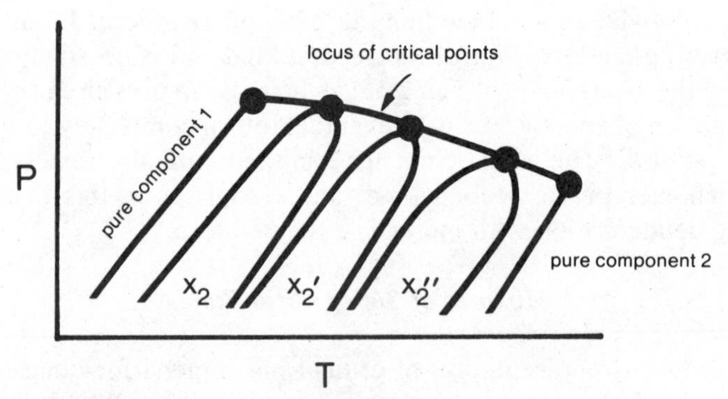

Figure 2.6. Critical *p-T* curves for two-component blends of gases showing the locus of critical points for various mole-fraction compositions.

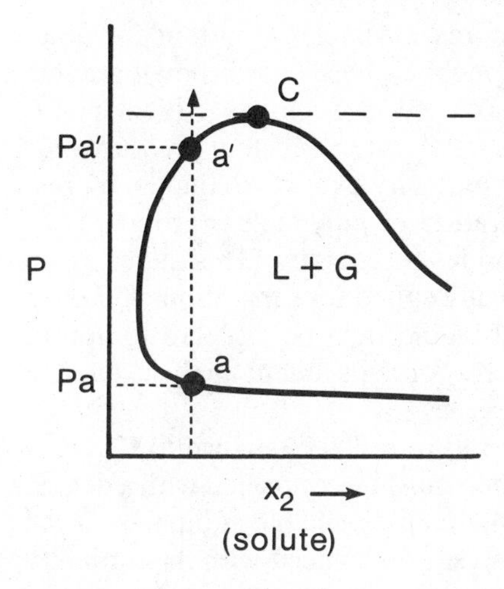

Figure 2.7. *p-x* plane for a two-component blend of gases at $T > T^c$.

only possibility, and in fact at least six major types of pressure-temperature behavior have been documented. These become quite complex when the binary liquids exhibit upper and/or lower consolute temperatures, as discussed in detail by Rowlinson and Swinton (12). From the standpoint of chromatographic separations, it is sufficient to note here that a dramatic increase in mutual solute-carrier solubility can be achieved by very small increments in p (13). Fig. 2.7 illustrates the situation for mixtures of CO_2 with *n*-alkanes (14) at $T > T^c$: as the pressure is increased to point a, a region of liquid + gas is encountered which persists up to point a'. Two states of matter are extant everywhere between these limits and solute migration will be slow. At pressures immediately above p'_a, however, solute and solvent become completely miscible once again as fluids, and the rate of solute migration through the chromatographic system thus increases significantly. The point p'_a explains the so-called threshold pressure observed experimentally by Giddings and his coworkers (15), and unfortunately portends a dramatic dependence of solute elution times on the system pressure. Moreover, curves shaped differently from that of Fig. 2.7 can also be expected, depending upon the chemical nature of solute and solvent. For example, Zlatkis and his coworkers Novotny and Bertsch (16), and Jentoft and Gouw (17) have shown that small amounts of additives in supercritical mobile phases can decrease *or* increase solute migration depending upon the additive. (The situation is somewhat analogous to the sensitivity of liquid-liquid consolute temperature where impurities in the parts-per-million range can offset $T_{consolute}$ by tens of degrees.)

Accurate pressure control is therefore extremely important for practical application of supercritical fluid mobile phases in chromatographic separations, particularly if the fluid is used close to its critical point. Moreover, different classes of solutes may elute at system pressures (or under pressure programming) that are mutually exclusive, that is, a pressure program that enhances the separation of polynuclear aromatic hydrocarbons may cause inorganic chlorides to coelute. (This suggests on the other hand that the technique is ideally suited for separation of *classes* of solutes, where the column effluent would then be directed to appropriate manifolded GC and/or LC systems for subsequent analysis of each member of each class.)

If for any reason the system temperature utilized is close to T^c, as is commonly the case in practice, then it too must be controlled with a degree of precision that exceeds that of commercially available equipment if the results are to be reproducible. The analyst is also faced with the difficulty that even if these practical requirements are met by the (of necessity laboratory-constructed) apparatus, solute retentions will be unpredictable as functions of p and T. Formulation of even the most rudimentary guidelines for selection of the column pressure (program) and temperature (program) must therefore be carried out empirically. For these reasons, supercritical fluid chromatography has remained little more than a novelty in separations science.

Nevertheless, and even in the face of the difficulties cited above, the technique does possess distinct advantages. These arise primarily as a result of the lower viscosity of supercritical fluids as compared with liquids (approximately a factor of 100), and the 200–500 times greater density over that of gases. The first of these factors permits higher mobile-phase linear velocities to be achieved at column pressure drops lower than is possible with liquids, while the second yields two important results. First, the solubilization of solutes is greatly increased over that exhibited by gases, which permits the chromatography of labile or high-molecular-weight materials at column temperatures considerably less than would ordinarily be required in gas chromatography. Secondly, solute diffusion coefficients are at least two orders of magnitude larger in supercritical fluids than in liquids. In principle, therefore, the kinetics of mass transport should be faster, which in turn should result in higher column "efficiency" (see later), that is, better separations. Hence, while it can be concluded that supercritical fluid chromatography is at best a difficult experimental technique to employ in practice, its inherent advantages hold the promise of routine application in at least some (albeit limited) areas. Supercritical fluid chromatographic separations of polystyrenes (18) and other polymers (19), as well as organometallic species (20,21) provide several illustrations; the interested reader is referred to refs. 22–24 for other examples.

2.2.2 Condensed States

The distinguishing feature of condensed states of matter is their incompressibility relative to that of gases. Moveover, liquids and solids can be distinguished from one another on the basis that under the influence of gravity the former will assume the shape of a container (although not occupy its volume fully) while the latter will not. The degree of ordering in liquids is, arguably, also much closer to that of solids than that extant in gases. Short-range ordering in liquids has in fact been recognized for many years; for example, Stewart (cf. ref. 1, Sec. 6.8) referred to the phenomenon as cybotaxis in the late 1920's, while polymorphs of triglycerides were recognized as long ago as 1849 (25). No such analogous behavior is known to exist either for pure or blended gases.

Equation of State for Liquids and Solids

An equation similar to eq. 2 can be written for liquids and solids:

$$V = V_0 + V_0\alpha_p t \tag{15}$$

where V_0 is the volume of the substance at 0°C, and t is in degrees Centigrade. The term α_p is called the coefficient of isobaric thermal expansion and is defined formally as

$$\alpha_p = \left(\frac{1}{V}\right)\left(\frac{\partial V}{\partial T}\right)_p \tag{16a}$$

It is conveniently measured from density data, where

$$\alpha_p = \left(-\frac{1}{\rho}\right)\left(\frac{\partial \rho}{\partial T}\right)_p \tag{16b}$$

Values of α_p are always positive for gases and solids, and are usually positive for liquids (liquid water expands upon cooling from 4 to 0°C). V_0 is a function of pressure and is given by the relation

$$V_0 = V_0^\circ[1 - \beta_T(p - 1)] \tag{17}$$

where V_0° is the volume of the substance at 0°C and 1 atm, and where β_T is called the coefficient of isothermal compressibility:

$$\beta_T = \left(-\frac{1}{V}\right)\left(\frac{\partial V}{\partial p}\right)_T \tag{18}$$

Values of β_T are very small for liquids and solids, typically 10^{-6} to 10^{-5} atm^{-1}, and are always positive.

An equation of state for condensed matter can now be written by combining eqs. 15 and 17:

$$V = V_0^\circ(1 + \alpha_p t)[1 - \beta_T(p - 1)] \tag{19}$$

(The equation is in fact valid for all forms of matter where, for ideal gases, $\alpha = RT/p$ and $\beta = RT/p^2V$.)

A derived (and often-ignored) property of condensed phases is the ratio α_p/β_T, which is called the coefficient of isochoric thermal pressure γ_V:

$$\gamma_V = \frac{\alpha_p}{\beta_T} = \left(\frac{\partial p}{\partial T}\right)_V \tag{20}$$

Table 2.4 lists values of α_p, β_T, and γ_V for various solids and liquids at 20°C, as well as those for carbon tetrachloride at 0 to 70°C, the latter providing illustration of the variation of these properties with temperature.

The usefulness of these data should not be underestimated. For example, the common experience of bursting a mercury thermometer is readily explained by the value of γ_V of 47 atm/deg for this substance. That is, a change of only 1° of a thermometer whose stem is completely filled causes an increase in pressure of 47 atm. Consider, on the other hand, the process of packing a high-performance liquid chromatography column: it is not uncommon that a 25-cm length of tube is attached to a slurry reservoir, and the entire apparatus then pressurized to 25,000 psig. The slurry is thereby forced into the column, where the packing is retained by a porous nickel frit at the column outlet. A large pressure drop occurs across the fitting containing the frit, which heats up noticeably during the packing process. Suppose that the packing liquid comprises a 50:50 mixture of methanol:carbon tetrachloride. The inverse of γ_V can then be expected to be approximately 9×10^{-2} deg/atm, for which a 25,000-psig pressure drop would cause a 153°C temperature increase. Fortunately, the column inlet/outlet pressure drop occurs across the entire column; even so, it is easy to visualize how the flash point of common LC solvents could be reached or exceeded at the column outlet and it is therefore wise to cool such end-fittings continuously while the column is being packed (26). Careful monitoring of the temperature at various points could, on the other hand, provide a detailed picture of the pressure drop along the column during the packing process (or for that matter, during column use). Since the variation of temperature along the column length should

Table 2.4 Thermal Expansion (α_p) Compression (β_T), and Thermal Pressure (γ_V) Coefficients

Substance	$10^4\ \alpha_p$ (deg^{-1})	$10^6\ \beta_T$ (atm^{-1})	γ_V (atm/deg)
Solids[a]			
Graphite	0.24	3.0	8.0
Ag	0.58	1.0	58.0
NaCl	1.21	4.2	29.0
Liquids[a]			
Hg	1.81	3.85	47.0
CCl_4	12.2	102.0	12.0
CH_3OH	12.0	120.0	10.0
C_2H_5OH	11.2	110.0	10.2
H_2O	2.0	45.3	4.42
CCl_4[b]			
T/°C			
0	11.77	87.0	13.5
10	11.98	94.2	12.7
20	12.20	101.9	12.0
30	12.41	110.2	11.3
40	12.66	119.7	10.6
50	12.96	130.5	9.93
60	13.27	141.8	9.36
70	13.49	154.3	8.74

[a] After G. W. Castellan, *Physical Chemistry*, Table 5-1. Copyright 1964 Addison-Wesley.
[b] After Rowlinson and Swinton (12), Table 2.7.

be a smooth curve (see later), any discontinuities would indicate sections that are packed nonuniformly (or have developed "void" space).

Heat Capacities

The molar heat capacity of a substance at constant volume is defined formally as

$$C_V = \left(\frac{\partial \bar{E}}{\partial T}\right)_V \tag{21}$$

This quantity is difficult to measure directly for liquids or solids, since such an experiment requires heating a completely filled container. The heat capacity at constant pressure generally is determined instead, where

$$C_p = \left(\frac{\partial \overline{H}}{\partial T}\right)_p \tag{22}$$

C_V can then be calculated from C_p via the relations

$$\begin{aligned} C_V &= C_p - T\overline{V}\alpha_p\gamma_V \\ &= C_p - T\overline{V}\beta_T\gamma_V^2 \\ &= C_p - T\overline{V}\alpha_p^2/\beta_T \end{aligned} \tag{23}$$

Over the ranges of temperatures most often employed in chromatography the heat capacity at constant pressure is constant, which means that enthalpies can be taken to be constant. This has the consequence that (van't Hoff) plots of log (solute retention) against inverse temperature will be linear when only one sorption mechanism contributes to elution times. When multiple mechanisms are operative, for example concurrent solution and gas-liquid interfacial adsorption in GC, such plots may or may not approach linearity depending upon the relative magnitudes of the contributions of each. Furthermore, in such instances solute peak shapes frequently are severely distorted. Even in the absence of such irregular shapes, however, nonlinear van't Hoff plots can be taken as firm evidence of multiple retention mechanisms.

Viscosity

When an incompressible liquid is forced through an empty tube at moderate pressure, a linear velocity gradient du/dr exists across the radius of the tube such that molecules closest to the wall move slower than those toward the center. The flow velocity profile is therefore that of a parabola proceeding down the tube, much like the extension of concentric cylinders of a telescope. The "dragging" or viscous force on layers nearest the tube wall is exactly balanced by the external applied force per unit area (pressure), which is expressed by Poiseuille's law:

$$\frac{\Delta V}{\Delta t} = \frac{\pi r^4(p_i - p_o)}{8L\eta_l} \tag{24a}$$

where $\Delta V/\Delta t$ is the volume flow rate per unit time t, r is the radius of the tube of length L, p_i and p_o are the inlet and outlet pressures, and η_l is the coefficient of viscosity, or simply the viscosity.

Equation 24(a) requires modification for use with gases because of compressibility effects along the length of the tube: below 1.1 atm, the average gas pressure $\bar{p}$ is $(p_i + p_o)/2$ and the relation becomes

$$\frac{\Delta V}{\Delta t} = \frac{\pi r^4(p_i^2 - p_o^2)}{16L\eta_g p_o} \tag{24b}$$

where it has been assumed that the gas volume is measured at the outlet pressure p_o. Above 1.3 atm or so the error in calculating the average pressure exceeds 2% and the relation

$$\bar{p} = \frac{2}{3}\left[\frac{(p_i/p_o)^3 - 1}{(p_i/p_o)^2 - 1}\right] p_o \tag{25}$$

must instead be employed.

η has units of g/cm/sec in the cgs system, the *poise*. Liquids commonly exhibit viscosities on the order of centipoise (cP), while gases are usually of the order of 100–200 μP.

There are several aspects of viscosity that bear directly upon chromatography, such as laminar vs. turbulent flow, relaxation of the parabolic flow profile, and so forth; these are taken up in a later section. For the time being, we consider the temperature- and density-dependence of η.

Gas viscosity increases with increasing temperature, which is explained on the basis of increasing molecular speed at higher T. The situation is weakly analogous to the vigor with which marbles are shaken out of a milk bottle: the more vigorous the shaking, the fewer is the number of marbles that emerge because of the increase in marble–marble collisions at the mouth (as well as throughout the rest) of the container. Viscosity also increases with molecular size, hence the density of gases for the same reason: the larger the molecules or the greater their number, the greater is the chance of intermolecular collisions (27). These two phenomena are observed very commonly in gas chromatography. At higher column temperature, a higher pressure is required to maintain a given volume flow rate. Mobile phases of increasing molecular weight also require a higher inlet pressure in order to achieve a certain flow rate. For example, the linear velocity of nitrogen mobile phase is easily half that of hydrogen through an open-tubular column at, say, 100°C and at 20 psig inlet pressure, and is halved again if the temperature is adjusted to 150°C. This has the practical consequence that elution times as well as column pressure drops will be lower (more favorable) with H_2 carrier than with N_2.

The situation is exactly reversed for the temperature coefficient of viscosity of liquids: η_l decreases with increasing temperature in approximate accordance with the relation

$$\log \eta_l = k_3 + k_4/RT \tag{26}$$

where k_3 and k_4 are empirical constants. The relation predicts that $\log \eta_l$ will vary linearly as the inverse of temperature, which holds for very many species over several tens of degrees. Water is an important exception, for which it is found empirically that (28), for 0 to 20°C,

$$\log \eta_l(t) = \frac{1301}{998.333 + 8.1855(t - 20) + 0.00585(t - 20)^2} - 3.30233$$

and for 20 to 100°C,

$$\log \frac{\eta_l(t)}{\eta_l(20)} = \frac{1.3272(20 - t) - 0.001053(t - 20)^2}{t + 105}$$

where t is in degrees Centigrade. The qualitative explanation of these phenomena is that molecules in the liquid state must overcome an energy barrier k_4 in order to pass from a layer of low linear velocity to a higher one. Higher temperatures result in more molecules in "states" of higher velocity, much like a Boltzmann energy distribution, such that larger numbers of them find their way nearer to the center of the tube. In the case of water, even more energy is required at lower temperatures in order to disrupt the short-range order arising from hydrogen bonds in this liquid (29). However, such a description fails to account for the increase of η_l with increasing density or pressure, for which we must resort again to the model of intermolecular collisions. The more closely packed are molecules, the more difficult it becomes for one to pass unhindered from one velocity layer to the next.

The consequences of the above phenomena are readily apparent in modern column liquid chromatography, where inlet pressures of several hundred atmospheres are frequently employed. Equation 24(a) indicates that halving η_l will double the volume flow rate at fixed inlet pressure or, conversely, will approximately halve the requisite inlet pressure for a given flow rate since $p_i \gg p_o$ is generally the case. For example, a temperature change of from 15 to 50°C alters the viscosity of pure water from 1.139 to 0.5468 cP, a very considerable change indeed.

For blended LC mobile phases, the isothermal viscosity of the mixture $\eta_l(1 + 2)$ can be expected to be a volume-fraction average of that of the pure components η_l^0:

$$\eta_l(1 + 2) = \sum_{i=1}^{n} \phi_i \eta_l^0(i) \tag{27}$$

The temperature dependence of viscosity of liquid mixtures is more difficult to forecast, but most likely conforms to some or other form of geometric combining rule:

$$\log \eta_l(1 + 2) = \{[k_3(1)k_3(2)]^{1/2} + [k_4(1)k_4(2)]^{1/2}\}/RT \tag{28}$$

Tables 2–5(a) and (b) provide viscosity data for a variety of gases and liquids, respectively, over the indicated ranges of temperature.

Surface Tension

Several consequences arise from surface-tension phenomena that are important in GC and LC, such as the capillary rise of liquids (as in TLC), the vapor pressure of solutes in pores (GC), the wetting of column walls and packings by liquids (LC), displacement of a sorbed liquid from a solid surface by a second liquid (LC), gas–solid and gas–liquid interfacial adsorption (GC), liquid–liquid interfacial adsorption (LC), and so forth. Specific considerations relevant to chromatography are deferred for the most part to later sections, while various fundamental aspects of surface tension are presented here.

Consider a molecule situated in the middle of a pure liquid: it is surrounded on all sides by other molecules and has the opportunity of colliding and interacting with them. The "influence" that one molecule exerts upon another can be considered to be an "energy of cohesion" (30) that results in liquefaction of the substance in the first place. However, *surface* molecules "feel" only a net pull inward from the surface since there are no molecules of the same kind above them. A droplet of mist suspended in air will therefore assume the shape of minimum surface area, namely a sphere, as illustrated conceptually in Fig. 2.8. Since all of the surface molecules are pulled toward the center of the sphere, energy must be supplied to overcome this cohesive force if the surface of a liquid is to be expanded (spread). In the same sense, one can view the surface as being held to the rest of the liquid by a *tension* (negative pressure). Since the tension occurs in one direction only (inward from the liquid surface),

Table 2.5(a) Viscosity Data for Listed Gases

Gas	η_g (μP) at T (°C)				
Air	$\frac{0}{170.8}$	$\frac{18}{182.7}$	$\frac{40}{190.4}$	$\frac{74}{210.2}$	$\frac{229}{263.8}$
NH_3	$\frac{0}{91.8}$	$\frac{20}{98.2}$	$\frac{50}{127.9}$	$\frac{150}{146.3}$	$\frac{200}{164.6}$
Ar	$\frac{0}{209.6}$	$\frac{20}{221.7}$	$\frac{100}{269.5}$	$\frac{200}{322.3}$	$\frac{302}{368.5}$
AsH_3	$\frac{0}{145.8}$	$\frac{15}{114.0}$	$\frac{100}{198.1}$		
Br_2(g)	$\frac{12.8}{151.}$	$\frac{65.7}{170.}$	$\frac{99.7}{188.}$	$\frac{139.7}{208.}$	$\frac{179.7}{227.}$
CO_2	$\frac{0}{139.0}$	$\frac{15}{148.0}$	$\frac{30}{153.}$	$\frac{40}{157.}$	$\frac{235}{241.5}$
CO	$\frac{0}{166.}$	$\frac{15}{172.}$	$\frac{21.7}{175.3}$	$\frac{126.7}{218.3}$	$\frac{227.0}{254.8}$
Cl_2	$\frac{20}{132.7}$	$\frac{50}{146.9}$	$\frac{100}{167.9}$	$\frac{150}{187.5}$	$\frac{200}{208.5}$
Kr	$\frac{0}{232.7}$	$\frac{15}{246.}$			
Hg(g)	$\frac{273}{494.}$	$\frac{313}{551.}$	$\frac{369}{641.}$	$\frac{380}{654.}$	
CH_4	$\frac{0}{102.6}$	$\frac{20}{108.7}$	$\frac{100}{133.1}$	$\frac{201}{160.5}$	$\frac{380}{202.6}$
Ne	$\frac{0}{297.3}$	$\frac{20}{311.1}$	$\frac{100}{364.6}$	$\frac{200}{424.8}$	$\frac{250}{453.2}$
H_2	$\frac{0}{83.5}$	$\frac{20.7}{87.6}$	$\frac{28.1}{89.2}$	$\frac{129.4}{108.6}$	$\frac{299}{138.1}$
He	$\frac{0}{186.}$	$\frac{20}{194.1}$	$\frac{100}{228.1}$	$\frac{200}{267.2}$	$\frac{282}{299.2}$
N_2	$\frac{10.9}{170.7}$	$\frac{27.4}{178.1}$	$\frac{127.2}{219.1}$	$\frac{226.7}{255.9}$	$\frac{299}{279.7}$
O_2	$\frac{0}{189.}$	$\frac{19.1}{201.8}$	$\frac{127.7}{256.8}$	$\frac{227}{301.7}$	$\frac{283}{323.3}$

Table 2.5(b) Viscosity Data for Listed Liquids

Liquid	η_l (cP) at T (°C)				
NH_3	$\frac{-69}{0.475}$	$\frac{-50}{0.317}$	$\frac{-40}{0.276}$	$\frac{-33.5}{0.255}$	
Bi(1)	$\frac{285}{1.61}$	$\frac{304}{1.66}$	$\frac{365}{1.46}$	$\frac{451}{1.28}$	$\frac{600}{1.00}$
Br_2(1)	$\frac{0}{1.241}$	$\frac{12.6}{1.07}$	$\frac{16.}{1.0}$	$\frac{19.5}{0.995}$	$\frac{28.9}{0.911}$
Acetone	$\frac{0}{0.399}$	$\frac{15}{0.337}$	$\frac{25}{0.316}$	$\frac{30}{0.295}$	$\frac{41}{0.280}$
n-Butanol	$\frac{0}{5.186}$	$\frac{15}{3.379}$	$\frac{30}{2.30}$	$\frac{50}{1.411}$	$\frac{70}{0.930}$
CCl_4	$\frac{0}{1.329}$	$\frac{15}{1.308}$	$\frac{30}{0.843}$	$\frac{50}{0.651}$	$\frac{70}{0.524}$
$CHCl_3$	$\frac{0}{0.700}$	$\frac{15}{0.596}$	$\frac{25}{0.542}$	$\frac{30}{0.514}$	$\frac{39}{0.500}$
Cyclohexane	$\frac{17}{1.02}$				
Cyclohexene	$\frac{13.5}{0.696}$	$\frac{20}{0.66}$			
Cyclohexanol	$\frac{20}{68\ (!)}$				
Diethyl ether	$\frac{0}{0.284}$	$\frac{17}{0.240}$	$\frac{20}{0.2332}$	$\frac{25}{0.222}$	
Ethyl acetate	$\frac{10}{0.512}$	$\frac{20}{0.455}$	$\frac{30}{0.400}$	$\frac{50}{0.345}$	
Ethanol	$\frac{0}{1.773}$	$\frac{10}{1.466}$	$\frac{30}{1.003}$	$\frac{50}{0.702}$	$\frac{70}{0.504}$
n-Heptane	$\frac{0}{0.524}$	$\frac{20}{0.409}$	$\frac{25}{0.386}$	$\frac{40}{0.341}$	$\frac{70}{0.262}$
Isopropyl alcohol	$\frac{15}{2.86}$	$\frac{30}{1.77}$			
Hg(1)	$\frac{-10}{1.764}$	$\frac{10}{1.615}$	$\frac{30}{1.499}$	$\frac{50}{1.407}$	$\frac{70}{1.331}$
Methanol	$\frac{0}{0.82}$	$\frac{15}{0.623}$	$\frac{30}{0.510}$	$\frac{40}{0.456}$	$\frac{50}{0.403}$
H_2O	$\frac{20}{1.002}$	$\frac{40}{0.653}$	$\frac{60}{0.467}$	$\frac{80}{0.355}$	$\frac{100}{0.282}$

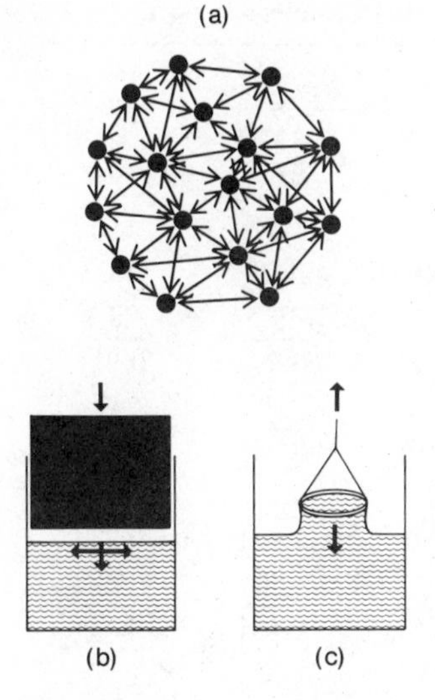

Figure 2.8. (*a*) Conceptual view of intermolecular interactions within a sphere of molecules suspended in air. The net "force" on the surface molecules is directed inward toward the center of the sphere. (*b*) Lines of force from an external applied pressure (force per unit area) within a liquid. (*c*) Tension (negative pressure) in one direction only causing surface molecules to resist being withdrawn from a liquid by a loop of wire. This is the method used by du Nouy for measurement of surface tension.

the applicable units are force per unit length as opposed to force per unit area, Figs. 2.8*b*,*c*. The unit commonly employed is the dyne/cm (N/m in the SI system).

The du Noüy tensiometer (31) has often been employed to measure the surface tension of liquids, where the force required to remove a ring from the surface is measured directly with a torsion balance or a quartz-spring apparatus such as that described by McBain and Bakr (32):

$$f = 4\pi r\gamma \tag{29}$$

where f is the force, r is the ring radius, and γ is the surface tension. The rise of a liquid in a capillary is also employed for surface-tension measurements. Fig. 2.9 illustrates the principle for a liquid that wets the walls of the container tube. The length r is taken to be a segment of the radius R of an imaginary circle formed from the arc represented by the meniscus of the liquid. The pressure difference across the meniscus ΔP is $(2\,\gamma\cos\theta)/r$, where the counterbalancing downward force applied by the weight of the column of liquid is $\pi r^2 gh(\rho - \rho')$; and where g is the gravity constant (980.665 cm/sec^2), h is the height of the column, and ρ and ρ' are the density of the liquid and the fluid above it (33). Converting the latter to force per unit area by division by πr^2 and setting the result equal to ΔP produces

$$\gamma = \frac{1}{2}\,gh(\rho - \rho')\,\frac{r}{\cos\theta} \tag{30}$$

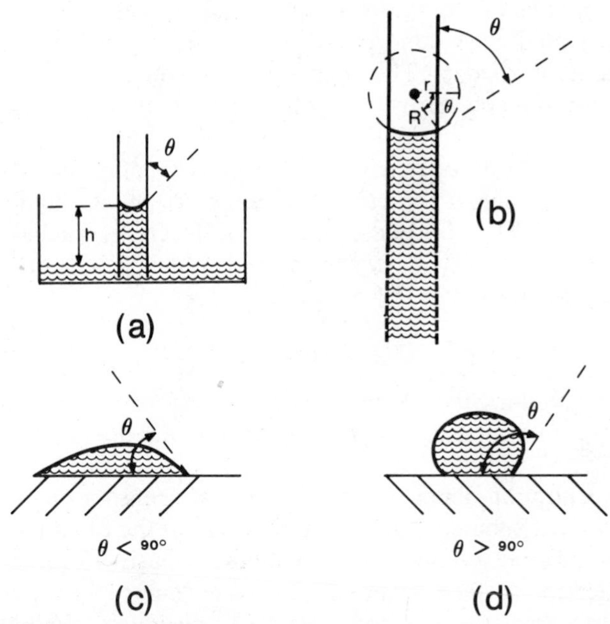

Figure 2.9. Rise in a capillary tube of a liquid which "wets" the walls (*a, b*). θ is the so-called contact angle. (*c, d*) Droplets of liquids on surfaces which exhibit good (*c*) and poor (*d*) wetting characteristics.

If the density of the liquid is large compared to the fluid above it and the contact angle θ is close to zero, eq. 30 reduces to

$$\gamma = \tfrac{1}{2}\, ghr\rho \tag{31}$$

The contact angle is less than 90° for liquids that wet the container material, is 90° for a planar (flat) meniscus, and is greater than 90° for materials that do not adhere well to the tube wall. Water/partially clean glass and mercury/glass are common examples of substances that exhibit convex and concave menisci. Since $\cos\theta \rightarrow 0$ as $\theta \rightarrow 90°$, materials that give close to planar menisci have surface tensions that are very much larger than those that exhibit convex or concave menisci. θ approaches zero for water and most organic substances with absolutely clean glass, Fig. 2.9*c*, whereas θ is about 105° for water with a paraffin wax surface and 140° for mercury/clean glass, as shown in Fig. 2.9*d*.

In 1925, Adam and Jessop (34) reported an ingenious device for the determination of contact angles that is particularly applicable to measurements of interest in modern chromatography. These include the wetting characteristics of GC stationary phases as well as LC mobile phases with glass, treated glass, amorphous fused silica, derivatized silica, and metals and their oxides. The recent contact-angle measurements by Voronkov and his coworkers (35) offer an example of studies of these kinds (36).

The surface tension of pure liquids decreases approximately linearly with temperature below the critical temperature of the liquid, and becomes zero at T^c. The McLeod equation (37) describes γ accurately over a wide range of T for very many substances, although some discrepancies arise with liquid metals:

$$\gamma = k_5(\rho_l - \rho_g)^4 \tag{32}$$

where k_5 is an empirical constant which differs from one substance to the next, and where ρ_l and ρ_g are the densities of the liquid and its vapor, respectively.

The concentration-dependence of the surface tension of solutions appears to be approximately linear, but the slope of the function is difficult to predict in advance on other than a qualitative basis. In general, solutes that decrease the surface tension of a solvent tend to accumulate at the solution surface and are said to be *adsorbed positively*. Conversely, solutes that cause an increase in γ are said to be *adsorbed negatively*. The interpretation of these observations is that positively adsorbed species are only poorly soluble in the solvent; hence, they tend to accumulate at the surface so as to avoid contact insofar as is possible with the bulk liquid. The energy required to remove them from the *solution* is therefore less than would be needed if they were situated somewhere within the bulk liquid. In contrast, solutes that interact strongly with the solvent increase the surface tension of the solution, since the surface molecules (solute or solvent) will be attracted into the bulk liquid more strongly than otherwise.

Tables 2.6(a),(b), and (c) present selected values of γ for pure and blended liquids at various temperatures. The water/ethanol data are particularly interesting because of the strong nonlinear dependence of γ upon composition, as shown in Fig. 2.10. The sharp increase toward the right-hand side of the plot is presumably due to hydrogen-bonding between water molecules that is of greater strength than that between water–ethanol or, weaker still, ethanol–ethanol.

Transitions Between Condensed States of Pure Compounds

Pure-compound phase transitions are for the most part straightforward in that the physical appearance of material is altered considerably, as in

Table 2.6(a) Surface Tensions of Liquid Elements

Element	Interfacial Fluid	T (°C)	γ (dyne/cm)
Bi	Vacuum	280.	392.
		320.	375.
		400.	374.
		500.	353.
Ga	H_2, CO_2	30.5	735.
In	Ar, He	170.	554.7
		200.	552.1
		250.	548.4
		300.	544.2
Pb	Vacuum	330.	450.
		400.	444.
Hg	Air	20.	435.5
	Vacuum	20.	474.
		200.	436.
		300.	405.
Cl_2(l)	Air	20.	18.4
Br_2	Air	20.	41.5
H_2(l)	Vapor	−255.	2.31
He(l)	Vapor	−270.	0.239
Ar(l)	Vapor	−188.	13.2
Ne(l)	Vapor	−248.	5.50
O_2	Vapor	−183.	13.2
N_2	Vapor	−183.	6.6

melting, boiling, or changes in crystalline form. These processes are called first-order transitions and are characterized by discontinuous changes in enthalpy and volume with temperature. Other first-order transitions are not nearly so easily recognized because the degree of change is much less readily apparent than, say, melting ice. For example, a "glassy state" is said to arise when a liquid undergoes supercooling with a concomitant marked increase in viscosity until the substance does indeed resemble glass (many polymers give more than one glass transition). However, the glassy state exhibits neither an ordered crystalline structure nor a sharp melting point. Upon warming, such materials soften with eventual liquefaction, although polydimethylsiloxanes commonly employed in GC decompose well before a true melting point is reached.

Liquid crystals, commonly called mesomorphs, also undergo first-order transitions. In contrast to polymers, these occur at sharply defined

Table 2.6(b) Surface Tensions of Liquid Compounds (Air or Vapor Interface)

Compound	T (°C)	γ (dyne/cm)
$CO_2(l)$	20	1.16
CS_2	20	32.33
$CO(l)$	−203	12.1
	−193	9.8
N_2H_4	25	91.5
N_2O_4	20	27.5
$HCN(l)$	17	18.2
$H_2O_2(l)$	18	76.1
H_2O	5	74.9
	10	74.22
	20	72.75
	40	69.56
	80	62.6
	100	58.9

Table 2.6(c) Surface Tensions of Liquid Solutions (Air Interface)

Solute	T (°C)	% (w/w) γ (dyne/cm)	Water Solvent			
HCl	20	%	1.78	12.81	35.29	
		γ	72.55	71.85	65.75	
NaCl	20	%	0.58	10.46	25.92	
		γ	72.92	76.05	82.55	
LiCl	25	%	5.46	7.37	13.95	
		γ	74.23	75.10	78.10	
			Ethanol Solvent			
LiCl	14	%	0.72	2.30	4.62	
		γ	22.90	23.17	23.26	
H_2O	40	%	20.	40.	90.	95.
		γ	23.43	26.18	46.77	53.35

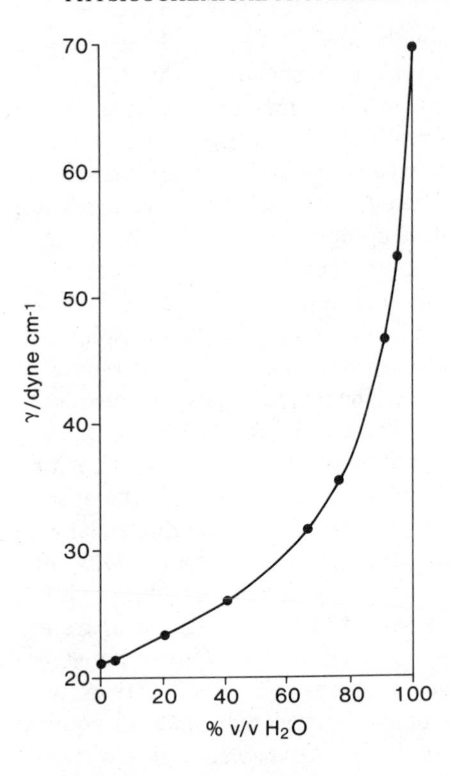

Figure 2.10. Variation of solution surface tension with composition for mixtures of water with ethanol (40°C).

temperatures. Several types of mesomorphs have been identified and are classified on the basis of ordering in one or two dimensions. *Smectics* are comprised of two-dimensional ordered layers, with no ordering between layers, and represent the closest proximity to solids. At least six variants of smectics are known, where each represents a slight variation of molecular alignment within layers. *Discotic* mesomorphs are a newly discovered class of compounds that form (planar) molecular columns, where the columns lay roughly parallel. *Nematic* liquid crystals are less ordered than smectics or discotics and invariably comprise rod-like molecules aligned in one direction only. *Chiral-nematic* mesomorphs form layered structures with a helical twist superposed on the layers such that the optical activity of the substance is greatly enhanced. It was thought at one time that chiral-nematics could be formed only from cholesteric-like species, but this has since been disproved.

Phase Diagrams for Mixtures of Condensed Substances

Many examples of types of phase diagrams are available in any undergraduate text on physical chemistry. Those involving metals are particularly intriguing, such as that for iron with carbon. We limit the discussion here to systems that are of direct interest in chromatography.

All of the pure-component phase transitions mentioned above can be represented by plots either of p against T as shown in Fig. 2.2 or, more commonly, of V against T, which dramatize transition temperatures. Plots of dV/dT against T aid in data interpretation when phase changes are indistinct. However, when two or more components are present in a system, the various forms of phase diagrams can become quite complex. Phase data generally are presented as plots of temperature (or reduced temperature) against composition for liquid–liquid and liquid–solid systems, and as pressure or temperature against composition for blends of gas + liquid and gas + solid substances. The former types are considered here while discussion of the latter is deferred to Secs. 2.2.5 and 2.2.6.

Fig. 2.11 presents the three possibilities of plots of T-x data for two liquids 1 and 2 that are only partially miscible. In Fig. 2.11*a*, the system beneath the curve is comprised of two phases. At the temperature labeled UCT, the phase boundary disappears and the liquids become fully miscible in all proportions. This value is called the upper consolute temperature, although it could equally be referred to as a critical temperature since it marks a phase transition. Fig. 2.11*b* illustrates a lower consolute temperature (LCT), while the system in (*c*) gives both a UCT and an LCT. All three types of curves need not be symmetric nor centered about a mole fraction of 0.5 The *International Critical Tables* series provides many examples of each type, where that shown in (*a*) appears to be the most common.

Interpretation of the phase diagrams is straightforward. Consider the line of constant composition (isopleth) $c'cx_c$ in Fig. 2.11*a*. At temperatures above the UCT, the system consists of a single phase of liquid of composition x_c. As the temperature is lowered toward the UCT, the solution may exhibit a pearly opalescence. This is thought to be due to incipient phase separation at the microscopic level. At the UCT the system breaks into two macroscopically immiscible phases. If the temperature is lowered further to T_a, the compositions of the two phases are given by x_b and x_e, that is, perpendiculars dropped from the points of intersection of the tie-line *bcde* with the two sides of the curve. That of greater density will be the lower phase. Now suppose that the overall composition of the system is changed from x_c to x_d by the addition of pure component 1. If the temperature is maintained at T_a, it is found that the compositions of the two phases do not change but the respective volumes of each are altered. The ratio of tie-line segment lengths *de*/*bd* gives the mole ratio of the two phases. Points of x,T in Figs. 2.11*b* and *c* are interpreted in the same way.

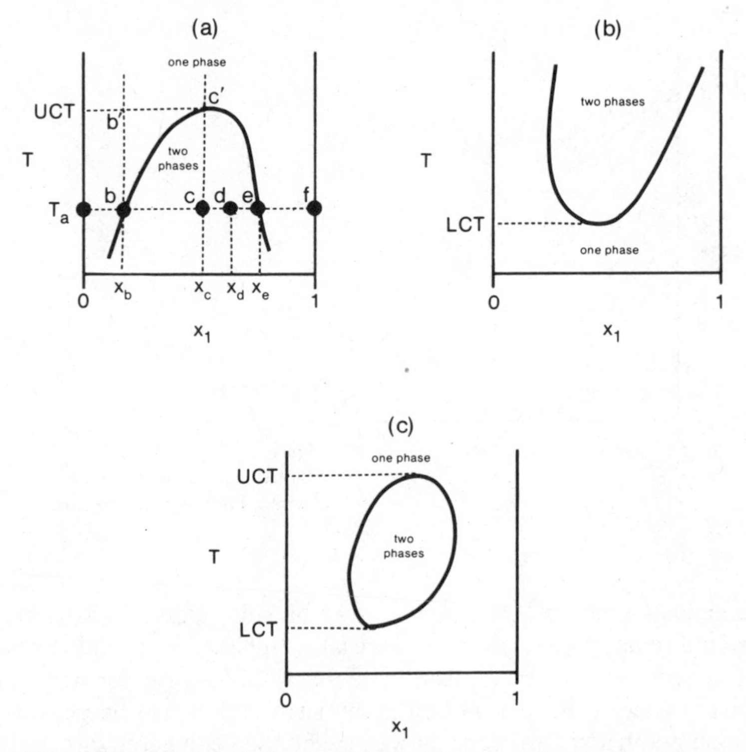

Figure 2.11. Plots of T against x for partially miscible pairs of liquids. An upper consolute temperature is exhibited in (a); a lower consolute temperature in (b); and both lower and upper consolute temperatures in (c).

Representation of partial miscibility of ternary-component liquids as a function of temperature is most easily done using triangular coordinates as the (composition) base of a prism, with T as its height. Data for quaternary, quinary, and higher-complexity systems can also be drawn in one plane, the details of which are provided in ref. 38.

Phase diagrams of systems containing a liquid-crystal component present several interesting situations. First, it has been known for many years that small amounts of impurities depress the mesomorphic/isotropic transition temperature (39). Secondly, when sufficient solute is added to a liquid crystal, two phases form. One is mesomorphic while the other is isotropic. Peterson and Martire (40) appear to be the first to represent this phenomenon in terms of a reduced temperature $T^* = T/T_{M \to I}$, where $T_{M \to I}$ is the transition temperature. A typical plot of T^* against x_1 is depicted in Fig. 2.12. The points (x_a, T_a^*) and (x_b, T_a^*) are interpreted in a

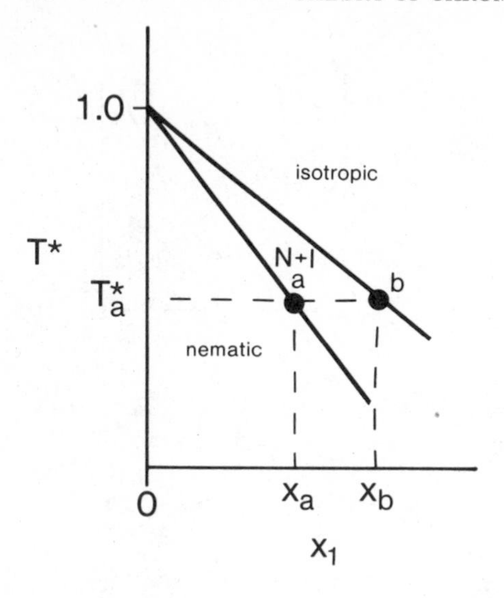

Figure 2.12. Plots of $T^*(= T/T_{M\to I})$ against x_1 for a nonmesomorphic solute 1 with a liquid-crystalline solvent.

manner analogous to those of Fig. 2.11 outside, along, and within the consolute-temperature curves. There also appears to be advantage in using a reduced temperature ordinate $T^* = T/T_{\text{consolute}}$ for liquids that are everywhere isotropic, in that many such curves can be placed on a common graph and variations between each then compared directly. Fig. 2.13 illustrates the plots of T^* against x for *n*-butanol/H_2O and phenol/H_2O.

Solid–liquid equilibria for which the liquid components are fully miscible and the solid components immiscible are described typically by Fig. 2.14*a*. The point *E* is called the eutectic point. Cooling the liquid mixture along the isopleth $aa'a''$ to point a' causes pure solid 1 to precipitate from the solution. At point a'', liquid solution is in equilibrium with solid 1 and solid 2; below this temperature the entire mixture freezes. In contrast, Fig. 2.14*b* presents the phase diagram for the situation in which the solid components are completely miscible. The upper and lower boundaries are called the liquidus (or freezing) and solidus (or melting) curves. Figure 2.14*c* illustrates the phase-boundary curves when species 1 and 2 form a solid complex. Six possible combinations of states of matter arise: liquid solution; solid 1 + solution; solid 2 + solution; solid complex + solution; solid complex + solid 1; and solid complex + solid 2. These diagrams are particularly useful for identifying complex formation as well as determining mole ratios of species within a complex. For that shown in (*c*), the ratio is one-to-one.

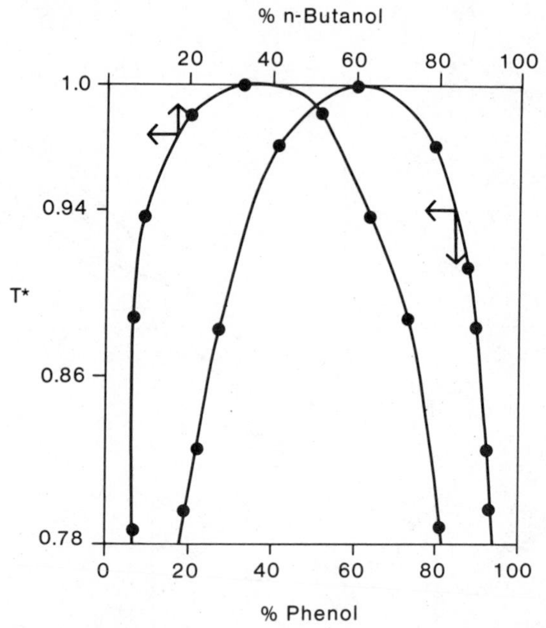

Figure 2.13. Reduced temperature $T^*(= T/T_{consolute})$ against composition for (left) *n*-butanol/H_2O and (right) phenol/H_2O.

Table 2.7 provides melting points and compositions for several metal-alloy systems that are of potential interest as stationary phases in gas chromatography.

2.2.3 Vaporization and Condensation of Liquids and Solids

Enthalpy of Vaporization and Heat Capacity

The change in the vapor pressure with temperature of a pure compound in equilibrium with its liquid (or solid in the case of sublimation) is given by

$$\frac{dp_1^0}{dT} = \frac{S_1^g - S_1^l}{V_1^g - V_1^l} \tag{33}$$

where S represents the entropy of the gas or liquid. At equilibrium the chemical potentials of the gas and liquid are equal:

$$u_1^{l,0} = u_1^{g,0} \tag{34}$$

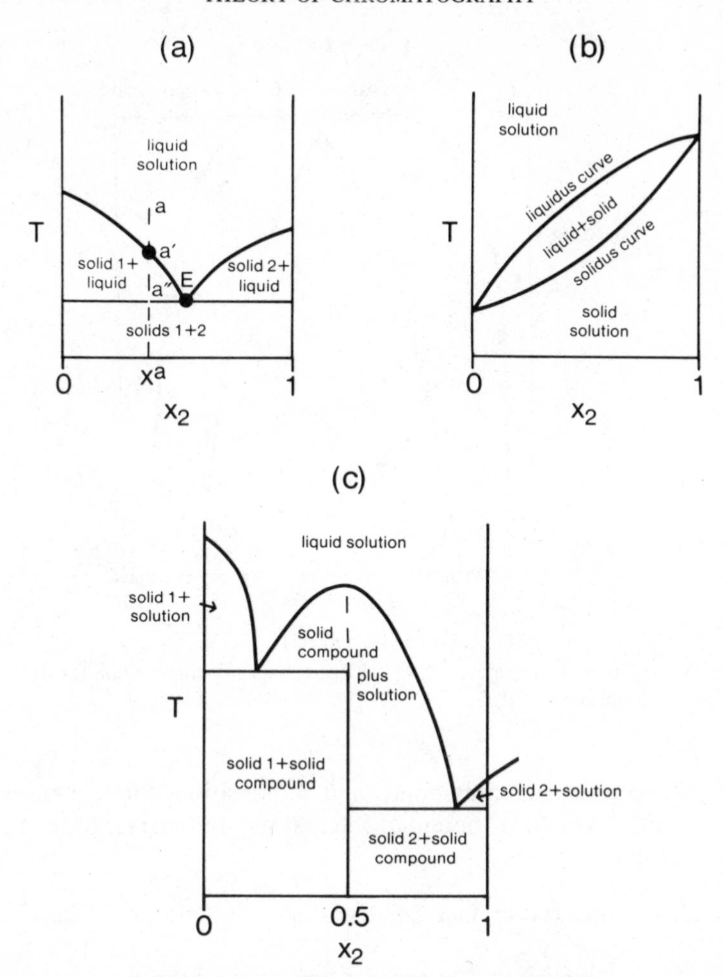

Figure 2.14. Phase diagrams for substances 1 and 2: (*a*) whose liquid forms are fully miscible but whose solid forms are immiscible; (*b*) fully miscible liquids and solids; (*c*) fully miscible liquids with solid-compound formation.

so that the change in entropy on passing from a liquid to a gas is related to the change in the enthalpy arising in the vaporizing process, $H^g - H^l$, via

$$\Delta S = \frac{\Delta H^v}{T} \tag{35}$$

Table 2.7 Melting Points of Pure Metals and Eutectic Alloys

Melting Point (°C)	% Composition (w/w)								
	Ga	In	Sn	Bi	Cd	Pb	Zn	Sb	Ag
29.78	100								
156.6		100							
231.88			100						
271.3				100					
320.9					100				
327.5						100			
419.4							100		
630.5								100	
960.8									100
3.0	61.0	25.0	13.0				1.0		
5.0	62.0	25.0	13.0						
13.0	67.0	29.0					4.0		
15.7	75.5	24.5							
20.0	92.0		8.0						
46.8		19.1	8.3	44.7	5.3	22.6			
58.0		21.0	12.0	49.0		18.0			
70.0			13.3	50.0	10.0	26.7			
91.5				51.6	8.2	40.2			
95.0			15.5	52.5		32.0			
102.5			26.0	54.0	20.0				
117.0		52.0	48.0						
124.0				55.5		44.5			
138.5			42.0	58.0					
142.0			51.2		18.2	30.6			
144.0				60.0	40.0				
177.0			67.8		32.2				
183.0			61.9			38.1			
199.0			91.0				9.0		
221.3			96.5						3.5
236.0					17.7	79.7		2.6	
247.0						87.0		13.0	
304.0						97.5			2.5

Substituting eq. 35 into eq. 33 produces

$$\frac{dp_1^0}{dT} = \frac{\Delta H^v}{T\Delta V_1} \tag{36}$$

where $\Delta V_1 = V_1^g - V_1^l$. Equation 36 is that of Clausius, and its derivation is brought out in full to point out the various approximations used in deriving a form of it that is very useful in gas chromatography, namely the Clausius–Clapeyron relation. We make the first approximation by discounting V_1^l as negligible in comparison with the volume of the gas:

$$\frac{dp_1^0}{dT} = \frac{\Delta H^v}{TV_1^g} \tag{37a}$$

Substituting the ideal gas relation for V_1^g next provides:

$$\frac{dp_1^0}{dT} = \frac{p_1^0 \Delta H^v}{RT^2} \tag{37b}$$

Rearranging eq. 37b to a form suitable for integration,

$$\frac{dp_1^0}{p_1^0} = \frac{\Delta H^v}{RT^2}\, dT \tag{37c}$$

In carrying out the integration we make the final approximation, that is, that ΔH^v is temperature-independent:

$$\ln p_1^0 = -\frac{\Delta H^v}{RT} + \text{const.} \tag{37d}$$

The first approximation, eq. 37a, appears to be valid so long as extraordinary accuracy for the vapor pressure is not of concern. For example, one mole of vapor of a substance ideally occupies 22.4 dm^3/mol at 0°C under 1 atm pressure. A reasonable value for the molar volume of its condensate is 0.2 dm^3/mol, which represents less than 0.9% of V_1^g. The second approximation, eq. 37b, is much less satisfactory and can lead to errors that approach 10%. Substituting instead the virial expression for p_1^0 in the form of eq. 8 yields

$$d \ln p_1^0 = \frac{\Delta H^v dT}{RT^2 + p_1^0 B_{11} T} \tag{38}$$

which brings vapor pressures calculated from heats of vaporization to well within experimental error.

The last approximation made in the derivation of eq. 37d is not so easily dealt with. The variation of the enthalpy of the gas with temperature is given by

$$\frac{dH^g}{dT} = C_p^g + \left(\frac{\partial H^g}{\partial p_1^0}\right)_T \left(\frac{\partial p_1^0}{\partial T}\right)_V = C_p^g + \gamma_V \left(\frac{\partial H^g}{\partial p_1^0}\right)_T \tag{39}$$

where γ_V is the isochoric thermal pressure coefficient defined earlier by eq. 20. The change in enthalpy with pressure at constant temperature is

$$\left(\frac{\partial H^g}{\partial p_1^0}\right)_T = -\left(\frac{\partial T}{\partial p_1^0}\right)_H \left(\frac{\partial H^g}{\partial T}\right)_p = -\left(\frac{\partial T}{\partial p_1^0}\right)_H C_p^g \tag{40}$$

The isenthalpic change of temperature with pressure is the Joule-Thompson coefficient μ_{JT}, so that

$$\frac{dH^g}{dT} = C_p^g(1 - \mu_{JT}\gamma_V) \tag{41}$$

Finally, the overall change of the heat of vaporization with temperature is given by

$$\frac{d(\Delta H^v)}{dT} = C_p^g(1 - \mu_{JT}\gamma_V) - C_p^l = \Delta C_p - \mu_{JT}\gamma_V C_p^g \tag{42}$$

Since the heat capacity of a liquid is very much larger than that of a gas, we see that, from this viewpoint, the enthalpy of vaporization appears to be invariant with temperature. We recall, however, that vapor pressures calculated with it will nevertheless be in error unless the virial correction given by eq. 38 is made.

Representation of Bulk Vapor Pressure Data

While the Clausius–Clapeyron relation (eqs. 37d or 38) is certainly appealing, others have frequently been cited as more convenient for purposes of data presentation. The most popular of these are relations of the Antoine type:

$$\log p_1^0 = A - \frac{B}{(t + C)} \tag{43}$$

Table 2.8(a) Comparison of Experimental with Calculated (Eq. 34) Vapor Pressure Data for 3-Methyl-1-Butene

T (°C)	$p°$ (torr)		
	Exptl.	Calc.	Dev.
0.218	355.25	355.25	0.00
5.112	433.53	433.54	0.01
10.053	525.86	525.86	0.00
15.033	633.94	633.94	0.00
20.061	760.00	759.99	−0.01
25.128	906.00	906.00	0.00
30.245	1074.60	1074.64	0.04
35.402	1268.10	1268.10	0.00
40.602	1489.20	1489.15	−0.05
45.847	1740.70	1740.72	0.02
51.139	2025.90	2025.93	0.03

From Boublík et al. (1973).
$A = 6.82643$; $B = 1013.605$; $C = 236.833$; $T_b = 20.061$°C

where A, B, and C are empirical constants and where t is in degrees Centigrade. Equation 43 corresponds exactly to eq. 37d when C is 273.15 and, more often often than not, C is in fact found to be less than 300 (the value for xenon has in contrast been reported to be 1651.838 in the region of its boiling point, − 108°C). Many tabulations of Antoine constants have appeared over the years, those by Boublík, Fried, and Hála (41) and Dreisbach (42), in addition to *The International Critical Tables*, being among the most frequently cited. Table 2.8(a) provides an example of the goodness of fit of eq. 43 to the vapor pressure data for 3-methyl-1-butene (41).

The Clausius–Clapeyron relation holds equally for the process of sublimation (where $\Delta\overline{H}^{v}$ is replaced by $\Delta\overline{H}^{s}$, a molar heat of sublimation); equations of the Antoine type hence can also be employed for solids, the vapor pressure data for iodine (43) being presented as an example in Table 2.8(b).

Equation 37d indicates that log p_1^0 will vary regularly with reciprocal temperature and in fact, such plots are invariably found to be linear over surprisingly large ranges of T. $\Delta\overline{H}^{v}$ (or $\Delta\overline{H}^{s}$) must therefore be very nearly constant as deduced above, and the several approximations made in the derivation of eqs. 37d and 38 either self-canceling or not seriously in error. We note as well that the heat of condensation must equal but be opposite in sign to that for vaporization (or sublimation).

Table 2.8(b) Comparison of Experimental with Calculated (Eq. 34) Vapor Pressure Data for Iodine

	$p°$ (torr)		
T (°C)	Exptl.	Calc.	Dev.
0.0	0.030	0.030	0.000
15.0	0.128	0.131	−0.003
30.0	0.467	0.469	−0.002
45.0	1.497	1.498	−0.001
60.0	4.291	4.285	0.006
75.0	11.18	11.21	−0.03
90.0	26.79	26.78	0.01

From Baxter and Grose (1915).
$A = 9.7522$; $B = 2863.54$; $C = 254.000$; $T_m = 113.5$°C.

Frequently, plots of boiling point against the number of repeating units are also linear for an homologous series of compounds, as are the Antoine constants. These results infer that the respective enthalpies of vaporization vary in a regular fashion, which implies in turn a relation between log p_1^0 and boiling point:

$$\log p_1^0 = k_6 t_b + k_7 \tag{44}$$

$$\log p_1^0 = k_8 n + k_9 \tag{45}$$

where k_i are empirical constants for a given series of related compounds, and n is the number of repeating units. The existence of such relations is often taken as evidence that intermolecular forces between neighboring molecules comprise simply the sum of interactions of their respective component segments. On this basis, the constants k_7, k_9 might be taken to represent the dominant form of interaction (hydrogen bonding, for example) while k_6, k_8 merely reflect a weighting of the numerical frequency of such interactions, these increasing as the number of homologous units in the series increases (44,45).

Variation of Bulk Vapor Pressure with External Applied Pressure

As noted above, the molar volume of the liquid is ignored in deriving the integrated form of the Clausius–Clapeyron relation, eq. 37d. We now consider inclusion of this term and the resultant effect of the external applied pressure on the vapor pressure of pure substances. The Gibbs free energy for a pure liquid is

$$dG_1^l = \overline{V}_1^l dP - S_1^l dt \tag{46}$$

where P is the external applied pressure and where $\overline{V}_1^l$ is the molar volume of bulk liquid. The analogous expression for the free energy of the vapor is

$$dG_1^g = \overline{V}_1^g dp_1^0 - S_1^g dT \tag{47}$$

Equations 46 and 47 may be equated if the system is at equilibrium so that, at constant temperature,

$$\frac{dp_1^0}{dP} = \frac{\overline{V}_1^l}{\overline{V}_1^g} \tag{48}$$

Substituting the ideal gas expression (!) for the volume of vapor, followed by rearrangement, yields

$$\frac{dp_1^0}{p_1^0} = \frac{\overline{V}_1^l dP}{RT} \tag{49}$$

Integrating between the limits P_2 and P_1,

$$\ln p_1^0(P_2) - \ln p_1^0(P_1) = \frac{\overline{V}_1^l}{RT}(P_2 - P_1) \tag{50}$$

We can now calculate the difference in vapor pressures for any given set of external pressures. Consider that of iodine [Table 2.8(b)] under its own pressure ($P_1 = p_1^0 = 11.18$ torr at 75°C; $\overline{V}_1^s = 51.48$ cm^3/mol) and at 2 and 50 atm applied pressure. At the latter, the vapor pressure of iodine is calculated to be 12.097 torr, while at 2 atm, a pressure encountered frequently in gas chromatography, $p_1^0 = 11.215$ torr. The relative vapor pressure of iodine under vacuum and at 2 atm external applied pressure is therefore 1.0031, that is, a difference of only 0.3%, which is considerably smaller than experimental errors that arise even with gas-chromatographic instrumentation and techniques of high precision and accuracy (46). However, no account was taken of virial effects in the derivation. These will result in an apparent increase of the (solute) vapor pressure with carriers other than hydrogen or helium, where the effects can be as much as 5% and which increase with increasing pressure. As a result, we can assume that the solute vapor pressure is invariant with external applied (column) pressure and that, in any event, what minor deviations arise will be negligible compared with virial interactions.

Kelvin Pressure

The vapor pressure of a solute dissolved in a solvent of concave surface geometry is less than that exerted by a plane surface. The relation that describes the vapor-pressure lowering is the Kelvin equation (47)

$$\frac{\Delta p_1}{p_1^0} = \frac{2\gamma \overline{V}_S}{rRT} \qquad (51)$$

where $\overline{V}_S$ is the molar volume of the solution (equal to that of the solvent at infinite dilution of solute), γ is the interfacial surface tension, and r is the radius of curvature of the solution surface. Conder and Young (48), among others, have discussed the effects in terms of solute liquid-gas partition coefficients and conclude that Kelvin vapor-pressure lowering will not have an appreciable effect upon K_R unless the support pore size is unusually small or the (liquid) stationary-phase loading particularly light. This can be expected to be true even in the case of polymeric stationary phases where $\overline{V}_S$ may in fact be very high, since if the support were wetted only poorly by stationary phase any change in p_1^0 due to the Kelvin effect would almost certainly be masked by gas-solid interfacial adsorption.

2.2.4 The Concepts of Activity and Fugacity

For a pure liquid 1 in equilibrium with its vapor at some fixed temperature, the chemical potentials of each phase are related by eq. 34:

$$u_1^{l,0} = u_1^{g,0} \qquad (34)$$

where the superscripts 0 indicate some chosen reference state, here the pure compound in equilibrium with its vapor. Rowlinson (49) has then argued that for component 1 of a two-component blend of liquids, each of which exerts a partial pressure corrected for non-ideal gas-phase effects,

$$u_1^g = u_1^{g,0} + RT \ln\left(\frac{y_1 P}{p_1^0}\right) + (P - p_1^0)B_{11} + 2PB_{12}^e y_2^2 \quad (52a)$$

$$u_1^l = u_1^{l,0} + RT \ln x_1 + (P - p_1^0)\overline{V}_1^0 \quad (52b)$$

where

$$B_{12}^e = B_{12} - \tfrac{1}{2}(B_{11} - B_{22}) \quad (53)$$

and where implicit in eq. 52 is the definition

$$f_1^0 = p_1^0 \exp\left[\frac{p_1^0(B_{11} - \overline{V}_1^0)}{RT}\right] \quad (54a)$$

The term f_1^0 is called the fugacity of the pure compound and as shown is related to the saturation vapor pressure via the second-interaction virial coefficient. The fugacity f for the substance at some pressure $p_1 \neq p_1^0$ is given by

$$f_1 = p_1 \exp\left[\frac{p_1(B_{11} - \overline{V}_1^0)}{RT}\right] \quad (54b)$$

Two new quantities, called the activity a and fugacity coefficient γ of the substance can now be defined:

$$a_1 = \frac{f_1}{f_1^0} \quad (55a)$$

$$f_1 = \gamma_1 p_1 \quad (55b)$$

where a approaches unity as the partial pressure approaches the saturation vapor pressure, while γ approaches unity as the gas approaches ideality.

The term B_{12}^e is referred to at times as the "excess" second virial coefficient (50), although it amounts in fact only to a convenient means of representation of the right-hand side of eq. 10.

The quotient $(y_1 P/p_1^0)$ can be considered as an "extent of reaction," where the reaction in this case is simply vaporization. The difference $(P - p_1^0)$, on the other hand, takes into account the change in molar volume of component 1 upon dilution with component 2.

The concepts of activity and fugacity are important ones, for they yield an exact description of the chemical potentials of solutes in mobile and stationary phases. Both are therefore utilized below in order to provide a coherent description of the chromatographic process.

2.2.5 Solubility of Gases in Liquids

Self-evidently, the solubility of solutes in liquids is of predominant importance in gas–liquid, liquid–liquid, and liquid–solid chromatography. In fact, as will be shown later, if it were possible to predict even approximately activity coefficients, the a priori forecasting of chromatographic separations would be reduced to a few simple manipulations with a hand calculator. Unfortunately, and despite occasional claims to the contrary, the ability to do so for species other than *n*-alkanes is still very far from realization even in theory. We proceed nevertheless to detail the activity, fugacity, and virial coefficients that come into play since in any case these lie at the heart of the chromatographic process and thereby indicate means whereby separations can be achieved, or, if already partially satisfactory, enhanced.

Partial and Total Vapor Pressures of Liquid Mixtures

Equating μ_1^g with μ_1^l, invoking eq. 52, and solving the result in terms of P provides

$$P = x_1 p_1^0 \exp\left[\frac{(P - p_1^0)(\overline{V}_1^0 - B_{11}) - 2PB_{12}^e y_2^2}{RT}\right] + x_2 p_2^0 \exp\left[\frac{(P - p_2^0)(\overline{V}_2^0 - B_{22}) - 2PB_{12}^e y_1^2}{RT}\right] \tag{56}$$

In the instance of negligible B_{12}^e, B_{11}, and B_{22}, eq. 56 reduces approximately to the more familiar form commonly celebrated as Raoult's law:

$$P = x_1 p_1^0 + x_2 p_2^0 \tag{57}$$

since the quotient $p_i \overline{V}_1^0/RT$ is likely to be small. We see that eq. 56 can also reduce to 57 if $B_{12}^e = 0$ and $B_{11} = B_{22}$. This is likely to be the case for closely related compounds such as ethylene bromide with propylene bromide, for which eq. 57 applies exactly. If it is found that $B_{12}^e = 0$ but that the pure-component virial coefficients are finite, eq. 56 then becomes (49):

$$P = x_1 p_1^0 + x_2 p_2^0 - x_1 x_2 (p_1^0 - p_2^0) \left[\frac{p_1^0(\overline{V}_1^0 - B_{11}) - p_2^0(\overline{V}_2^0 - B_{22})}{RT} \right] \quad (58)$$

which is tantamount to assuming that the cross virial coefficient B_{12} is given by the arithmetic average of the virial terms of the pure components. This is at considerable odds with experiment in most instances, and so the full relation, eq. 56, is to be recommended.

Representation of Partial Pressures: the Solute Activity Coefficient

We now rearrange eq. 57 to

$$P = x_1(p_1^0 - p_2^0) + p_2^0 \quad (59)$$

for which plots of P against x_1 (so-called Raoult's-law plots) are expected to be linear with ordinate intercepts of p_1^0 and p_2^0 at $x_2 = 0$ and $x_1 = 0$, respectively. (We note in passing that linearity will in fact be obtained only with fortuitous reduction of eq. 56 to 57 as indicated above, since it is hardly likely that the second-interaction virial coefficients of any of the compounds of interest here will be zero at any temperature relevant to gas chromatography.)

The first comprehensive test of eq. 59 was that by von Zawidski (51) at the turn of this Century. More often than not the relation was found not to hold and, more particularly, did not even approximate the behavior of solutes at infinite dilution in solvents, that is, the situation of immediate concern in the normal elution mode of GC. Plots typical of negative and positive deviations, the latter being by far the most common, are shown in Figs. 2.15–2.17.

However, eq. 59 is obeyed approximately near an axis of pure solvent, and also approaches linearity in the limit of infinite dilution of solute. This suggests for the latter a reformulation of the relation in terms of the slope of the curve tangent:

$$p_1 = (\text{const.})\, x_1 \quad (60)$$

where eq. 60 is called Henry's law. It is important to note that the reference state for the solution is thereby taken to be that of pure solute that

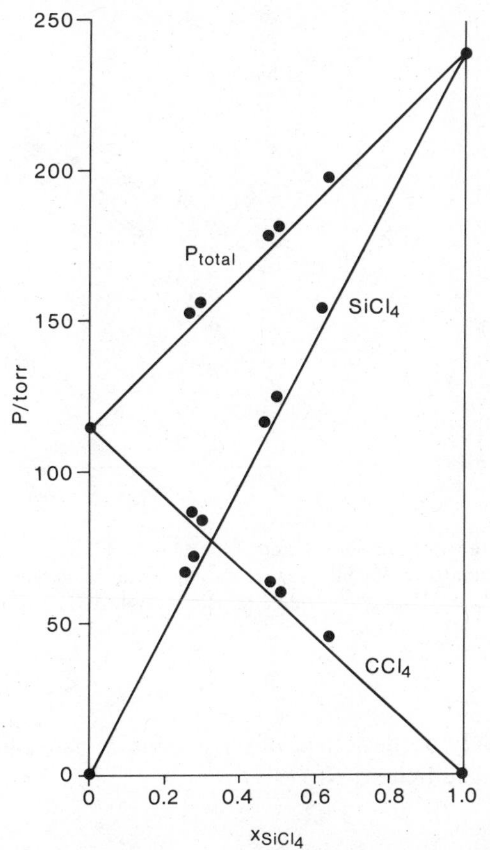

Figure 2.15. Plots of partial and total vapor pressures against mole fraction of silicon tetrachloride in admixture with carbon tetrachloride at 29°C. Straight lines have been drawn between the end-point pressures, where the mixture data [S.E. Wood, *J. Am. Chem. Soc.*, **59,** 1510 (1937)] deviate very slightly positively from Raoult's law.

(hypothetically) exhibits properties pertaining to infinite dilution in the solvent. Our choice of a reference state of pure solute in Sec. 2.2.4 is therefore internally consistent.

We now redefine Henry's law in the more usual terms of the mass m_1^S of solute gas dissolved in a volume V_S of solvent,

$$\frac{m_1^S}{V_S} = k_1 p_1 \tag{61}$$

We must therefore give the Henry's constant k_1 units of mass per unit volume per unit pressure, conveniently g/cm³/atm.

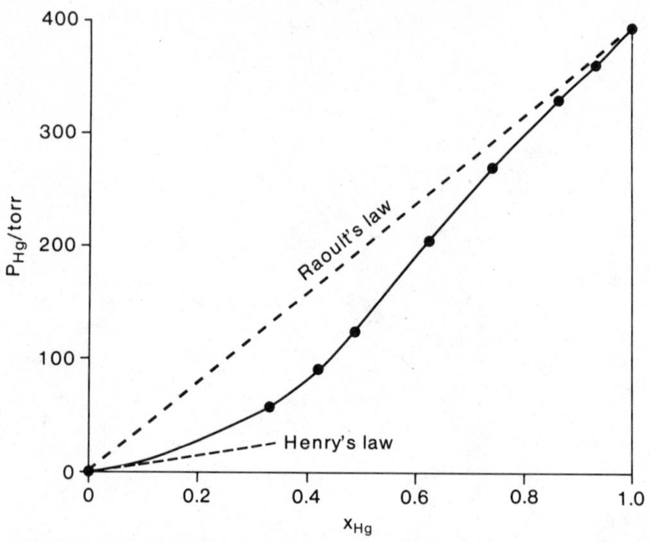

Figure 2.16. Plot of vapor pressure of mercury in admixture with cadmium at 322°C, showing negative deviation from Raoult's law. The tangent to the curve at low mole fraction represents Henry's law. Data taken from J.H. Hildebrand, A.H. Foster, and C.W. Beebe, *J. Am. Chem. Soc.*, **42,** 545 (1920).

Another widely used method of expression of gas solubility in liquids is the Ostwald coefficient, B:

$$B = \frac{V_1^S}{V_S} \tag{62}$$

where k_1 and B are evidently related by

$$B = \frac{k_1 RT}{M_1} \tag{63}$$

and where V_1^S is the volume of dissolved solute gas and M_1 is its molecular weight.

We now recast Raoult's law in terms of the solute partial pressure as a function of the mole fraction x_1 of solute dissolved in the solution:

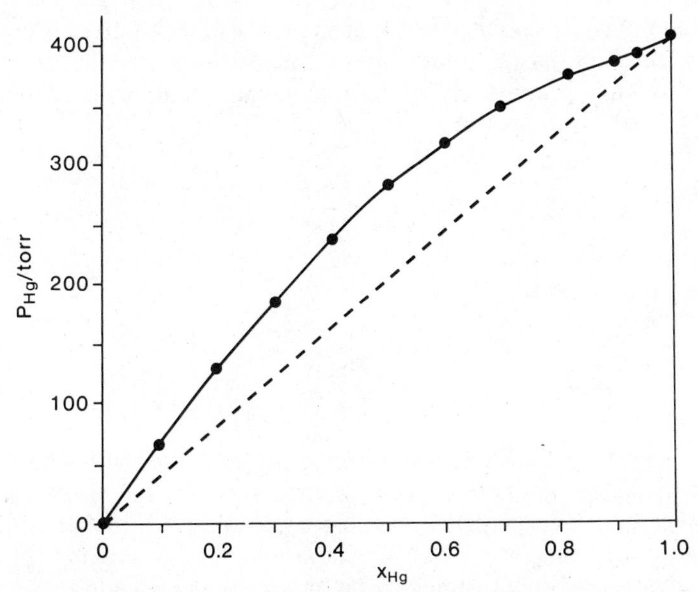

Figure 2.17. Plot of vapor pressure of mercury in admixture with tin at 324°C, showing positive deviation from Raoult's law. Data taken from J.H. Hildebrand, A.H. Foster, and C.W. Beebe, *J. Am. Chem. Soc.*, **42,** 545 (1920).

$$p_1 = \gamma_p^{\infty} p_1^0 x_1 \tag{64}$$

where γ_p^{∞} is the Raoult's-law solute activity coefficient at effective infinite dilution in the solvent, and where the product $\gamma_p^{\infty} p_1^0$ has been identified as the constant in eq. 60. It is thereby possible to establish a relationship between k_1 and γ_p^{∞}. We first rearrange eq. 64 and note that $C_1^S \overline{V}_S = x_1$ since the solute is at effective infinite dilution in the liquid phase:

$$\gamma_p^{\infty} = \frac{p_1}{C_1^S p_1^0 \overline{V}_S} \tag{65}$$

where C_1^S is the molar concentration of solute in the solvent. Utilizing next the identity $C_1^S = m_1^S / M_1 V_S$, where m_1^S is the weight of dissolved solute vapor,

$$\gamma_p^{\infty} = \frac{1}{k_1} \frac{M_1}{p_1^0 \overline{V}_S} \tag{66}$$

Purnell (45) has also pointed out that the Ostwald coefficient is equal to the concentration-based solute liquid–gas partition coefficient, $K_R = C_1^S/C_1^M$ (which is proportional directly to solute retentions in chromatography) so that, in the particular instance of GLC,

$$K_R = \frac{C_1^S}{C_1^M} = \frac{k_1 RT}{M_1} = B \tag{67}$$

whence we find

$$K_R = \frac{RT}{\gamma_p^\infty p_1^0 \overline{V}_S} \tag{68}$$

Thus, the Henry's constant and Raoult's-law activity coefficient are related to the liquid–gas Ostwald (i.e., partition) coefficient via eqs. 67 and 68; conversely, Henry's-law constants can always be converted to K_R values and vice versa. Moreover, the importance of the activity coefficient is now clear: if γ_p^∞ could somehow be predicted or estimated, the solute partition coefficient would immediately be calculable. In addition, we could then predict gas–liquid chromatographic separations, since the relative retention $\alpha_{i/j}$ of two solutes, the parameter of ultimate concern in analysis, is

$$\alpha_{i/j} = \frac{K_{R(i)}}{K_{R(j)}} = \frac{\gamma_p^\infty p_1^0 \text{ (solute } j)}{\gamma_p^\infty p_1^0 \text{ (solute } i)} \tag{69}$$

It is for this reason if no other (and given in particular the simplicity of eqs. 68 and 69), that the present-day inability to do so even to an approximate degree is so profoundly disappointing.

It is worth pointing out that the demand placed upon the accuracy of calculated values of γ_p^∞ is high. For example, the separation of two solutes exhibiting an alpha value of 1.01 represents at the present time the limit of capability of separation with commercially available GLC columns and instrument systems. Therefore, for compounds of equal vapor pressure, the prediction of activity coefficients must be accurate to no worse than $\pm 1\%$, a severe requirement indeed.

Because of the obvious importance of the matter, we briefly examine below the most popular approach currently being employed for the interpretation of activity coefficient data. However, before doing so, we must first take into account fugacity and virial effects in order to provide an exact description of γ_p^∞.

Fugacity and Virial Corrections to γ_p^∞

The full correction to be applied to K_R is derived by considering the chemical potentials of the solute in the gas and liquid phases. Without belaboring the mathematics (52), we set $u_1^g = u_1^l = 0$ to satisfy the general condition of equilibrium and then subtract $u_1^{l,0}$ (eq. 52b) from $u_1^{g,0}$ (eq. 52a). Substitution of the product $\gamma_p^\infty p_1^0 x_1$ for the solute partial pressure p_1 into the result thereby produces after suitable approximation and with 1 atm chosen as the standard-state pressure:

$$\ln K_R = \ln K_R^0 + \beta P \tag{70}$$

where P is the total pressure as before, and where β is given by

$$\beta = \frac{2B_{12} - \overline{V}_1^\infty}{RT} \tag{71}$$

$\overline{V}_1^\infty$ is the molar volume of the solute at infinite dilution in the solvent which is commonly replaced with negligible error by $\overline{V}_1^0$, the solute bulk molar volume at the system temperature. The term K_R^0 represents the solute partition coefficient for the hypothetical instance of zero system pressure drop. Recalling eq. 68, which relates K_R to γ_p^∞, we then arrive at an expression for the fully corrected solute infinite-dilution activity coefficient γ_1^∞:

$$\ln \gamma_1^\infty = \ln \gamma_p^\infty - \frac{p_1^0(B_{11} - \overline{V}_1^0)}{RT} + \frac{P(2B_{12} - \overline{V}_1^0)}{RT} \tag{72}$$

Making the stipulation that the pressure-based activity coefficient is that derived from the zero-pressure limit of the solute partition coefficient K_R^0,

$$\ln \gamma_1^\infty = \ln \gamma_p^\infty - \frac{p_1^0(B_{11} - \overline{V}_1^0)}{RT} \tag{73}$$

Equation 70 can be viewed as application essentially of a virial correction that takes account of the cross term B_{12}. Applying in turn eq. 73 then amounts to making the fugacity correction to γ_p^∞, for which we use the symbol γ_1^∞ to indicate that the correction has in fact been made.

There are two important consequences of eq. 70–73. First, we note that since B_{12} will be negative for all combinations of solutes and (gaseous) mobile phases of concern here, plots of K_R against P will be linear and of finite negative slope. Moreover, ln K_R cannot be invariant with pressure since B_{12} is negative while $\overline{V}_1^0$ must be positive. Even at some temperature presumably close to the Boyle temperature of the solute ($B_{11} = 0$) where B_{12} could conceivably be zero, the product $P\overline{V}_1^0/RT$ will not be, as pointed out some time ago by Cruickshank, Windsor, and Young (53).

Secondly, we note that given the magnitudes of the virial coefficients provided in Tables 2.1 and 2.2, there might well be advantage in using pure or blended carrier gases other than hydrogen or helium in analytical GLC. Mixed mobile phases have heretofore been employed only infrequently in gas chromatography; indeed, the variation of retentions (i.e., K_R) from one pure carrier to another has been regarded as more of a nuisance than anything else. Laub (54) has nevertheless shown recently that in fact, the use of blended mobile phases offers the prospect of considerable and quantitative control over GC retentions and should therefore not be overlooked as an alternative means of enhancing separations.

Mention is required at this point also of the potential effects of gas-phase solubility in the liquid solvent. Bearing in mind that gas–liquid chromatography comprises a ternary system of solute, solvent (liquid stationary phase), and carrier (gaseous mobile phase), it is conceivable that even at low pressures the latter may be somewhat soluble in the stationary liquid. This would then alter the extent of solubilization of solute, that is, change γ_1^∞ hence retentions. Cruickshank and his coworkers (55) were the first to consider suitable modifications to eq. 72 in order to take this possibility into account. Laub (54) has since argued that there is advantage in recasting the β term as follows:

$$\beta' = \frac{2B_{12} - \overline{V}_1^0}{RT} + \left(\frac{\partial x_2^S}{\partial x_2^M}\right)_\infty \left[1 - \left(\frac{\partial \Gamma_R^0}{d x_2^M}\right)_\infty\right] \tag{74}$$

where the subscript ∞ indicates infinite dilution of the solute. The first partial differential expresses the change in the carrier solubility with mole fraction in the gas phase, and is used instead of the parameter λ employed by Cruickshank and colleagues, where λ is the first coefficient of a power series expansion for the solubility of carrier in the stationary phase:

$$x_2 = \lambda p_2 + \phi p_2^2 + \mu p_2^3 + \ldots \tag{75}$$

The right-hand partial differential of eq. 74 represents the change in the solute sorption isotherm with carrier mole fraction in the liquid solution and is given the symbol Γ_R^0 to distinguish it from K_R and K_R^0. Either

Table 2.9 Corrected and Uncorrected Activity Coefficients for Metal Chloride Solutes[a]

Solute	Stationary Phase	T (°C)	p_1^0 (torr)[a]	γ_P^∞	γ_1^∞
$SbCl_5$	$NaFeCl_4$	470	444	2.6	2.5
$SnCl_4$	Squalane	423	2060	0.81	0.84
$TiCl_4$	Squalane	423	1064	0.55	0.56
$NbCl_5$	Squalane	473	221	0.31	0.30

[a] From Zado and Juvet (56).

differential can also be expanded as necessary or convenient in order to take into account the carrier pressure:

$$\left(\frac{\partial x_2^S}{\partial x_2^M}\right)_\infty = -\left(\frac{\partial x_2^S}{\partial p_2}\right)_{x_2^M}\left(\frac{\partial p_2}{\partial x_2^M}\right)_{x_2^S} \tag{76a}$$

$$\left(\frac{\partial \Gamma_R^0}{\partial x_2^M}\right)_\infty = -\left(\frac{\partial p_2}{\partial x_2^M}\right)_{\Gamma_R^0}\left(\frac{\partial \Gamma_R^0}{\partial p_2}\right)_{x_2^M} \tag{76b}$$

Equations 74 and 76 can in addition be modified with the appropriate units to account for surface-adsorbed carrier (54), and will also be utilized below to describe retentions in liquid chromatography. Meanwhile, we see that the relations reduce immediately to eq. 71 when the carrier is insoluble in the stationary phase.

The importance of activity coefficients cannot be underestimated. For example, suppose that one were faced with the separation of two organometallic solutes that are closely related in chemical structure and vapor pressure, yet where one contains an olefin bond while the other does not. An appropriate stationary phase would then be one that interacts selectively with double bonds, for example, a charge-transfer reagent such as Ag(I) or Tl(I) in polyethylene glycol. The activity coefficient of the olefin-containing solute would thereby be decreased relative to that of the other, and the alpha ratio (cf. eq. 69) hence would show an increase (improved separation). The activity coefficient ratio, along with that of the solute vapor pressures, thus governs entirely the *selectivity* of gas–liquid chromatographic systems.

The values of γ_P^∞ and γ_1^∞ reported by Zado and Juvet (7,56) for several metal chloride solutes are presented in Table 2.9. The differences are obviously not large for the compounds listed but may well prove to be so for other systems, particularly with the more exotic mobile phases described by Laub (54) in combination with stationary phases comprising the metal alloys listed in Table 2.7.

Interpretation of Activity Coefficient Data

The qualitative view of activity coefficients provides that, for values of $\gamma_1^\infty > 1$, the solute and solvent interact unfavorably such that the latter tends to "push" the former out of solution. Thus, the partial vapor pressure of the solute above the solution is higher than would otherwise be calculated from the form of Raoult's law given by eq. 59. The activity-coefficient multiplier must therefore be greater than unity in order for the right side of the relation to conform to the left, that is, eq. 64. Conversely, for instances in which $\gamma_1^\infty < 1$, the solute and solvent interact favorably, that is, solution is encouraged. γ_1^∞ must therefore be less than unity in order that eq. 59 be obeyed. Henry's and Raoult's laws become equivalent in those situations where $\gamma_1^\infty = 1$, and the respective partial vapor pressures of solute and solvent correspond precisely to those calculated directly from the products of the individual bulk vapor pressures and mole-fractional compositions.

A more quantitative view of activity coefficients is derived by consideration of the thermodynamic quantities pertaining to mixing. Since the solute is at infinite dilution in the solvent, the excess Gibbs free energy of solution is

$$G^e = RT \ln \gamma_1^\infty \tag{77}$$

(The absolute excess free energy, rather than a difference in excess free energies, is the appropriate term, since the reference-state excess is zero.) Two situations commonly arise with regard to the excess enthalpy and entropy of solution, where

$$G^e = H^e - TS^e \tag{78}$$

When the excess enthalpy H^e is approximately zero, eqs. 77 and 78 provide

$$\ln \gamma_1^\infty \text{ (athermal)} = -\frac{S^e}{R} \tag{79}$$

where the activity coefficient is determined solely by the entropy of mixing, that is, the so-called athermal activity coefficient. These will be temperature-invariant.

Solutions for which eq. 79 holds are termed athermal, and are encountered only infrequently in gas–liquid chromatography. This may be somewhat surprising, since solvents pertinent to GLC are necessarily of negligible volatility when compared with solutes, which usually dictates that the molecular weight of the former exceed the latter by a considerable degree. Polydimethylsiloxane stationary phases (M_2 on the order of 10^5 daltons) offer an example (57), while Laub, Martire, and Purnell (58) have presented and discussed the situations arising with *n*-alkane solvents.

Considerable efforts have been spent over the years on the theoretical evaluation of athermal solutions, since the seemingly simple situation is presented where deviations from ideal solution behavior arise as a consequence solely of disparities in molecular size. One of the more common treatments assumes that the atoms of solute and solvent molecules occupy the sites of a lattice, where interactions occur in regions where the lattice-points of each are adjacent. This is the so-called point-contact model, and serves well for systems of small noninteractive solutes + large (polymer) inert solvents. A refinement is to consider that whole segments of molecules, rather than individual atoms, lie at and interact between lattice sites. This is the segment-contact model of athermal solutions; the interested reader is referred to the excellent description of lattice models offered by Littlewood (59), and to ref. 58 for further quantitative details.

Even less likely to be encountered in GLC are so-called thermal solutions, for which H^e is finite but S^e is zero. The activity coefficient in these instances is given by

$$\ln \gamma_1^\infty \text{ (thermal)} = \frac{H^e}{RT} \tag{80}$$

and is referred to as the thermal activity coefficient. The excess enthalpy can be positive (heat absorbed) or negative (heat evolved from the system). The former corresponds to activity coefficients larger than unity and the same reasoning regarding poor mixing applies as used above. The latter situation, H^e negative, indicates that solution is favored in excess of random dispersion and that the solute and solvent must therefore interact, however fleetingly, in some specific fashion, for example, dipole–dipole interaction, hydrogen-bonding, charge-transfer, and so forth.

By far the most common situation in gas–liquid chromatography is that neither the athermal nor the thermal activity coefficients are negligible, and so we write a summation of the two:

$$\ln \gamma_1^\infty = \ln \gamma_1^\infty \text{ (athermal)} + \ln \gamma_1^\infty \text{ (thermal)} \tag{81}$$

Recent refinements of eq. 81 by, among others, Patterson and his coworkers (60) include further subdivision of the athermal contribution to γ_1^∞ into combinatorial (comb.), free-volume (f.v.), and order/disorder (o./d.) activity coefficients.

The first of these is commonly assumed to be given by an expression involving only the solute and solvent molar volume ratio:

$$\ln \gamma_1^\infty \text{ (comb.)} = \ln \frac{1}{r} + \left(1 - \frac{1}{r}\right) \tag{82}$$

where $r = \overline{V}_S/\overline{V}_1$. Its interpretation follows closely the lattice models mentioned above.

The free-volume contribution is said to arise from actual contraction or expansion of solvent in the immediate vicinity of dissolved solute: the free space between molecules of like kind (that is, in bulk) will most likely be different from that extant when the molecules are inserted into a different environment (i.e., into solution). Hence, upon mixing solute with solvent, the resultant volume of the solution may be enlarged or condensed over that predicted from the relation

$$\overline{V}_{1+S} = x_1\overline{V}_1 + x_S\overline{V}_S \tag{83}$$

Nor is the contribution necessarily a trivial one, since the concentration of solute may be high within the length of a GC column over which the solute band is spread. That is, x_1 and x_S refer only to the column region occupied by the solute, not the entire column.

The order/disorder contribution to γ_1^∞ can be of two types. Consider first a solvent (e.g., a long-chain *n*-alkane) within which some ordering persists. Placing a bulky solute into solution causes the local order to be somewhat disrupted which in turn results in an increase in the entropy and a corresponding decrease in the free energy. Anomalous as it may first appear, solution is therefore favored in this instance. In contrast, suppose that a long-chain alkane solvent contains a butyl branch at its midpoint. Since there is free rotation about all skeleton carbons, the positional loci described by the butyl end-carbon will approximate a sphere. However, this free rotation will be hindered if another molecule is brought into the vicinity. The reduction in rotation amounts to an increase in the system order, a decrease in the entropy, and an increase in the free energy; anomalous as it again seems, solution in this case is therefore disfavored.

It must be borne in mind that enthalpic contributions often accompany the entropies associated with free-volume and order/disorder effects. The distinction between thermal and athermal activity coefficients is therefore somewhat blurred. (As a result, it might in fact be suggested that the terms, however convenient, no longer merit consideration.) Thus, the overall free energy may not follow the trends predicted solely on the basis of molecular size or geometry (60). Moreover, reliable data are available thus far only for mixtures of alkanes, and it remains to be seen what role these factors play in determining experimental activity coefficients for systems comprised of moderate to strong specific interactions.

An alternative interpretation of the above formulation distinguishes only the combinatorial contribution to the observed activity coefficient, the remainder being taken as the "energetic" contribution:

$$\ln \gamma_1^\infty \text{ (energetic)} = \ln \gamma_1^\infty \text{ (obs.)} - \ln \gamma_1^\infty \text{ (comb.)} = \chi \tag{84}$$

where χ is called the interaction parameter (61). (It can, of course, be argued that the free-volume and order/disorder contributions are energetic interactions also, since as noted above each may give rise to an excess enthalpy.) The brunt of interpretation of activity coefficients today is placed upon interaction parameters, since prediction of these immediately provides the activity coefficient, hence (and of particular importance in GLC) partition coefficients and retentions. To date, however, the results in either case amount at best only to a qualitative description of solutions, the most notable failure being solubility-parameter theory (cf. ref. 1, Ch. 5). Nevertheless, eq. 84 provides at least some insight into the energetics of solutions, as well as a convenient means of data presentation in the form of tabulations of values of chi.

Surface Excess Concentration

Of compelling interest in many branches of science is the sorption of solutes at surfaces, be these from the gas phase onto a liquid or solid, or from a liquid phase onto the interface with a second immiscible liquid or onto a solid. For example, the transport properties of nutrients in systems relevant to molecular biology are now known to depend largely upon the surface features of cell and organelle boundaries extant at the interface with the surrounding medium. Broadly speaking, it is in fact true to say also that, irrespective of whether the solvent is aqueous or organic, solutions of electrolytes comprising at least some organic character are invariably inhomogeneous, that is, a concentration gradient of solute exists on passing from bulk liquid to its surface. Moreover, plots of solute concentration against depth from the surface are in some instances described very nearly by a step function: virtually all of the solute is concentrated at the surface and the concentration gradient verges on infinite. The matter is also relevant to chromatography and, in particular, liquid–solid chromatography (retentions there being dictated almost entirely by the support surface and associated mobile-phase constituents sorbed on it).

We examine at this point relations that describe the situation pertinent to GLC, namely, the solution of (gaseous) solutes in liquid solvents, taking into account in addition the potential for adsorption at the liquid–gas interface. The treatment is completely general, however, and so will evidently pertain also to gas–solid, liquid–liquid, and liquid–solid chromatography (62).

Consider that a certain gaseous solute is adsorbed at the liquid–gas interface as well as dissolved in bulk liquid. Since the surface is presumed to have no volume (i.e., is taken to be infinitely thin), the concentration of solute at the surface over and above that from the bulk liquid (the "excess" surface concentration), Γ_1, is given units of moles per unit area. Dividing Γ_1 by the molar concentration of solute in the gas phase gives the solute liquid–surface/gas partition coefficient:

$$K_I^0 = \frac{\Gamma_1}{C_1^M} \tag{85}$$

where a superscript zero is added to indicate full application of the fugacity and virial corrections. The overall partition coefficient is now the sum of two terms, which Laub (54) chose to represent with the symbol Γ_R^0:

$$\Gamma_R^0 = K_R^0 + K_I^0 \tag{86}$$

where K_R^0 is retained to indicate the bulk-liquid/gas partition coefficient.

The surface excess concentration of solute can also be described by the Gibbs sorption equation (62):

$$\Gamma_1 = -\frac{1}{RT}\frac{d\gamma}{d \ln a_1} \tag{87}$$

where a_1 is the bulk-liquid solute activity and γ is the liquid surface tension. Replacing the former with the product $\gamma_1^\infty x_1$, where γ_1^∞ is the bulk-liquid solute activity coefficient at infinite dilution, and noting that $d \ln \gamma_1^\infty / dx_1 \rightarrow 0$ as $x_1 \rightarrow 0$,

$$\Gamma_1 = -\frac{x_1}{RT}\left(\frac{\partial \gamma}{\partial x_1}\right)_\infty \tag{88}$$

It can be shown that

$$\frac{K_I^0}{K_R^0} = -\frac{RT\rho_S}{M_S}\left(\frac{\partial\gamma}{\partial x_1}\right)_\infty \tag{89a}$$

or, inserting eq. 73 and solving for K_I^0,

$$K_I^0 = -\frac{1}{\gamma_1^\infty f_1^0}\left(\frac{\partial\gamma}{dx_1}\right)_\infty \tag{89b}$$

Equations 89a and 89b provide two alternative methods of measurement of Γ_1, hence K_I^0. Martin (63) used essentially the former in 1963, and determined in separate experiments K_R^0 (gas chromatography) and the right-hand side of eq. 89a (tensiometry). He was then able to calculate K_I^0. Martire, Pecsok, and Purnell (64) employed the latter relation the following year, and measured $(\partial\gamma/\partial x_1)$ and γ_1^∞ with a McBain balance and a du Noüy tensiometer, thence K_I^0 directly. The results from each were mutually consistent so long as the measured surface area used in the calculations corresponded to that available to the solute (65).

The amount of solute vapor sorbed at the surface of a liquid relative to that dissolved in its bulk, as noted above, can be very substantial indeed. Martire (64) has provided tabulations of what few data are available and, as might be presumed (although the notion was at one time quite controversial), "polar" solutes tend to sorb onto (rather than dissolve in) "nonpolar" solvents as well as the converse. Unfortunately, data have yet to be provided for inorganic and organometallic solutes with various liquid solvents, although it can be presumed from the GLC peak shapes often found with such systems that surface adsorption does in fact play a very prominent role in their retentions. Moreover, and irrespective of the type of solute, this is certain to be the case when the metal alloys of Table 2.7 are employed as GLC stationary phases.

Blended Solvents

Given the lack of quantitative descriptions of solutions comprising only two components (solute + solvent), the problems inherent with multicomponent systems would seem to be entirely intractable. There is nevertheless merit in considering these, since multicomponent solvents can be used to adjust in a quantitative way the selectivity of chromatographic systems.

Rowlinson and Swinton (ref. 12, Ch. 5) have reviewed the various methods of representation of solute activity coefficients as a function of blended-solvent composition. The most useful of these is that provided by Martire and his colleagues (66,67) where, for binary mixtures of solvents B and C,

$$\ln \gamma_{1(M)}^{\infty} = \ln \frac{\overline{V}_1}{\overline{V}_M} + \left(1 - \frac{\overline{V}_1}{\overline{V}_M}\right) + \chi_{1(B)} + (\chi_{1(C)} - \chi_{1(B)})\,\phi_C - \frac{\overline{V}_1}{\overline{V}_C}\,\chi_{C(B)}\phi_C(1 - \phi_C) \tag{90}$$

where ϕ represents a volume fraction and $\overline{V}_M$ is the molar volume of the liquid solution (care should be taken not to confuse this with V_M, the volume of mobile phase). The pure-solvent interaction parameters $\chi_{1(i)}$ ($i = B$ or C) are derived from eq. 84, following which the solvent–solvent interaction parameter $\chi_{C(B)}$ is calculated with eq. 90.

It is straightforward to recast the relation in terms of partition coefficients:

$$\ln K_{R(M)}^{0} = \ln K_{R(B)}^{0} + \left[\left(\frac{\overline{V}_1}{\overline{V}_C} - \frac{\overline{V}_1}{\overline{V}_B}\right) + \left(\chi_{1(B)} - \chi_{1(C)} + \frac{\overline{V}_1}{\overline{V}_C}\chi_{C(B)}\right)\right]\phi_C - \frac{\overline{V}_1}{\overline{V}_C}\,\chi_{C(B)}\phi_C^2 \tag{91}$$

Alternatively,

$$\ln K_{R(M)}^{0} = \phi_B \ln K_{R(B)}^{0} + \phi_C \ln K_{R(C)}^{0} + \frac{\overline{V}_1}{\overline{V}_C}\,\chi_{C(B)}\phi_B\phi_C \tag{92}$$

The importance of these formidable-looking expressions lies in their prediction that neither $\ln \gamma_{1(M)}^{\infty}$ nor $\ln K_{R(M)}^{0}$ will vary linearly as ϕ_C ($=1 - \phi_B$). However, considerable controversy has arisen over this point following the work of Purnell and his colleagues, Vargas de Andrade (68) and Laub (69). Moreover, the matter is far from merely academic since, if the pure-phase solute partition coefficients were known or could be measured, those arising with mixed phases could then be predicted and alpha values, hence separations, could in turn be calculated directly with eq. 69. The selectivity spectrum available to the analyst would thereby be expanded many fold. Thus, the prospect is presented of quantitative adjustment of separations by alteration of the composition of blended phases. The traditional alternative, of course, has been (and remains) random trial-and-error utilization of many individual pure phases, guidelines for choices of which, in the absence of reliable methods of prediction of activity coefficients, are virtually nonexistent beyond the personal experience of the analyst. Therefore, because the matter bears directly both upon analytical liquid and gas chromatography, we explore it here in some detail.

Consider the Raoult's-law (eq. 64) partial pressure of a solute in the three solvents B, C, and $B + C$ (64):

$$\frac{p_1}{p_1^0} = \gamma_{1(B)}^{\infty} x_1^B = \gamma_{1(C)}^{\infty} x_1^C = \gamma_{1(M)}^{\infty} x_1^M \tag{93}$$

where as before $M = B + C$, and where the appropriate superscripts and subscripts have been used to identify the pertinent solvent. Since the solute is at infinite dilution in each solvent, the pure-phase mole-fraction terms are given by

$$x_1^B = \frac{n_1^B}{n_B} \tag{94}$$

$$x_1^C = \frac{n_1^C}{n_C} \tag{95}$$

We next assume that the solvent blend M corresponds in all ways to complete ideality. Thus

$$x_1^M = \frac{n_1^M}{n_B + n_C} = \frac{n_1^B + n_1^C}{n_B + n_C} \tag{96}$$

Next, from eqs. 94–96,

$$\frac{1}{\gamma_{1(M)}^{\infty}} = x_1^M \frac{p_1}{p_1^0} = \frac{p_1}{p_1^0}\left(\frac{n_1^B}{n_B + n_C} + \frac{n_1^C}{n_B + n_C}\right) \tag{97}$$

Recognizing that $n_1^i = n_i x_1^i$, followed by substitution of the appropriate terms for x_1^i from eqs. 94 and 95 into 97, thereby yields an expression for $\gamma_{1(M)}^{\infty}$ with mixed solvents:

$$\frac{1}{\gamma_{1(M)}^{\infty}} = \frac{x_B}{\gamma_{1(B)}^{\infty}} + \frac{x_C}{\gamma_{1(C)}^{\infty}} \tag{98}$$

where x_i indicates the mole fraction of the ith solvent in the solution. Equation 98 can also be cast in terms of partition coefficients by substitution from eq. 68:

$$K^0_{R(M)} = \phi_B K^0_{R(B)} + \phi_C K^0_{R(C)} \tag{99}$$

where ϕ_i indicates a volume fraction (70).

As it happens, this relation corresponds also to the instance where B and C are completely immiscible. It is as if two separate determinations of solute liquid-gas partitioning were carried out with beakers of the pure solvents (assumed to have negligible vapor pressure) standing side by side, each having simultaneous access to the same volume of solute vapor. The total mole fraction of dissolved solute gas would then be taken as the sum of moles of solute found in each beaker, divided by the total number of moles of the two solvents, that is., eq. 96 thence 98 and 99.

The surprising feature of eqs. 98 and 99 is that they describe nearly exactly a very large number of solute/mixed-solvent systems (such as diethyl maleate/quinoline) for which it is difficult to imagine that B and C do indeed form ideal solutions. Nor was macroscopic phase separation observed with any of the several hundred systems shown to conform to the relations by Laub, Purnell, and their coworkers (68,69,71). Moreover, and despite predictions to the contrary (72), the equations were also obeyed to well within experimental error when tested against intimately blended solvents that do exhibit macroscopic immiscibility (73), these including polymer systems (74,75) and mixed-bed GLC packings (76). Laub and Purnell (69) hence christened all solutions described by eqs. 98 and 99 as diachoric, since the relations make no distinction between immiscibility and ideality.

It has also been shown (77) that eq. 99 can be expanded to the form

$$K^0_{R(M)} = \sum_{i=1}^{n} \phi_i K^0_{R(i)} \tag{100}$$

which holds for n at least 5. In addition, one (or more) of the pure-phase partition coefficients may comprise sorption onto the liquid surface(s) in addition to exhibiting complexation or other specific interactions. Thus the summation is completely general (78) and makes no attempt to identify the source of the $K^0_{R(i)}$; rather, emphasis is placed upon utilizing them to predict a volume-fraction weighted average, $K^0_{R(M)}$.

However, the diachoric solutions hypothesis and relations, as written, are by no means consistently obeyed. For example, the solvent system squalane/dinonyl phthalate gives solute activity and partition coefficients that deviate from eq. 99 by upward of 9% at a volume fraction of 0.5 (67). Moreover, the latter data were described to well within experimental error by eqs. 86 and 87, albeit with judicious choice of $\chi_{C(B)}$. On the other hand, when $\chi_{C(B)}$ is set to zero, eq. 87 reduces to

$$\ln K^0_{R(M)} = \phi_B \ln K^0_{R(B)} + \phi_C \ln K^0_{R(C)} \tag{101}$$

That is, plots of $K^0_{R(M)}$ against volume fraction are predicted to be curved and it therefore fails to account for the very many systems that in fact regress linearly. Laub, Martire, and Purnell (58) found in subsequent studies of *n*-alkane solutes with blended *n*-alkane solvents that eqs. 99 and 101 were indistinguishable (all $\chi_{i(j)}$ set to zero), while Laub and Wellington (71), and Laub (79), in an extension of earlier work (67), showed that incorporation of terms into the former to take account of solvent–solvent dimerization and solute–solvent specific interaction leads to plots of K^0_R against ϕ that can be curved positively or negatively, depending upon the interactions as well as the relative magnitude of the pure-phase partition coefficients.

Generally speaking, the situation remains at this point at the present time: the prediction of solute activity and partition coefficients is at least approximately possible with mixed solvents, and is without question notably more advanced than is the case with pure solvents. However, there remains some disagreement about appropriate expressions with which to do so, that is, eq. 92 vs. eq. 99. Moreover, even the most carefully designed of experiments has thus far failed to resolve the controversy. Nevertheless, we encourage further and comprehensive work in this area since, as shown later, the use of mixed solvents places in the hands of the analyst an extraordinarily powerful *quantitative* means of adjustment of system selectivity.

Partially Miscible Solvents

Partial (or complete) immiscibility introduces essentially a third component into two-component solvent mixtures, namely, the liquid–liquid interface. It is usually considered in terms of area, however, since only infrequently is the interface more than a few molecules thick. Nevertheless, solvent molecules in this region are generally ordered to a very high degree, much like the liquid-crystalline state (80). Phase diagrams for the interface may therefore resemble those typified in Fig. 2.12. The importance of this lies in the manner in which the GLC experiment is carried out. Stationary phase is dispersed as a thin film onto particles of solid support in packed-column gas chromatography. And, while it is generally true that effects due to the radius of curvature of the film can be ignored, the local concentration of solute, as noted above, may well be high. As a result, dissolved solute may cause local disruption in the ordering of the interface, which may in turn give rise to a local shift in the conjugate-solution compositions. In such instances, retentions would be dependent upon the amount of solute injected (81). On the other hand, the degree of disruption of the interface will in all probability vary from one solute to the next, which portends the enhancement of separations as a function of solute concentration. Little else can be said at this time, however, since there are no corroborative data for organic solutes, let alone inorganic or organometallic species.

Polymer Solvents

In order that the amount of stationary phase remain constant in a gas-chromatographic system, the solvent must be of negligible volatility. A general rule-of-thumb requirement for its vapor pressure is in fact ~0.01 torr or less. Few liquids of molecular weight 200 daltons or so meet this criterion at temperatures of in excess of 100°C, and so polymeric stationary phases have over the years been preferred to small-molecule solvents. This is particularly true for open-tubular column systems, where the ratio of stationary/mobile-phase volumes is small. However, eq. 68 requires knowledge of the stationary-phase molar volume in order to assess partition coefficients. Of equal importance, a rearranged form of the relation indicates that as this quantity becomes large, the activity coefficient tends to zero:

$$\gamma_P^\infty = \frac{RT}{K_R p_1^0 \overline{V}_S} \tag{102}$$

The difficulty of measuring densities of polymer solvents at high temperatures, while not impossible, is overcome by redefinition of the partition coefficient in terms of the mass m_S of stationary phase

$$V_g^T = \frac{RT}{\gamma_P^\infty p_1^0 M_S} \tag{103}$$

where V_g^T is referred to as the "specific retention volume" of the solute at the system temperature, a value directly calculable from gas-chromatographic data, and where M_S is the solvent molecular weight. A more convenient parameter is V_g^0, the "specific retention volume" corrected to zero degrees C:

$$V_g^0 = V_g^T \frac{273}{T} \tag{104}$$

In either case, the quantities are merely new forms of the equilibrium partition coefficient K_R that obviate determination of the stationary-phase molar volume.

Nevertheless, there remains the difficulty of the stationary-phase molecular weight. Frequently, a number- (hence, mole-) average value is available for use with eqs. 103 and 104; however, the activity coefficient will still tend to zero as this becomes large. To circumvent the problem, we must at this point redefine the activity coefficient (82). Substituting eq. 54 into 55a provides

$$a_1 = \frac{f_1}{f_1^0} \approx \frac{p_1}{p_1^0} \tag{105}$$

Recalling next Raoult's law, eq. 64,

$$a_1 = \gamma_P^\infty x_1 \tag{106}$$

Accordingly, we may write

$$a_1 = {}^x\gamma_1^\infty x_1 = {}^w\gamma_1^\infty w_1 = {}^\phi\gamma_1^\infty \phi_1 = \ldots \tag{107}$$

where the symbols represent mole-, weight-, and volume-fraction-based units. That is, the solute activity (hence activity coefficient) may be defined in any suitable units. The following relations are useful in converting activity coefficients from one compositional base to another:

$${}^w\gamma_1^\infty = {}^x\gamma_1^\infty \frac{M_S}{M_1} \tag{108a}$$

$${}^w\gamma_1^\infty = {}^\phi\gamma_1^\infty \frac{\rho_s}{\rho_1} \tag{108b}$$

Those of weight fraction are convenient in the present instance, whence eq. 104 becomes

$$V_g^0 = \frac{273\,R}{{}^w\gamma_1^\infty p_1^0 M_1}$$

where M_1 is the solute molecular weight. Thus, V_g^0 and ${}^w\gamma_1^\infty$ asymptotically approach lower and upper limits respectively as the stationary-phase molecular weight is increased to infinity.

We note in passing that gas-chromatographic derived activity coefficients with polymer solvents were thought at one time to be at variance with those arising with static (i.e., nonchromatographic) apparatus and techniques (83); this has since been disproved (84).

2.2.6 Adsorption at the Gas-Solid Interface

Historically, the study of gas–solid adsorption parallels to a large degree the development of all of chemistry. For example, it has been recognized for at least two hundred years that a porous solid can sorb gases in amounts several times its own volume. Furthermore, adsorption phenomena would seem, at least in principle, to be amenable to exact description, since the only apparent interactions are those of the adsorbate with the adsorbent surface, that is (and in contrast to the situation with liquid solutions), in two dimensions only. This has been borne out over the years to a large degree, and it is in fact fair to say that the *process* of adsorption, while perhaps not fully understood, is nevertheless well characterized (85–89).

Forces of Adsorption

In the simplest instance of two atoms isolated in space yet near enough to each other to interact, the potential energy of attraction $\epsilon(r)$ arising as a result of their proximity is given by London's relation:

$$\epsilon(r) = -Cr^{-6}$$

where r is the distance between the atoms and where C is a constant. To this must be added a second term that reflects repulsive forces arising when the electron clouds of the two atoms become intermingled, which takes the form Br^{-12}. The total potential energy is then given by

$$\epsilon(r) = -Cr^{-6} + Br^{-12}$$

which is often referred to as the Lennard-Jones potential.

In the particular case of a gaseous atom or molecule interacting with many ions or molecules comprising an adsorbent, the individual interactions must be summed to obtain the potential energy of attraction:

$$\phi(z) = \sum \epsilon_{ij}(r_{ij}) \tag{109}$$

where subscripts i and j have been added to distinguish different kinds of atoms or molecules, and where z indicates that the distance being measured extends from the center of the adsorbate to a hypothetical plane (i.e., surface) formed by the locus of centers of the adsorbent atoms.

Specification of particular adsorbate-adsorbent interactions generally is carried out in terms of ϕ. One such expression follows (89):

$$\phi(z) = \phi_D + \phi_P + \phi_{F\mu} + \phi_{FQ} + \phi_R + \ldots \tag{110}$$

where the subscripts indicate terms in r^{-6} and r^{-12} (D and R); induced-dipole (P) and dipole–dipole ($F\mu$) effects; and quadrupole-moment/surface-field (FQ) interactions. Clearly, additional terms could be added as appropriate.

All of the interactions indicated above are collectively termed van der Waals or physical adsorption, since they arise from physical (i.e., bulk) properties of the interacting species. Physical adsorption is characterized by relatively small (~5 kcal/mol) heats of adsorption that are on the order of heats of vaporization; rapid and reversible attainment of equilibrium at low temperatures; and, hence, low heats of activation. All adsorption systems exhibit some manner or other of physical adsorption, which is the phenomenon of immediate concern in most forms of gas-solid and liquid-solid chromatography.

Clathrate adsorption (sometimes referred to as "persorption") is a special case of physical adsorption wherein adsorbate molecules fit into, and are retained by, vacant spaces within the adsorbent. Zeolite "molecular sieves" represent an example. The phenomenon is said to differ from true solid solution only in the respect of heterogeneity of distribution of (adsorbate) "solute" throughout the (adsorbent) "solvent". (It is of some interest in this regard that heats of clathrate adsorption are generally higher than those corresponding to true physical adsorption, and yet appreciably smaller than values usually associated with the solution process.) In any event, molecular sieves are widely used as GC stationary phases for the separation of permanent gases, as illustrated by the chromatogram shown in Fig. 2.18.

In contrast to the above, chemical adsorption is said to ensue when the forces of adsorbate-adsorbent interactions are chemical in nature, that

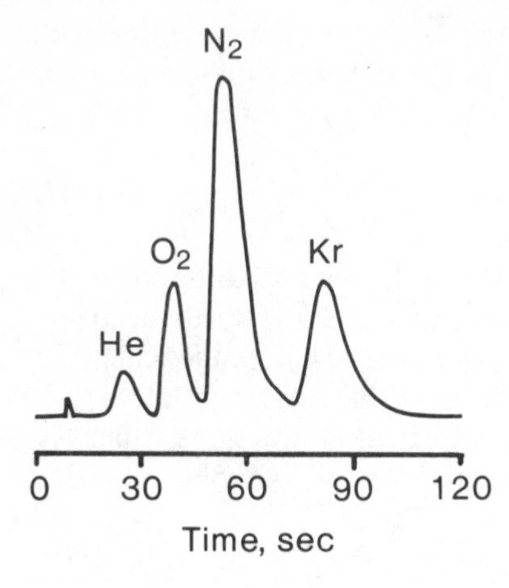

Figure 2.18. Chromatogram of fixed gases with molecular sieve adsorbent stationary phase (90).

is, involve the formation and/or destruction of chemical bonds. They are characterized by high heats of interaction (on the order of 20 to 100 kcal/mol); irreversibility; and high (often 5–20 kcal/mol) heats of activation. Thus, such interactions generally are favored at high temperature, the irreversible adsorption of hydrogen on tungsten being a notable exception.

The Adsorption Isotherm

An adsorbent suspended in a closed thermostated vessel containing some gas will adsorb some of the gas, such that at equilibrium the pressure is reduced and the weight of adsorbent increased. Plots of such data are conveniently presented as weight gain by the adsorbent (more usefully, moles of adsorbed gas) against the equilibrium pressure of the gas, and are called adsorption isotherms. In the simplest case, the data describe a straight line that passes through the origin and that extends up to the saturation pressure of the adsorbate, p^0. The slope of the line is then a kind of partition coefficient.

Langmuir, shortly after the turn of this Century, was the first to study such behavior from a quantitative standpoint (91). The matter bears directly on solute retentions both in GC and in LC and so, we briefly recollect his treatment in what follows.

To begin, let S be the total number of sites available for adsorption on the surface as well as within the pores of an adsorbent. Upon exposure to an adsorbate, S_1 sites become occupied by individual molecules, while S_0 sites are left free. Once equilibrium has been established, the rate of evaporation of adsorbate species from the surface (the number of sites vacated per unit time), k_1S_1, will be equal to the rate of condensation of new adsorbate molecules onto the surface, $k_2p_1S_0$:

$$k_1S_1 = k_2p_1S_0 = k_2p_1(S - S_1) \tag{111}$$

where k_i are rate constants. We next let Θ equal the fraction of sites that are occupied:

$$\Theta = \frac{S_1}{S} \tag{112}$$

From eq. 111,

$$\frac{S_1}{S - S_1} = \frac{k_2p_1}{k_1} \tag{113}$$

Rearrangement, followed by substitution of b for k_2/k_1, then produces

$$\Theta = \frac{p_1b}{1 + p_1b} \tag{114}$$

Alternatively, the numbers of vacant and occupied sites can be replaced by volumes of adsorbate: let v_m be the volume corresponding to complete monolayer coverage of the adsorbent and v_1 to the volume of partial coverage. Equation 114 then becomes

$$v_1 = \frac{v_mp_1b}{1 + p_1b} \tag{115}$$

where, at sufficiently low adsorbate pressure, $v_1 = v_mp_1b$, that is, v_1 becomes a linear function of p_1.

Equation 115 is known as the Langmuir isotherm relation, where plots of the lhs against the rhs can either be linear or curved concave to the abscissa, as shown in Fig. 2.19. Very many systems conform to the relation, particularly at very low adsorbate pressure. However, linearity does not ensue even at the limit of pressure measurement for other adsorbate-adsorbent systems (92). For example, Hobson (93) found deviations from linearity for argon, krypton, and xenon with a porous silver adsorbent at fractional pressures of as low as 10^{-13}, although variations in porosity may have contributed also to these results.

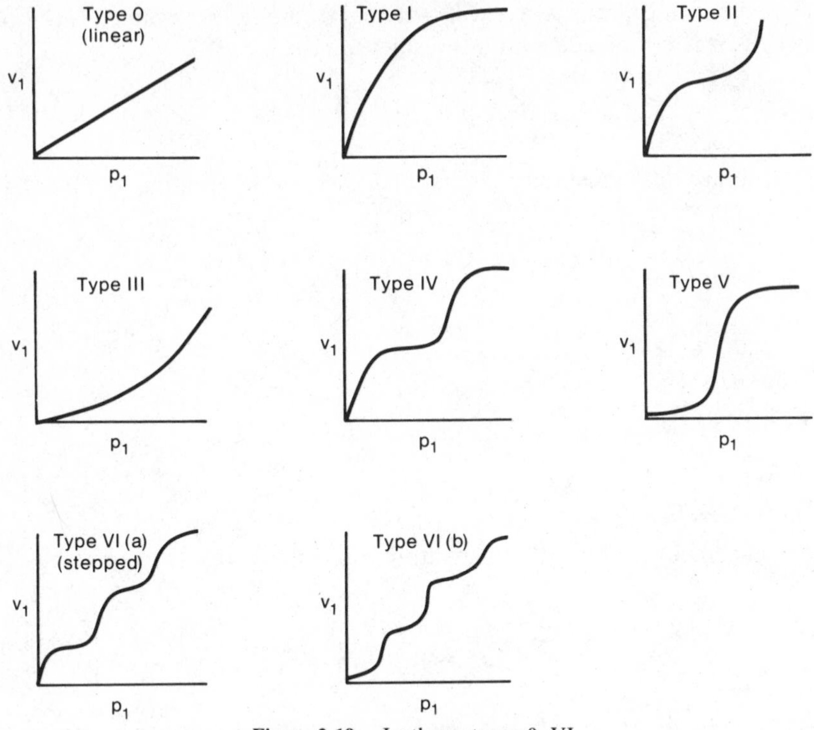

Figure 2.19. Isotherm types 0–VI.

Alternative rearrangement followed by inversion of eq. 115 produces a linear form of the relation

$$\frac{p_1}{v_1} = \frac{1}{v_m} + \frac{p_1 b}{v_m} \tag{116}$$

Plots of p_1/v_1 against p_1 may then be used to derive b/v_m from the slope and $1/v_m$ from the intercept. The slope/intercept quotient then yields b. If the cross-sectional area σ_1 of the adsorbate is known, the surface area A_S/cm^2 and specific surface area $S/\text{cm}^2/g$ of the adsorbent can also be calculated:

$$A_S = \frac{N_0}{22{,}414} \sigma_1 v_m \tag{117a}$$

$$S = \frac{A_S}{w_S} \tag{117b}$$

where w_S is the mass of adsorbent material.

While the Langmuir approach is certainly highly successful in describing linear and concave isotherms, some systems exhibit curvature that is convex to the abscissa while still others show step changes in plots of v_1 against p_1, as illustrated in Fig. 2.19. The curve shapes are classified as Types I–VI (94,95) as shown, linear isotherms being the zeroth case. In order to explain the phenomena, Brunauer, Emmett, and Teller (94) modified the Langmuir kinetic approach so as to allow for multilayer formation. Their treatment also assumes that heats of adsorption are relevant only for the first layer of adsorbate, all other layers corresponding essentially to liquefaction. Then, proceeding essentially as described above for monolayers, they derived the relation known today as the BET equation:

$$\frac{v_1}{v_m} = \frac{c(p_1/p_1^0)}{(1 - p_1/p_1^0)[1 - (1 - c)(p_1/p_1^0)]} \tag{118a}$$

where c is a constant. This relation, too, can be rearranged to yield linear plots:

$$\frac{p_1}{v_1(p_1^0 - p_1)} = \frac{1}{v_m c} + \frac{(c - 1)}{v_m c}\frac{p_1}{p_1^0} \tag{118b}$$

where regressing the lhs against p_1/p_1^0 yields $(c - 1)/v_m c$ as the slope and $1/v_m c$ as the intercept, from which $(c - 1)$ is found as the slope/intercept quotient.

The importance of sorption isotherms in chromatography lies in their shapes, and in the magnitudes of slopes of lines drawn through points corresponding to the amounts of adsorbate (solute) introduced into the system. In the "normal" (i.e., high-dilution) elution mode of chromatography, only Types 0, I, and III are of importance, as shown in Fig. 2.20. In the simplest instance (Type 0) the isotherm is a straight line, the slope of which (the partition coefficient) must of course be constant. Chromatographic peaks will therefore be nearly symmetric (96) and retention times will be independent of the amount injected. In contrast, for isotherms of Type I, the peak will "tail". Moreover, depending upon the amount injected, the peak maximum (i.e., retention time) will vary. For example, the slope of line 1 is greater than that of 2 and so, the retention time of the solute peak maximum in the former instance will be longer than in the latter. That is, the retention time of the peak maximum decreases as the amount injected is increased. Solutes giving rise to Type III isotherms will "front", and the retention time will increase as the amount injected is increased.

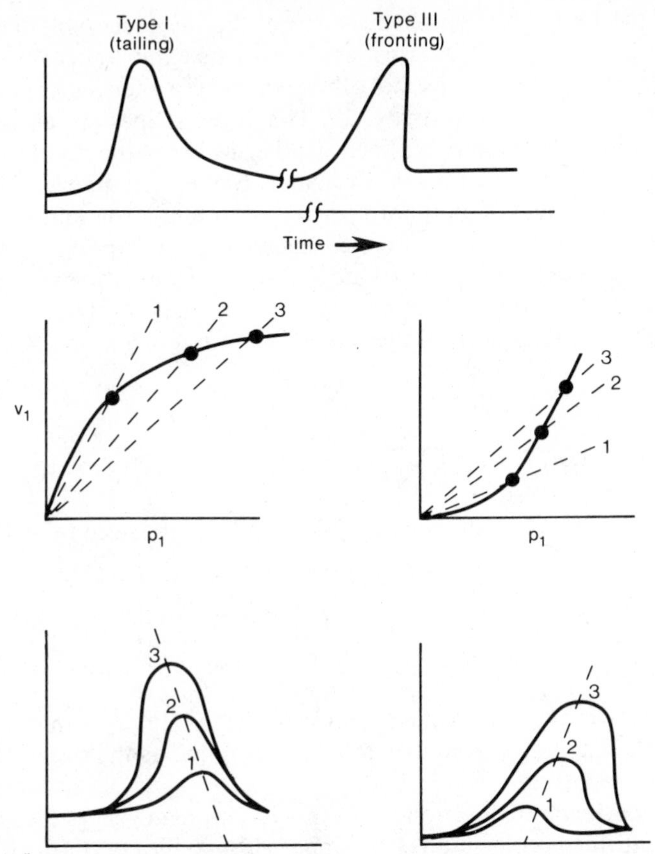

Figure 2.20. Chromatographic band shapes and peak emergence times corresponding to various amounts of injected solute with isotherm types I and III.

Peak tailing is extremely common in gas–solid chromatography because of multilayer surface adsorption by solutes as well as surface heterogeneity, that is, persistence of a multiple retention mechanism even at very high solute dilution in the carrier phase. The same is also true of liquid chromatography, although tailing there is due, as discussed below, to the peculiarity of the retention mechanism inherent in the technique. Tailing in gas–liquid chromatography arises most commonly as a result of gas–liquid interfacial adsorption, that is, when adsorption (in addition to solution) contributes substantially to retention. In contrast, fronting in GLC is almost exclusively an indication of solute "overload", where the (liquid) stationary phase has become saturated with solute.

2.2.7 Sorption at the Liquid–Solid Interface

Initial reflection would seem to indicate that the adsorption of liquids on solids would differ little from that of adsorbate gases; in fact, the properties of liquid–solid systems are as different from gas–solid systems as are liquids from gases. For example, when a vapor is adsorbed onto a solid surface, molecular motions of gaseous molecules that happen to pass in the vicinity immediately above the adsorbate-adsorbent layer hardly appear to be affected. In contrast, it is now well established that finite ordering occurs within even simple liquids in the bulk state. The introduction of an adsorbent then appears to catalyze a kind of nucleation, that is, ordering is markedly increased. Moreover, a concentration gradient will be established in the manner of surface-excess interfacial adsorption as discussed in Sec. 2.2.5. The appropriate liquid-solid isotherm expression is therefore that of an "excess" concentration of the adsorbate, where the excess is taken as that at the surface over that found in bulk liquid. Thus, the radial distribution of molecules near an adsorbent is quite different from that far removed from its surface. Water provides an excellent example (97), where its density varies from 1.14–2.55 g/cm^3 near the surfaces of nylon, dacron, glass fibers, and glass powders (98).

The manner of representation of surface-excess isotherms for liquid–solid systems has been reviewed and discussed comprehensively by Everett and his coworkers (99,100), and as noted above closely parallels that of functions presented earlier. Köster and Findenegg (101) have also recently presented a liquid-chromatographic technique with which such data can be measured directly. The importance of the matter lies in the ease with which system selectivity can be adjusted in liquid chromatography (102–109): since mobile-phase components can be made to sorb onto the surface of the (sorbent) column packing, virtually any manner of stationary phase (including ion-exchange) can be fabricated in situ simply by doping the mobile phase with the desired material. However, there are some important differences in this technique vs. classical liquid-liquid chromatography. First, since the amount of sorbed mobile-phase component is small, it is relatively easy to "overload" the column with solute. Secondly, strongly sorbed solutes may well displace the mobile-phase component in a kind of competition for surface (or pore-interior) sites of adsorption. Thirdly, the establishment of equilibrium between the mobile and stationary phases very often is slow, such that many column volumes of liquid must be passed through the system before solute retention times become reproducible. This is particularly annoying when gradient elution must be employed for a given separation (110,111), since re-equilibration of the column from the final mobile phase to the starting composition may require several hours. Even so, there clearly remain the overwhelming advantages of flexibility and convenience in judicious choice of multicomponent mobile phases for particular separations as opposed, for example, to fabrication of specific stationary phases permanently attached to the support.

Nevertheless, it must be recognized that the liquid–solid sorption process is as yet only poorly characterized overall, and that considerable clarification is required before any quantitative predictions can be made regarding the efficacy of this or that model for multicomponent mobile phases, let alone solute/sorbed mobile-phase interactions. The current popularity of column liquid chromatography should ensure a continuation of what has become, over the past five or so years, an area of active and lively research.

2.3 DIFFUSION

Diffusional phenomena are of major importance both in gas and liquid chromatography, solute band-spreading being but one of the several consequences that enter into the elution process. In general, diffusion is one of several of what are known as "transport properties", that is, measures of the rate of transport of matter or energy across real or hypothetical boundary planes. For example, the rate of transport of thermal energy is related to the thermal conductivity of the medium; the coefficient of transport of electrical energy (charge) is the electrical conductivity; and viscosity arises as a result of the transfer of momentum.

There are two phenomena related to the transport of mass: effusion and diffusion, where the former pertains to particles passing through a pinhole leak of dimensions close to the size of the particle (the result being called Knudsen flow), while the latter is related to the relaxation of concentration gradients, the property of concern here. It is also usual to speak of self-diffusion D_{11} and interdiffusion D_{12}, where the former is diffusion of like kinds of particles (atoms or molecules) into and among one another ($1 \rightarrow 1$) while the latter, pertinent to chromatography, is diffusion of one kind of particle into and among those of a different kind ($1 \rightarrow 2$). Self-diffusion is considered first in what follows, the results then being applied to interdiffusion, since the formulations for each are very nearly identical.

2.3.1 Self-Diffusion

We begin by defining the net flux of matter, J_x, as the number of particles (or moles) of species 1, N_1, contained in a volume V that pass linearly with an average velocity $\bar{v}_x$ (= $\bar{v}/6$) in a direction x through a plane of unit area in time t:

$$J_x = \frac{\bar{v}}{6}\frac{N_1}{V} = \frac{1}{A}\left(\frac{dN_1}{dt}\right) \tag{119}$$

The cgs units of flux are therefore g/cm²/sec. If we now let $N_1/V = C_1$, a concentration, the flux can be related to the concentration gradient of particles along the path, but in the opposite direction, over which diffusion occurs:

$$J_x = \frac{1}{A}\left(\frac{dN_1}{dt}\right) = -D_{11}\left(\frac{dC_1}{dx}\right) \tag{120}$$

Where D_{11}, the self-diffusion coefficient, is simply a constant of proportionality (units of cm²/sec), and where the negative sign has been introduced to indicate that the concentration gradient is negative, that is, the flow of matter occurs in the direction x of from high to low concentration.

Equation 120 is known as Fick's first law of diffusion: the flux of matter is equal to a negative concentration gradient multiplied by a proportionality constant, the diffusion coefficient (112). It is of a form that, intuitively, should be correct: matter passes from one region of a system to another so as to relax a concentration imbalance (113). Fick's second law, restricted to one direction, relates the change in concentration as a function of time that arises because of diffusion to the change in the concentration as a function of distance in the direction in which diffusion occurs:

$$\frac{dC_1}{dt} = D_{11}\left(\frac{d^2C_1}{dx^2}\right) \tag{121}$$

The steady-state condition is $dC_1/dt = 0$, whence dC_1/dx = a constant, that is, the change in concentration varies linearly with distance in the x direction. For example, consider a box of cross-sectional area 1 cm² and length $x + dx$, as shown in Fig. 2.21, into and out of which matter diffuses in a direction parallel with x. The number of moles of material that enter the box from the left is the flux at x multiplied by the time interval, $J_x dt$, while the number of moles leaving the box to the right is the flux at x + dx again multiplied by the time interval, $J_{x+dx}dt$. If there is a number of moles flowing into the box that is in excess of that flowing out, the difference dN_1 will be given by

$$\frac{dN_1}{A} = \frac{(J_x - J_{x+dx})\,dx\,dt}{dx} \tag{122}$$

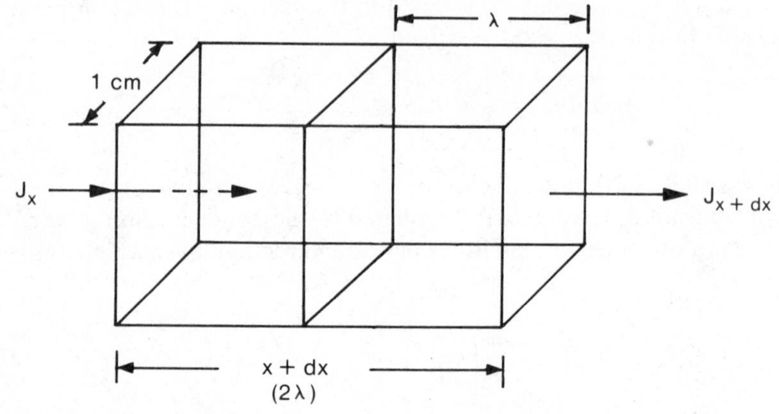

Figure 2.21. Non–steady-state flux of matter passing through a box of dimensions 1 cm × 1 cm × $(x + dx)$, where $x + dx = 2\lambda$, twice the mean free path of the particles. The flux entering at the left is J_x, while that leaving to the right is J_{x+dx}.

Since the change in flux over the distance $x + dx$ is negative, that is, decreases with distance on passing from left to right through the box,

$$\frac{(J_x - J_{x+dx})}{dx} = -\left(\frac{dJ_x}{dx}\right) \tag{123}$$

Equation 122 therefore becomes

$$\frac{1}{A}\left(\frac{dN_1}{dx}\right) = dC_1 = -\left(\frac{dJ_x}{dx}\right) dt \tag{124}$$

or

$$\frac{dC_1}{dt} = -\left(\frac{dJ_x}{dx}\right) \tag{125}$$

Substitution of $-D_{11}(dC_1/dx)$ from eq. 120 for J_x in eq. 125 then produces eq. 121.

The importance of this result lies in the choice of the distance $x + dx$, which we take at this point to be equal to twice the mean free path λ of particles within and outside of the container:

$$\lambda = \frac{\overline{v}}{Z_1} = \frac{V}{2^{1/2}\pi(d_1)^2 N_1} \tag{126}$$

where $\overline{v}$ is the average velocity of a particle, Z_1 is the number of collisions per unit time that a single particle engages in with all other particles, d_1 is the particle diameter, and N_1 is the number of particles (or moles) of species 1 as before (the units of λ are therefore distance per collision). We now let the concentrations of particles at the left, middle, and right-hand planes of Fig. 2.21 be $(C_1 + \lambda dC_1/d\lambda)$, (C_1), and $(C_1 - \lambda dC_1/d\lambda)$, respectively. The net number of particles dN_1 moving from left to right in a time interval dt, that is, the flux, is then given by analogy with eq. 122: recalling from eq. 119 that $(1/A)(dN_1/dt) = \overline{v}_x(N_1/V) = \overline{v}_x C_1$ ($\overline{v}_x$ again being the average velocity in direction x), we find

$$\frac{1}{A}\left(\frac{dN_1}{dt}\right) = \frac{\overline{v}\left(C_1 + \lambda \dfrac{dC_1}{d\lambda}\right) - \left(C_1 - \lambda \dfrac{dC_1}{d\lambda}\right) d\lambda}{6\, d\lambda}$$

$$= -\frac{\overline{v}}{3}\lambda\left(\frac{dC_1}{d\lambda}\right) \tag{127}$$

Comparison of this result with eq. 120 then provides

$$D_{11} = \frac{\overline{v}\lambda}{3} \tag{128}$$

The average velocity of a particle is given by the kinetic theory of gases as

$$\overline{v} = \left(\frac{8RT}{\pi M_1}\right)^{1/2} \tag{129}$$

M_1 being the molar mass of the particle, so that

$$D_{11} = \frac{V(8RT/\pi M_1)^{1/2}}{3(2)^{1/2}\pi(d_1)^2 N_1} \tag{130}$$

Combining numerical constants and replacing N_1/V by p_1/RT then yields

$$D_{11} = \frac{2(RT)^{3/2}}{3\pi^{3/2}(d_1)^2 p_1 (M_1)^{1/2}} \tag{131}$$

D_{11} is therefore said to be inversely pressure-dependent, directly proportional to $T^{3/2}$, and inversely proportional to $M^{1/2}$. Hence, diffusion will apparently be fastest for small molecules of low molecular weight at high temperature (114).

2.3.2 Interdiffusion

The phenomenon of interest in chromatography is of course the diffusion of a solute in a solvent, be this a liquid (LC mobile phase; GC stationary phase) or a gas (GC mobile phase). An expression involving the interdiffusion coefficient D_{12} is derived in exactly the same manner as used to arrive at D_{11}, the result being known as the Stefan-Maxwell relation:

$$D_{12} = \frac{V(8RT/M_{12})^{3/2}}{3(2)^{1/2}\pi^{3/2}(d_{12})^2 N_{12}} = \frac{2(RT)^{3/2}}{3\pi^{3/2}(d_{12})^2 P_t (M_{12})^{1/2}} \qquad (132)$$

where M_{12}, d_{12}, and N_{12} are, respectively, the effective molar mass, particle diameter, and number of particles per unit volume V. The former of these is often taken to be approximately equivalent to a reduced molecular weight $(1/M_1 + 1/M_2)$, while the arithmetic average is used for d_{12}. For gas-phase solutions, N_{12}/V is simply the sum of the partial pressures, $P_{total} = p_1 + p_2$ while, for liquid solutions, N_{12}/V is the sum of the concentrations C_1 and C_2.

Equation 132, like eq. 131, predicts that interdiffusion coefficients are directly proportional to $T^{3/2}$, yet will vary as the inverse pressure of the gas phase. In the instance that the solute (species 1) is highly dilute, the partial pressure of component 2 is equal to the total pressure. In highly dilute liquid solutions, the total concentration of species 1 + 2 is very close to the inverse molar volume of the latter. Thus, at least for dilute solutions, the interdiffusion coefficient is predicted to be independent of the composition of the solution, the experimental verification of which was one of the great triumphs of the kinetic theory.

Tables 2.10 and 2.11 present a few examples of interdiffusion coefficients for gaseous and liquid species drawn from the *International Critical Tables,* where the temperature dependence of D_{12} is seen to correspond only approximately to the three-halves power of temperature.

2.3.3 Viscosity and Diffusion

We suspect that diffusion should somehow be related inversely to viscosity, that is, the more viscous the medium the slower will be the rate of mass transport. For hard spheres, the relation is

$$D_{12} = \frac{RT}{3\pi\eta\, d} \qquad (133)$$

Table 2.10 Interdiffusion Coefficients for Gases

Gas A	Gas B	D_0 (cm²/sec)
H_2	CO	0.651
	CO_2	0.550
	N_2	0.674
	Air	0.611
	SO_2	0.480
	O_2	0.697
He	Ar	0.641
O_2	CO	0.185
	CO_2	0.139[a]
	Air	0.178
	N_2	0.181
N_2O	CO_2	0.096[a]
CO	CO_2	0.137
CO_2	Air	0.138[a]
H_2O	Air	0.220
	CO_2	0.139[a]
	H_2	0.752

[a] $m = 2.00$ (else 1.75) in the relation: $D = D_0(T/T_0)^m(p_0/p)$, where D_0 is the diffusion coefficient at 273K and $p_0 = 1$ atm, D being the diffusion coefficient at pressure p and temperature T.

Equation 133 is known as the Stokes—Einstein relation, where η is the viscosity and d the particle diameter, these referring to the pure solvent for highly dilute solutions. While only approximately obeyed, eq. 133 nevertheless confirms that D_{12} varies inversely as the viscosity, and also the particle diameter: the larger the diffusing species, the more numerous will be interparticle collisions and, hence, the slower will be the net progress made in any one direction.

2.3.4 Diffusional Spreading

We now suppose that, instead of diffusion in one direction only, we introduce a certain concentration of solute at a point along the longitudinal axis of a system. The solute will spread in all directions, whereby its point-source concentration will be diminished. Further, the increase in concentration at some distance x from the point source will be related

Table 2.11 Interdiffusion Coefficients for Gases and Liquids in Water

Solute	T (°C)	10^5D (cm²/sec)
H_2	10	4.3
	16	4.7
	21	5.2
N_2	19	1.9
	22	2.0
O_2	18	2.0
CO_2	10	1.46
	15	1.60
	20	1.77
CH_3OH	15	1.28
CH_3CN	15	1.26
CH_3CH_2OH	15	1.00

exponentially to the time allowed for spreading. Fick's single-dimensioned second law, eq. 121, provides the link between the concentration, time, distance along the longitudinal axis, and diffusion coefficient, for which Purnell (ref. 45, p. 53) has provided a solution

$$\frac{C_1}{C_1^0} = \frac{1}{2(\pi D_{12}t)^{1/2}} \exp(-x^2/4D_{12}t) \tag{134}$$

where C_1 is the concentration found after time t at a distance x in either direction from the point source, and where C_1^0 is the concentration at the point source. Equation 134 describes a Gaussian distribution, for which the standard deviation σ_x is one-fourth the baseline width of the diffusing band:

$$\sigma_x = (2D_{12}t)^{1/2} \tag{135}$$

where eq. 135 is known as the Einstein relation. It predicts that for a given time allowed for diffusion, the larger the diffusion coefficient, the further will the solute band have spread.

The importance of the Einstein equation is twofold. First, we see from Tables 2.10 and 2.11 that diffusion is roughly 10^4 times faster in gases than in liquids. Therefore, in gas chromatography, the diffusional spreading of a solute dissolved in a liquid phase is likely to be a factor of 10^2 smaller than that in the gas phase. Thus, spreading as a result of gas-phase diffusion likely represents the limiting minimum bandwidth that can be achieved in GC. Secondly, as will be shown later (Sec. 2.5.4), the major source of peak spreading arises as a result, not of diffusion, but of flow heterogeneities (due to mass-transfer nonequilibrium). Diffusion helps to relax these, and so higher column efficiencies are in fact obtained for low-molecular-weight solutes that are retained little in stationary phases of low viscosity.

The same nonequilibrium effects are extant in liquid chromatography, but because of the small magnitude of liquid-phase diffusion coefficients, diffusional relaxation of solute band-spreading is much less efficient than in gas chromatography. As a result, the latter technique offers inherently greater system efficiency. Higher temperature will help to enhance D_{12} in the absence of multiple retention mechanisms (which may in fact be rare in LC) and, hence, yield narrower solute bands as a result of the increased extent of relaxation of mass-transfer nonequilibrium.

2.4 FLOW THROUGH POROUS MEDIA

In general, there are three real modes and one ideal mode whereby a fluid (here taken to be the solute) can be made to flow through a tube, irrespective of whether the latter contains a packing, as shown diagrammatically in Fig. 2.22. The idealized situation is called plug flow, where the moving boundary (particles of solute flowing along with those comprising the mobile phase) lies everywhere orthogonal to the longitudinal axis of the system. Plug flow is of course the most desirable situation, and is often referred to as ideal chromatography since there is no band-spreading as the solute peak moves through the column. In the first real mode, called diffusional or Knudsen (115) flow, the mean free path of the solute particles is very much greater than the radius of the tube, where the tube can be defined variously as an orifice, pore through a particle, interparticle space, and so forth. Knudsen flow is therefore applicable only in instances of very low pressure. The second mode of mass transport is known as laminar or stream flow, which takes place at moderate pressure and moderate solute linear velocity. Here, the solute particles move in well-defined "streamlines" parallel to the longitudinal axis of the system. The third mode of mass transport is called turbulent flow, and takes place at high pressures and high linear velocities: turbulence arises in the streaming patterns of the solute particles such that "eddying" (swirling) occurs. The result is a flow profile that resembles plug flow more closely than do either Knudsen or laminar flow.

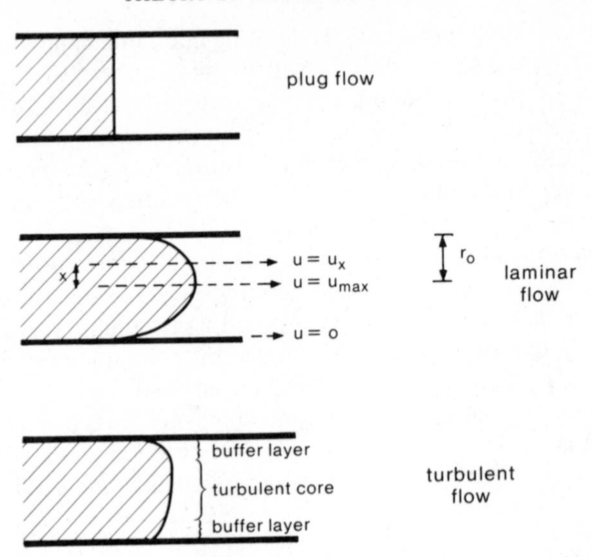

Figure 2.22 Schematic representation of plug, laminar, and turbulent flow.

The difference between laminar and turbulent flow, as well as the transition from one to the other, is easily seen in the smoke rising from a cigarette burning in absolutely calm air. Gases and particulate matter at the temperature of the smouldering end (~600–900°C), along with entrained air that is heated upon contact with the burning tobacco, rise in what appear to be near-straight (i.e., laminar) filaments to a height of a foot or so. Thereafter, the gases and particulates have cooled sufficiently such that intermolecular collisions cause the streamlines to begin to diverge into disperse swirls, that is, the vertical flow becomes "turbulent".

A useful measure for estimation of the onset of turbulence in laminar-flow open-tubular systems is given by the Reynolds number, Re:

$$\mathrm{Re} = \frac{\rho u\, d}{\eta} \tag{136}$$

where ρ is the density of the fluid of viscosity η; u is its linear velocity; and d is the tube diameter. Equation 136 actually represents the ratio of inertial forces, ρu^2, and viscous forces, $\eta u/d$. When viscous forces predominate (Re ~2000 or less), the flow is laminar. On the other hand, when inertial forces predominate, that is, when Re exceeds ~3000, the flow stream is comprised almost entirely of swirls and eddies. At intermediate values of the Reynolds number, the flow corresponds to a kind of transition state between streaming and turbulence, and contains elements of both.

For packed tubes, for which d is replaced by the particle diameter d_p in eq. 136, the onset of turbulence may well take place at values of Re ~100 or so as a result of obstructions (i.e., the packing) encountered by the flowing particles (116).

Only laminar flow will be considered here, since the conditions required to induce turbulent flow are beyond the practical limits of most available instrumentation, and since the interparticle space (even for LC packings of 3 μ diameter) is much larger than the mean free path of any solute likely to be encountered in the chromatography of inorganic and organometallic species.

2.4.1 Flow Profile in Empty Tubes

The hydrostatic force required to cause a fluid to pass through an empty tube is equal to the viscous drag exerted on the fluid:

$$\pi r^2 \Delta p = 2\pi r L \eta \left(\frac{du}{dr}\right) \tag{137}$$

where r is the radius of the tube of length L, and where Δp is the pressure drop resulting in a linear velocity of u. Solving this relation in terms of (du/dr),

$$\frac{du}{dr} = \frac{r\Delta p}{2\eta L} \tag{138}$$

Thus, the velocity is zero at the column wall ($r = 0$), and has a maximum value at the center of the tube, that is, at a distance r from the wall. If we now integrate eq. 138 from the wall to the center of the pipe, we find

$$u_x = \Delta p \frac{(r^2 - x^2)}{4\eta L} \tag{139}$$

where u_x is the linear velocity at some distance x from the center toward the wall; cf. Fig. 2.22.

The linear velocity will clearly be at a maximum when $x = 0$, and so

$$u_{\max} = \Delta p \frac{r^2}{4\eta L} \tag{140}$$

Combining eqs. 139 and 140 then yields the relation

$$u_x = u_{\max}\left(1 - \frac{x^2}{r^2}\right) \tag{141}$$

It is also true that the *average* linear velocity is equal to $(u_{max} - u_x)/2$, where u_x is that at $r = 0$, i.e., $\overline{u} = (u_{max})/2$. Thus, $u_{max} = 2\,\overline{u}$, and we find that the average velocity is related to that at any distance x from the center of the tube in the manner

$$u_x = 2\overline{u}\left(1 - \frac{x^2}{r^2}\right) \tag{142}$$

2.4.2 Flow Through Empty Tubes

We can now combine eqs. 139 and 142 to provide

$$\overline{u} = \Delta p \frac{r^2}{8\eta L} \tag{143}$$

and, recognizing that the volume flow rate $F = \overline{u}\pi r^2$,

$$F = \Delta p \frac{\pi r^4}{8\eta L} \tag{24}$$

which is Poiseuille's law, given earlier in Sec. 2.2.2, and which can be expected to hold over virtually the entire range of laminar flow, that is, that falling between diffusional flow and the point of onset of turbulence.

2.4.3 Flow Through Packed Beds

The relation that describes the linear velocity of a fluid through a packed bed is known as Darcy's law (117):

$$\overline{u}_D = B_1 \frac{\Delta p}{L} \tag{144a}$$

where B_1 is called the permeability coefficient of the packed bed, and where $\overline{u}_D$ is the linear (Darcy) velocity of the fluid at the tube outlet, measured by dividing the volume of fluid collected in a given time t by the cross-sectional area of the pipe: $\overline{u}_D = V/At$.

The permeability coefficient is often multiplied by the viscosity of the medium to give the specific permeability coefficient: $B_0 = \eta B_1$, whence eq. 144a becomes

$$\overline{u}_D = \frac{\Delta p B_0}{\eta L} \tag{144b}$$

Equations 144a and 144b are applicable to empty tubes, for which comparison with eq. 143 shows that $B_0 = r^2/8$. For packed columns, however, as in gas and liquid chromatography, the true average linear velocity $\overline{u}$ must actually be higher than $\overline{u}_D$ since the flowing medium must pass around the granular matter in the tube, that is, it cannot flow directly through the column in a straight line. Since the resultant path length L_e is longer than that corresponding to an empty tube, L, so also must the linear velocity be higher in order to achieve the same volume flow rate. The ratio L_e/L is given the symbol λ and is called the "tortuosity." Calculations with spherical nonporous particles indicate that λ is likely to be on the order of $2^{1/2}$ and, at the very least, lies between 0.5 and 1.5 (cf. Sec. 2.5.4).

Stated another way, the free volume that is available to flowing solute is greater in an empty tube than in one of the same dimensions which contains a packing. Then, if it is to be true that

$$\frac{V_{\text{packed}}}{At} = \frac{V_{\text{empty}}}{At}$$

and, since $V_{\text{packed}} < V_{\text{empty}}$, $\overline{u}_{\text{packed}}$ (or, simply, $\overline{u}$) must be greater than $\overline{u}_{\text{empty}}$, that is, what we have symbolized as $\overline{u}_D$, since it is true in general that $V/At = \overline{u}$.

The two velocities, $\overline{u}$ and $\overline{u}_D$, differ by an amount equal to the fractional free (interparticle) space ϵ $(0 < \epsilon < 1)$ in the tube:

$$\overline{u} = \frac{\overline{u}_D}{\epsilon} \tag{145}$$

so that the true average linear velocity in packed columns is given by

$$\overline{u} = \Delta p \frac{B_0}{\epsilon \eta L} \tag{146}$$

ϵ has been shown generally to fall between 0.38 and 0.42 for a wide range of materials in well-packed systems (118–120), and is therefore usually taken as 0.40.

The Hydraulic Radius

It is often useful to visualize the channels available for flow through a packed-bed column in terms of a large number of empty, twisted capillaries that have been bundled within the tube. It is then possible to define an average effective or "hydraulic" radius r_H that is available within the tube through which fluid may pass. This is taken as the volume available for flow, divided by the surface area of the packing with which the fluid is in contact. For empty tubes, the hydraulic radius is just $(\pi r^2 L)/(2\pi r L) = r/2$, that is, the volume of the tube divided by the surface area of its interior. For packed columns, we let S be the total surface area of particles that occupy a unit volume of the column and that are in contact with the flowing medium, and thereby find

$$r_H = \frac{\epsilon}{S} \tag{147}$$

A more directly accessible quantity is the specific surface area S_0, the surface area of particles in contact with the flowing fluid divided by the volume occupied by the particles:

$$S_0 = \frac{S}{(1 - \epsilon)} \tag{148}$$

where, for spherical packings, $S_0 = (3/4)(4\pi r^2)/(\pi r^3) = 6/d_p$. Equation 147 then becomes

$$r_H = \frac{\epsilon}{S_0(1 - \epsilon)} \tag{149}$$

The significance of r_H is thereby brought out clearly: it is the actual radius through which the fluid can flow.

The Kozeny-Carman Equation

We now utilize eqs. 147–149 with 143 to derive, after some rearrangements and combination of constants,

$$\overline{u} = \frac{(d_p)^2 \epsilon^2 \Delta p}{180\eta(1 - \epsilon)^2 L} \tag{150}$$

which is known as the Kozeny-Carman relation (121–123), and which appears to be borne out, at least approximately, in practice. Further, recalling eq. 146, the specific permeability coefficient B_0 can now be identified in terms of readily measurable quantities:

$$B_0 = \frac{(d_p)^2\epsilon^3}{180(1 - \epsilon)^2} \tag{151}$$

Equations 150 and 151 also verify what might well be surmised from intuitive considerations of the flow of liquids or gases through packed beds. The linear velocity depends directly upon the pressure drop and, of course, inversely as the column length, but also varies as the square of the particle diameter. In addition, higher linear velocities can be achieved at lower viscosity, which portends, for liquid chromatography, the use of elevated column temperature (cf. Sec. 2.2.2). In contrast, in gas chromatography, since the viscosity of gases increases with temperature, higher operating temperature will require a higher inlet pressure in order to maintain a constant flow rate.

Compressibility Effects

We have heretofore made use of the average velocity of the flowing fluid, $\bar{u}$. There is no difficulty in measuring this quantity for liquids, since they are assumed to be incompressible below ~10,000 psig. Thus the velocity measured at the column outlet $u_0 = \bar{u}$. However, this is not so for gases since they are compressible. The factor that accounts for compressibility was first derived by James and Martin (124) in their inaugural paper on gas chromatography, and was provided earlier in eq. 25:

$$\frac{p_o}{\bar{p}} = \frac{3}{2}\left[\frac{(p_i/p_o)^2 - 1}{(p_i/p_o)^3 - 1}\right] \tag{25}$$

where p_i and p_o are the inlet and outlet pressures, respectively. In addition, Laub and Pecsok (1) have shown that

$$\frac{p_i}{\bar{p}} = \frac{3}{2}\left[\frac{1 - (p_o/p_i)^2}{1 - (p_o/p_i)^3}\right] \tag{152}$$

The quantity on the right-hand side of eq. 25 is called the compressibility correction factor, and is commonly given the symbol j. It can assume values of from zero to unity as can readily be verified by substitution of hypothetical p_i and p_o. In contrast, the compressibility correction factor given in eq. 152 ranges from unity to 1.5.

Because various forms of compressibility correction factors appear in virial expressions, Everett (125) has suggested that equations of the form of 25 and 152 be represented generally by

$$(J_n^m) = \frac{n}{m}\left[\frac{(p_i/p_o)^m - 1}{(p_i/p_o)^n - 1}\right] \tag{153}$$

In this nomenclature, the compressibility correction factor in eq. 25 is represented as $(J_3^2)_o$, where we have appended a subscript o to indicate that the average pressure is calculated from that measured at the outlet. For eq. 152, then, the appropriate symbol is $(J_3^2)_i$, the subscript i indicating that the average pressure is calculated from that at the inlet.

We can now relate the average velocity of a gas flowing through a packed bed to that measured either at the column inlet or outlet; choosing the latter,

$$\bar{u} = (J_3^2)_o \, u_o \tag{154}$$

It is also possible to show that the pressure p_x at any distance x from the packed-column inlet is related to the overall column length L by

$$p_x^2 = \left(\frac{x}{L}\right)(p_i^2 - p_o^2) + p_o^2 \tag{155a}$$

or

$$\left(\frac{p_x}{p_o}\right)^2 = \left(\frac{x}{L}\right)[(p_i/p_o)^2 - 1] + 1 \tag{155b}$$

Similarly, the velocity at a distance x from the inlet, u_x (not to be confused with the cross-sectional velocity profile) is given by $u_o(p_o/p_x)$, or

$$\left(\frac{u_x}{u_o}\right)^2 = \frac{1}{\left(\frac{x}{L}\right)\left[1 - \left(\frac{p_i}{p_o}\right)^2\right] + \left(\frac{p_i}{p_o}\right)^2} \tag{156}$$

Figures 2.23 and 2.24 illustrate the behavior of plots of p_x/p_o and of u_x/u_o at values of x/L at various ratios of p_i/p_o. The curves are in each case a consequence of the compressibility of gases, and would otherwise be straight lines. It is also of interest that the curves steepen toward the end of the column, such that the pressure drops markedly while the linear velocity increases. The consequence of this is that, for columns operated

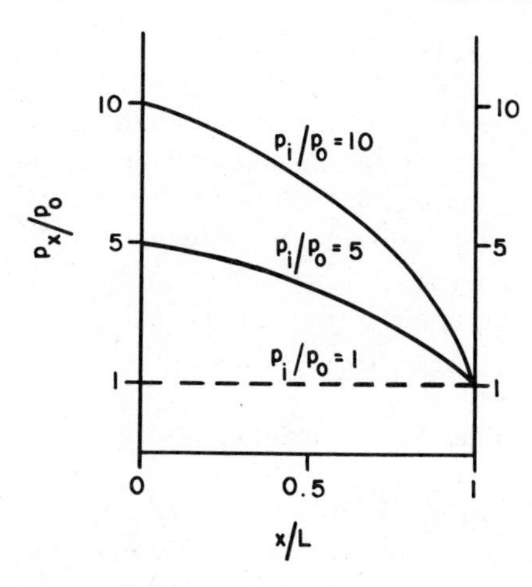

Figure 2.23. Plot of p_x/p_o against distance along a column, x/L, for various ratios of p_i/p_o.

at a high pressure drop, u_x is very nearly linear over the entire length of the packed bed.

It has occasionally been argued that the variation of $\bar{u}$ along a column length, that is, operating gas-chromatographic systems at a high inlet/outlet pressure ratio (as is required for long packed columns containing small particles) is detrimental to efficiency. However, Giddings and coworkers (126, 127) and Sternberg and Poulson (128,129) have demonstrated that this is not so: the column efficiency measured at its outlet represents the average performance achieved and, as such, is independent of the length so long as the pressure drop is not inordinately severe.

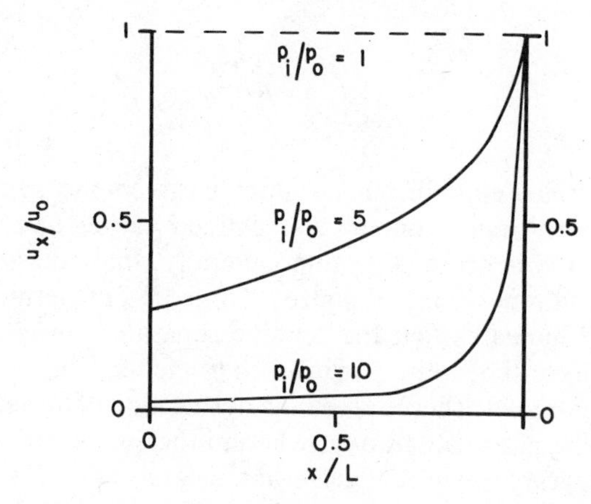

Figure 2.24. Plot of u_x/u_o against distance along a column, x/L, for various ratios of p_i/p_o.

2.5 DYNAMICAL RECTIFICATION

Virtually all of the physicochemical properties of pure and blended substances as discussed in the preceding sections can be taken advantage of in providing rectification of mixture components. Alternatively, the situation may be viewed from the standpoint that rectification *depends fundamentally upon* these properties. In either case, exposition of the manner in which each contributes to chromatographic separations forms the remainder of this chapter (130).

2.5.1 Solute Distribution Between Phases: the Partition Coefficient

For any given system comprised of two immiscible phases at constant temperature, a solute introduced into the system will distribute itself between the phases such that the ratio of its concentrations in each will be a constant. This postulate was first formulated by Berthelot and Jungfleisch (1872), although it has since been attributed to Nernst (1891), and is the fundamental premise upon which all chromatographic techniques are based (131). The demonstration of this is as follows. Stated formally, the Gibbs free energy of the system must be zero at equilibrium, whence the chemical potentials of solute in each phase must be equal:

$$\mu_S^0 + RT \ln a_1^S = \mu_M^0 + RT \ln a_1^M \tag{157}$$

where μ_i^0 is the standard-state chemical potential of solute in the *i*th phase, and where a_i is its activity. Since in the normal elution mode of chromatography the solute is highly dilute, we can replace activities by molar concentrations and find

$$\frac{C_1^S}{C_1^M} = e^{(-\Delta\mu^0/RT)} = K_R \tag{158}$$

where the relevant equilibrium constant is given the symbol K_R, and which is precisely equivalent to that defined earlier in eqs. 67 et seq. However, eq. 158 is presented quite generally, that is, without defining the identity or compositions of phases S and M. Furthermore, reference back to eq. 69 indicates that the relative separation of two solutes α is simply the ratio of the solute partition coefficients. Thus eq. 158 does in fact provide a complete thermodynamic description of all separations processes, including chromatography, wherein the solutes are highly dilute.

In chromatography, we recognize immediately that S is of course the stationary phase and M is the mobile phase. However, these might equally be identified as the lower and upper phases in a separatory funnel or countercurrent distribution apparatus; the refreezing and molten zones in a zone-refining apparatus; the liquid and vapor phases in a distillation column; and so forth. As a corollary, relations derived from consideration of any one of these separations techniques must evidently apply also to chromatography.

2.5.2 Plate Theory: the Linear Ideal Model of Chromatography

The simplest view of the chromatographic process draws an analogy with bubble-cap distillation columns or tandem-arranged separatory funnels, wherein the equilibration of solute takes place in distinctly defined chambers or regions within the respective apparatus. Summing over all equilibrations then mimicks the results (separations) derived from gas–liquid or liquid–liquid chromatographic systems. The treatment is known as "plate" or "stage" theory since each equilibration corresponds to a given region (stage) within the system (132–134).

Single-Stage Liquid–Liquid Extraction

Consider a solute distributed between two immiscible phases S and M in a separatory funnel. Expressing the amounts of solute in each phase in terms of mole fractions,

$$x_1^S = \frac{n_1^S}{n_1^S + n_1^M} \tag{159a}$$

$$x_1^M = \frac{n_1^M}{n_1^S + n_1^M} \tag{159b}$$

where n is the solute mole number. In the absence of excess volumes of mixing, it is true generally that $n_1^i = C_1^i V_i$, where C_1^i and V_i are the molar concentration of solute in the ith phase of volume V. Substituting the appropriate identities for mole numbers into eq. 142 thereby produces

$$x_1^S = \frac{C_1^S V_S}{C_1^S V_S + C_1^M V_M} \tag{160a}$$

$$x_1^M = \frac{C_1^M V_M}{C_1^S V_S + C_1^M V_M} \tag{160b}$$

Dividing through each equation by $C_1^M V_M$ yields the relations

$$x_1^S = \frac{(C_1^S/C_1^M)(V_S/V_M)}{(C_1^S/C_1^M)(V_S/V_M) + 1} \tag{161a}$$

$$x_1^M = \frac{1}{(C_1^S/C_1^M)(V_S/V_M) + 1} \tag{161b}$$

Replacing the concentration ratios by the partition coefficient, and defining the ratio of phases as β (appropriately called the phase ratio), we find

$$x_1^S = \frac{K_R/\beta}{K_R/\beta + 1} \tag{162a}$$

$$x_1^M = \frac{1}{K_R/\beta + 1} \tag{162b}$$

The ratio x_1^S/x_1^M is also of interest in chromatography, and is equal to

$$\frac{x_1^S}{x_1^M} = \frac{n_1^S}{n_1^M} = \frac{w_1^S}{w_1^M} \tag{163}$$

Since the relation expresses the ratio of amounts (either in terms of moles or weights) of solute in each phase, it is called the capacity factor k'.

Combining eqs. 162 and 163 gives

$$k' = K_R/\beta \tag{164}$$

This is one form of what is often called the fundamental retention equation of chromatography, since the fundamental parameter K_R is related to the system variables V_S and V_M via the experimentally-observed quantity k'.

Multiple-Stage Liquid–Liquid Extraction

Consider next that the solute in the separatory funnel has a partition coefficient of unity, and that the volumes of each phase are equal ($\beta = 1$). Upon reaching equilibrium, the mole fraction of solute will be $\frac{1}{2}$ in each phase. Now let us suppose that the M phase is drawn off and replaced in the funnel with an equal volume of fresh M. After partitioning has occurred, there will remain $\frac{1}{2} \times \frac{1}{2} = \frac{1}{4}$ mole fraction of solute in the S phase, that is,

$${}^2x_1^S = ({}^1x_1^S)^2$$

while the amount extracted by the fresh portion of M is

$${}^2x_1^M = {}^1x_1^S\,{}^1x_1^M$$

where a superscript has been added to the left-hand side of each mole-

Table 2.12 Solute Distribution in a Separatory Funnel after R Extractions of Phase S by Phase M

	K_R/β					
	0.1		1.0		10	
R	${}^{R}x_1^S$	$\sum {}^{R}x_1^M$	${}^{R}x_1^S$	$\sum {}^{R}x_1^M$	${}^{R}x_1^S$	$\sum {}^{R}x_1^M$
1	0.0909	0.9091	0.500	0.500	0.9091	0.0909
3	7.51×10^{-4}	0.9992	0.125	0.875	0.7513	0.2487
6	5.65×10^{-7}	~1	0.016	0.904	0.5645	0.4355
10	3.85×10^{-11}	~1	0.001	0.999	0.3855	0.6145

fraction symbol to indicate the extraction number. A third extraction will leave $\frac{1}{8}$ of the solute in S:

$${}^{3}x_1^S = ({}^{1}x_1^S)^3$$

while the amount taken into M will be

$${}^{3}x_1^M = {}^{1}x_1^S\,{}^{2}x_1^M = {}^{1}x_1^S({}^{1}x_1^M\,{}^{1}x_1^S) = {}^{1}x_1^M({}^{1}x_1^S)^2$$

In general, the fraction of solute left in the S phase following the Rth extraction will be (135)

$${}^{R}x_1^S = ({}^{1}x_1^S)^R \tag{165a}$$

while that removed from the system by the Rth portion of M is given by

$${}^{R}x_1^M = {}^{1}x_1^M({}^{1}x_1^S)^{R-1} = \left(\frac{1}{K_R/\beta + 1}\right)\left(\frac{K_R/\beta}{K_R/\beta + 1}\right)^{R-1} \tag{165b}$$

Table 2-12 presents values of ${}^{R}x_1^S$ and $\sum x_1^M$ $(= 1 - {}^{R}x_1^S)$ for different values of K_R/β after 1, 3, 6, and 10 extractions. It is immediately obvious upon going across any row that for fixed β, the larger the partition coefficient the more material will remain in S, that is, more M phase is required to reach some fixed level of solute removal. Upon going down any column, we find that the more extractions, the more complete is the removal of solute from S into M. However, the degree of extraction varies inversely as K_R/β: the higher this ratio, the higher is the number of required extractions.

The data also bear out the intuitive notion that the extractant phase (here, M) should be split into smaller portions, with each portion being used for an equilibration, rather than carrying out a single extraction with the total amount of M. For example, assume that a solute of unit partition coefficient is distributed between equal 100-cm^3 volumes of S and M. The mole fraction of solute extracted into M (i.e., after one extraction) is 0.50. Now suppose that M were instead split into ten 10-cm^3 portions and that ten extractions were performed. The ratio K_R/β would in each case be 10 and, according to eq. 165, the fractional amount left in S would be 0.3855 with 0.6145 transferred into M, that is, 30% superior to the use of the entire amount of M at one time.

It is of interest that there exists an upper limit to the amount of solute that can be extracted out of S:

$$^{\infty}x_1^M = e^{-\Sigma V_M / K_R V_S} \tag{166}$$

The Craig Apparatus

In 1943, an apparatus that mimicks tandem-arranged separatory funnels was developed by L. C. Craig for the study of multiple-stage extractions (hence chromatography), but which soon found success in its own right as a bulk-scale preparative separations technique (136). A Craig apparatus consisting of five tubes with $\beta = 1$ is illustrated schematically in Fig. 2.25 for a solute of unit K_R. Initially (first row), the solute is dissolved in M and is contained in a separate box, while only stationary (lower) phase is present in each tube. The tubes N are labeled as shown, with the first being the zeroth for reasons of mathematical convenience. The second row of boxes illustrates the injection process: the contents of the injection box are transferred to the first tube ($N = 0$), thus filling its upper half with M. An equilibration also takes place such that half the solute deposits in S, while half remains in M. The symbol $T_{R,N}$ is used for the fraction of solute in each tube rather than in each phase, and of course must be equal to unity as shown. (The major distinction between multiple extractions from a single separatory funnel and the Craig apparatus thus amounts to description of the solute in terms either of the fractions in each phase or those in each tube. However, the conclusions drawn from each technique are entirely consistent.) We now perform the first ($R = 1$) *intertube* displacement (row three) by transferring the entire amount of M phase from $N = 0$ to $N = 1$ while leaving behind the S phase. Simultaneously, fresh M is transferred from the injection box into the tube labeled $N = 0$. After equilibration, the fractions of solute in each tube are $T_{1,0} = \frac{8}{16}$

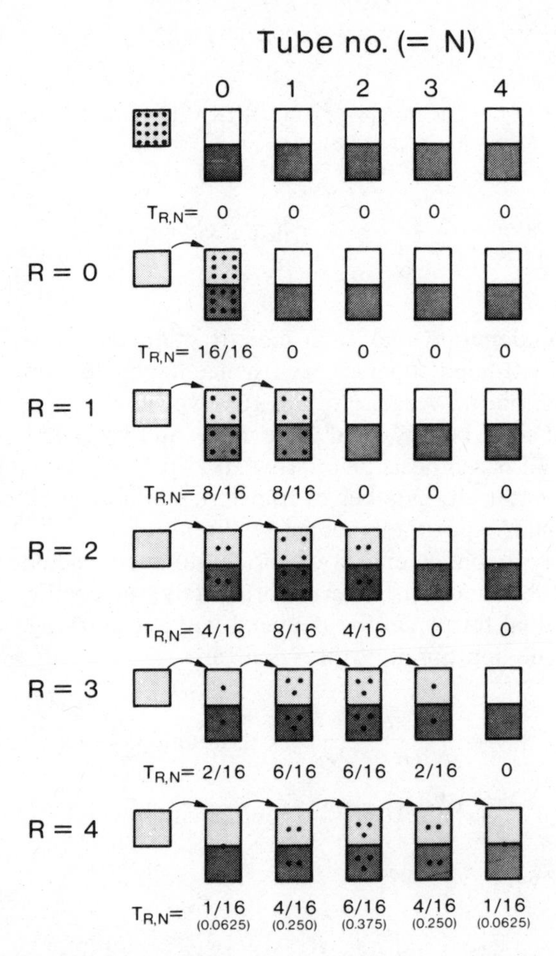

Figure 2.25. Diagrammatic illustration of the operation of a five-tube Craig apparatus.

and $T_{1,1} = \frac{8}{16}$. The process is repeated in rows 4–6 where, in each case, only phase M is moved and replenished. The seventh transfer thus will cause a volume of M + solute to be expelled from the system.

The distribution that results after completing R intertube transfers is conveniently expressed in terms of the mole fractions of solute present initially in each phase in the first tube, ${}^1x_1^S$ and ${}^1x_1^M$, that is, as if we had in fact arranged separatory funnels in tandem:

$$T_{R,N} = ({}^1x_1^S + {}^1x_1^M)^R \tag{167}$$

This is the form of a binomial expansion which, for $R = 4$, yields the terms

$$\begin{aligned} T_{R,N} = \underset{(N=0)}{(x_1^S)^4} &+ \underset{(N=1)}{4x_1^M(x_1^S)^3} + \underset{(N=2)}{6(x_1^M)^2(x_1^S)^2} \\ &+ \underset{(N=3)}{4(x_1^M)^3x_1^S} + \underset{(N=4)}{(x_1^M)^4} \end{aligned}$$

where it is understood that each mole-fraction term refers to the first tube and the left-hand superscripts 1 hence have been omitted.

As a check on the veracity of eq. 167, we substitute $\frac{1}{2}$ for each mole-fraction term since both K_R and β are unity, and proceed to calculate the values given in parenthesis below row six. That is, the amount of solute in each tube after any number of transfers can indeed be calculated directly by expansion of the binomial relation.

It is clearly inconvenient, however, to do such calculations for large numbers of tubes and/or transfers. Fortunately, the coefficients and powers of individual terms can be deduced without carrying out the entire expansion. The appropriate expressions are

$$C_{R,N} = \frac{R!}{N!(R-N)!} \qquad \text{(binomial coefficients)}$$

$$(x_1^S)^{R-N}(x_1^M)^N \qquad \text{(binomial powers)}$$

Thus overall we have

$$T_{R,N} = \left[\frac{R!}{N!(R-N)!}\right](x_1^S)^{R-N}(x_1^M)^N \tag{168}$$

For example, the term describing the third tube ($N = 2$) after four transfers ($R = 4$) is

$$\left[\frac{4!}{2!(4-2)!}\right](x_1^S)^{4-2}(x_1^M)^2 = 6(x_1^S)^2(x_1^M)^2$$

It is comforting to note that attempts at calculating impossible results yield meaningless answers, or answers of zero. For example, the coefficient for the fifth tube ($N = 4$) after only 3 transfers is $3!/4!(3 - 4)! = 1/4(-1)!$ which is undefined physically as well as mathematically.

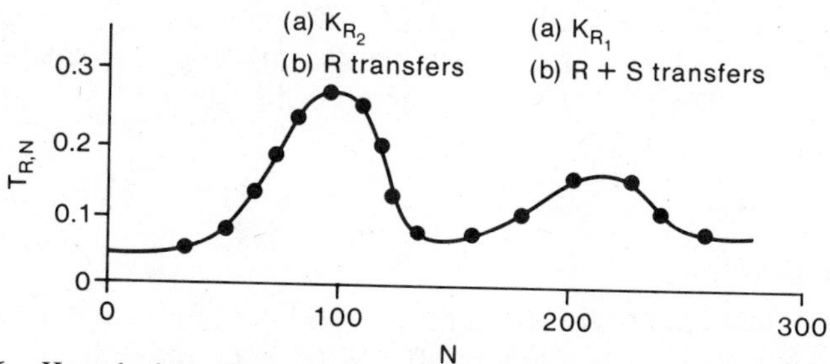

Figure 2.26. Hypothetical distribution in a Craig apparatus (*a*) for two solutes after *R* transfers, where $K_{R2} > K_{R1}$; (*b*) for a single solute after *R* and *R* + *S* transfers.

Figure 2.26 illustrates the hypothetical distribution of two solutes in a Craig apparatus after some number of transfers, where the partition coefficient of the leftmost peak is greater than that to the right. Alternatively, Fig. 2.26 might equally represent the distribution of a single solute after *R* transfers and after *R* + *S* transfers. From a purely qualitative standpoint, we can therefore say that the larger the solute partition coefficient the longer it is retained in the system, that is, the larger is the number of intertube transfers required to move any portion of it from the first to the last tube. Moreover, the further a solute progresses through the system, the broader (more spread out) its band-shape becomes. It is also observed experimentally that the solute distribution rapidly assumes a Poisson shape and then, as *R* and *N* become large (greater than ~100), a Gaussian profile.

We examine first the movement of the peak maximum in terms of *R* and *N*. Consider the fraction of solute in any given (*N*th) tube following two successive intertube transfers, *R* and *R* + 1. The ratio of the fractions remaining in the tube is given by (cf. eq. 168):

$$\frac{T_{R+1,N}}{T_{R,N}} = \frac{\left[\dfrac{(R+1)!}{N!(R+1-N)!}\right](x_1^S)^{R+1-N}(x_1^M)^N}{\left[\dfrac{R!}{N!(R-N)!}\right](x_1^S)^{R-N}(x_1^M)^N}$$

$$= \frac{R+1}{R+1-N}\,x_1^S \qquad (169)$$

However, if the *N*th tube happens to contain the peak maximum, the ratio will be close to unity. That is, the fraction in tube N_{max} (where the subscript max indicates the peak maximum) changes very little upon two successive transfers (i.e., the slope of a tangent-line drawn to the peak maximum is negligible). If R is sufficiently large, we can also make the approximation that $R + 1 \approx R$. Then, recalling that $x_1^S = 1 - x_1^M$, eq. 169 becomes

$$N_{\text{max}} = Rx_1^M \tag{170a}$$

Equation 170a thus provides means of calculating the number of the tube that contains the peak maximum after R transfers have been performed (137). Alternatively, we can write

$$R_{\text{max}} = \frac{N}{x_1^M} \tag{170b}$$

where R_{max} is the number of transfers required to move the peak maximum into any tube N. It is also of importance that the standard deviation of the (binomial) distribution obtained about N_{max} after R transfers in the limit of large R is given by (138):

$$\sigma = (Rx_1^M x_1^S)^{1/2} \tag{170c}$$

In principle, it is possible to calculate the fraction of solute in each tube with eq. 168, and then plot the data in the manner shown in Fig. 2.26 irrespective of the number of tubes, transfers, phase ratio, and solute partition coefficients. However, it would be much more convenient if the distribution curve could somehow be forecast in advance without resorting to computer or other techniques in expanding the relation. To do so requires that we examine the curve shape in the limit of a large number of tubes after an equally large (or even larger) number of intertube transfers, that is, the Gaussian distribution.

Consider that a large number of measurements, r, of the value of a single quantity, n, have been made, and that the average of all the measured values is $\bar{n}$. Each of the measurements will most likely deviate somewhat either positively or negatively from $\bar{n}$ by an amount $n - \bar{n}$. A plot of the number of times a particular value of $n - \bar{n}$ occurs in the data set (the *frequency* of $n - \bar{n}$) against $n - \bar{n}$ resembles a symmetric bell-shaped curve that is centered about the origin if the errors in measurement were in fact truly random. It is called the normal (Gaussian) error curve. Alternatively, we could simply plot the height of the curve y (in arbitrary frequency units) corresponding to each value of $n - \bar{n}$ against $n - \bar{n}$, for which the curve shape would be the same as that described previously.

One form of this is expressed in terms of the fraction y/y_0, where y_0 is the *maximum* peak height at $n - \bar{n} = 0$:

$$y = y_0 e^{-n^2/2\sigma^2} \tag{171}$$

where σ is the standard deviation of the curve. However, in chromatography the peaks (elution profiles) are located at some distance removed from the origin. Therefore, we need to carry out a coordinates transformation from y_0 centered about $n - \bar{n} = 0$ to y_0 centered about $\bar{n}$, that is, utilize an abscissa of zero to infinity rather than one of $-\infty$ to $+\infty$. In doing so, eq. 154 becomes

$$y = y_0 e^{-(n-\bar{n})^2/2\sigma^2} \tag{172}$$

where the curve will be of exactly the same shape as that of the previous relation, except that a perpendicular dropped from its maximum y_0 will now be located along the abscissa at $\bar{n}$.

Curves generated from eq. 172 have several well-defined properties. First, the total area under the curve in terms of its maximum height y_0 and standard deviation σ is given by

$$A = y_0\sigma(2\pi)^{1/2} \tag{173a}$$

The points of inflection n_i are found at

$$n = \bar{n} \pm \sigma \tag{173b}$$

while the baseline distance w_b between points of intersection of tangents drawn to the curve through the inflection points is

$$w_b = 4\sigma \tag{173c}$$

More often than not, the peak baseline breadth rather than its area is of prime concern, and so it is common practice to normalize eq. 172 to unit area by dividing by eq. 173a:

$$y' = \frac{1}{\sigma(2\pi)^{1/2}} e^{-(n-\bar{n})^2/2\sigma^2} \tag{173d}$$

In the particular instance of solute distributions in a Craig apparatus (or, for that matter, in an LC column) we set $n = N + 1 \approx N$, the number of tubes; $\bar{n} = N_{max}$, the tube containing the peak maximum; and $y' =$

$T_{R,N}$, the fraction of solute in each tube. Equations 173 then provide the identities

$$A = T_{R,N_{\max}}\sigma(2\pi)^{1/2} \tag{174a}$$

$$N_i = N_{\max} \pm \sigma \tag{174b}$$

$$\sigma = \frac{w_b}{4} = (N_{\max})^{1/2} = (Rx_1^M x_1^S)^{1/2} \tag{174c}$$

Equation 172 then takes the form

$$T_{R,N} = \frac{1}{(2\pi N_{\max})^{1/2}}\, e^{-(N-N_{\max})^2/2N_{\max}} \tag{175}$$

Moreover, for $N = N_{\max}$, that is, for the tube containing the peak maximum,

$$T_{R,N_{\max}} = (2\pi N_{\max})^{-1/2} = (2\pi R x_1^M x_1^S)^{-1/2} \tag{176}$$

These relations will be used shortly in identifying the number of transfers required to move the peak maximum into the last tube. Meanwhile, we note that eq. 175 describes completely the solute distribution only in terms of the tube number containing the peak maximum after R intertube transfers have been performed (cf. eq. 170).

We now set $N = N_{\text{last}}$ (that is, the last tube) and enquire about the number of transfers $R_{\text{last}}^{\max}$ required to get the peak maximum to the end of the system, that is, into N_{last}. Utilizing eq. 170,

$$R_{\text{last}}^{\max} = N_{\text{last}}/x_1^M \tag{177}$$

For the five-tube example presented earlier, $R_{\text{last}}^{\max} = 4/05. = 8$ transfers required to move the solute peak maximum into the final tube. The next transfer will carry it out of the system (139).

$R_{\text{last}}^{\max}$ is of great practical concern in the use of Craig apparatus, since it involves the actual volume of M phase that must be passed through the system in order to elute (get into the final tube thence out of the system) the solute(s) of interest. Let such a volume of M (corresponding to $R_{\text{last}}^{\max}$ for a particular solute) be designated as V_R^0, the elution volume. In addition, let v_M be the volume of M phase in each tube. The total volume of M in the system at any one time after sufficient transfers have occurred such that the last tube is filled must then be $N_{last}v_M$. We can therefore write

$$V_R^0 = R_{last}^{max} v_M = \frac{N_{last} v_M}{x_1^M} = N_{last} v_M (1 + k') \tag{178a}$$

and

$$V_M = v_M N_{last} \tag{178b}$$

It therefore follows that

$$V_R^0 = V_M(1 + k') \tag{179a}$$

and that (140):

$$k' = \frac{V_R^0 - V_M}{V_M} \tag{179b}$$

In addition, recalling that $\beta = V_M/V_S$ and that $k' = K_R/\beta$,

$$V_R^0 = N_{last}(v_M + K_R v_S) = V_M + K_R V_S \tag{179c}$$

Equations 179 are alternative forms of the fundamental retention equation of chromatography to be derived solely from consideration of liquid-liquid partitioning (141).

We now seek to convert the Gaussian function in terms of N_{max} into one involving N_{last} and V_R^0, that is, in terms of the amount of solute passing into the last tube (thence out of the system) and the solute elution volume, since these are the quantities pertinent to chromatography. First, we recast eq. 175 in terms of σ:

$$T_{R,N} = \frac{1}{\sigma(2\pi)^{1/2}} e^{-(N - N_{max})^2/2\sigma^2} \tag{180}$$

If we employ the mobile-phase volume $v_M R$ which has passed through the last tube as the abscissa instead of one comprised of the tube serial number, eq. 180 becomes

$$T_{R,N_{last}} = \frac{1}{\sigma(2\pi)^{1/2}} e^{-(V_R^0 - v_M R)^2/2\sigma^2} \tag{181}$$

The standard deviation *of the means* of *sets* of data (the *mean* standard deviation) is calculated from the standard deviation of each set of measurements divided by the square root of the total number of sets of measurements (142). In the particular instance of peak maxima passing through the last tube of a Craig apparatus, we choose each set of measurements to correspond to the volume transfer required to get the peak maximum into the ith tube, $v_M R_i^{\max}$, while the total number of such sets of measurements is just the total number of tubes, N_{last}. Then, according to eq. 178a,

$$\sigma_m = \left[\frac{\sum_{i=1}^{N_{\text{last}}} (v_M R_i^{\max})^2}{N_{\text{last}}} \right]^{1/2} = \frac{V_R^0}{(N_{\text{last}})^{1/2}} \approx \frac{V_R^0}{(R_{\text{last}}^{\max} x_1^M x_1^S)^{1/2}} \tag{182}$$

where use has been made in the approximation (closely obeyed for $K_R > 10$) that $x_1^S = (1 - x_1^M) \approx 1$. Substituting eq. 182 into 181 then yields the desired result:

$$T_{R,N_{\text{last}}} = \frac{(N_{\text{last}})^{1/2}}{V_R^0 (2\pi)^{1/2}} \exp \left\{ -(N_{\text{last}}/2) \left(1 - \frac{v_M R}{V_R^0} \right)^2 \right\} \tag{183}$$

Several significant conclusions can be drawn from eq. 166. Most importantly, we see that if the number of tubes is increased, the mean standard deviation increases only as the square root. For example, if the number of tubes is tripled, σ_m is increased only by $3^{1/2}$. Thus the *percentage* of tubes occupied by the solute actually *decreases* as the number of tubes is increased. This has the consequence that a solute band emerging from an apparatus containing many tubes in fact appears to be narrower than if eluted from one of a smaller number of stages, as illustrated in Fig. 2.27.

Next, we note that in the particular instance of chromatography solute partition coefficients typically are large (>50), and that the number of stages ("plates") is usually high (>500). Therefore, as a solute band ($K_R = 50$, say) approaches the last tube of such a Craig apparatus, R begins to approximate $R_{\max}$. That is, 96% ($\pm 2\sigma_m$) of the solute band is spread out over a comparatively few tubes with respect to the total number of tubes, which is a consequence of σ_m increasing only as the square root of N_{last} as noted above. To state the matter in another way, the number of transfers required to move any portion of the solute band into any tube at or near the end of a system comprised of a large (fixed) number of tubes is very nearly the same as the number of transfers required to move the peak *maximum* into that tube:

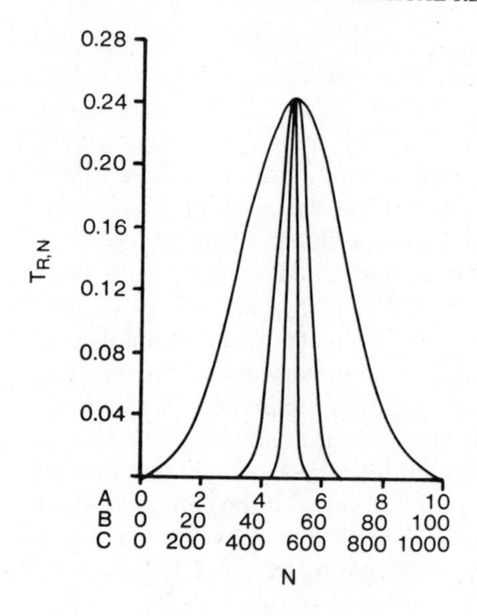

Figure 2.27. Distribution of solute for Craig systems containing different numbers of tubes.

$$R \rightarrow R_{\max} \text{ as } N \rightarrow N_{\text{last}}$$

or

$$R \approx R_{\max} = \frac{N}{x_1^M} \tag{184}$$

We now divide this result by eq. 178a, which expressed the exact number of transfers to get the solute band maximum into the last tube as a function of N_{last}. The result is

$$\frac{R}{R_{\text{last}}^{\max}} \approx \frac{N}{N_{\text{last}}} = \frac{v_M R}{V_R^0} \tag{185}$$

For example, assume that $v_M = v_S = 1$ cm^3. We then calculate, for $K_R = 50$ and $N_{\text{last}} = 500$, that 25,500 transfers (eq. 178a) will be required to move the peak maximum into the last tube and that the elution volume will therefore be 25,500 cm^3. When the peak maximum is actually present in the last tube, half of it has passed from the system while half remains in the system. The mean standard deviation of the peak (exact form of eq. 182) is 1140 cm^3 on an abscissa of $v_M R$ volume units, while the baseline width w_b is four times this ($4\sigma_m$), here, since $v_M = 1$ cm^3, 4560 cm^3. The half of the peak remaining in the system is therefore spread out over 2280 cm^3, that is, $2\sigma_m$. According to eq. 185, this amounts to 45 tubes (143).

We now assume that instead of $K_R = 50$, we have a solute of $K_R = 1$. Proceeding as above, we calculate that 1000 transfers are required to move the peak maximum into the last plate, and that the peak will have a mean standard deviation of 44.7 cm^3. Thus we confirm an intuitive suspicion that solutes with large partition coefficients will emerge from the system as low broad bands, while those of equal area but with small K_R will be relatively tall and sharp. Similarly, if the identities of the solvents were reversed, that is, if species S had instead been employed as mobile phase while species M were used as stationary phase, the partition coefficient of the first example above would be the inverse of its former value, $1/50 = 0.02$ (144). This would then require only $500(1 + 0.02) = 510$ transfers to move the peak maximum into the last tube with a mean standard deviation of 22.8 cm^3. That is, it appears as if the choice of which solvent should be used as which phase is determined by the magnitude of the solute partition coefficient: the combination of S and M that yields the lowest K_R will require the smallest number of intertube transfers to elute the solute through the system. However, there is a price to pay for arranging matters in this fashion, as we shall soon discover.

Resolution

The classical measure of the resolution between two wave-forms 1 and 2 is given by the relation

$$R_s = \frac{\Delta \bar{x}}{1/2(w_{b_1} + w_{b_2})} \tag{186}$$

where $\Delta \bar{x}$ is the difference between the positions of the peak maxima in units corresponding to those of the abscissa, and where w_b is as before the baseline width between tangents drawn through the inflection points of each band. If the two peaks are in fact due to closely eluting solutes emerging from a Craig apparatus, then the abscissa units corresponding to $\bar{x}$ could equally be taken as transfer numbers $R_{\text{last}}^{\text{max}}$ or as elution volumes V_R^0. In either event, the baseline widths of the bands will be very similar for large values of $R_{\text{last}}^{\text{max}}$ and N_{last}, and so it is common practice to replace the average in the denominator of eq. 186 by the value exhibited by either peak, that is, $4\sigma_m$.

We now choose to define satisfactory resolution between two peaks as that which corresponds to a separation of six standard deviations between their maxima (ca. 100% separation): $\Delta V_R^0 = 6\sigma_m$. Equation 186 therefore takes the form

$$R_s = \frac{6\sigma_m}{4\sigma_m} = 1.5$$

If we had instead defined adequate resolution as four peak standard deviations (~96% separation), the value of R_s would then be unity:

$$R_s = \frac{4\sigma_m}{4\sigma_m} = 1.0$$

The physical interpretation of these results is that, for closely eluting substances (i.e., nearly identical baseline widths), $(R_s - 1)$ solute bands can be placed between those of solutes 1 and 2. In practice, the baseline widths of solute bands can diverge quite considerably depending upon the chromatographic technique and conditions. However, in the special instances of temperature programming in GC and LC, as well as mobile-phase programming in LC, it can be arranged that very many solutes elute with near-identical bandwidths. The factor $(R_s - 1)$, where R_s is calculated from the first- and last-eluting compounds, then provides an exact measure of the number of solutes that can be placed in between the two.

We now solve eq. 182 in terms of the number of tubes in the system:

$$N = \left(\frac{V_R^0}{\sigma_m}\right)^2 \tag{187}$$

where it is understood that N refers to N_{last} and where, for convenience, the subscript has been dropped. But from eqs. 174c and 186, for solutes 1 and 2,

$$\sigma_m = \frac{V_R^0(2) - V_R^0(1)}{4R_s} \tag{188}$$

Thus, we arrive at a description of resolution in terms of the solute elution volumes and the number of tubes:

$$R_s = \frac{V_R^0(2) - V_R^0(1)}{4V_R^0(2)} N^{1/2} \tag{189}$$

Without proof,

$$\frac{V_R^0(2)}{V_R^0(2) - V_R^0(1)} = \left(\frac{\alpha}{\alpha - 1}\right)\left(\frac{k_2' + 1}{k_2'}\right) \tag{190}$$

The resolution relation, eq. 189, therefore becomes

$$R_s = \frac{1}{4}\left(\frac{\alpha - 1}{\alpha}\right)\left(\frac{k_2'}{k_2' + 1}\right)N^{1/2} = \frac{1}{4}\left(\frac{\alpha - 1}{\alpha}\right)\left(\frac{K_R}{K_R + \beta}\right)N^{1/2} \tag{191}$$

Equation 191 is a kind of summary fundamental retention equation of chromatography, since it relates separations (R_s, α) to measurable quantities (k'), the system variables (N and β), and the solute partition coefficients (K_R). For example, we see immediately that resolution increases only as the square root of the number of tubes. Thus a twofold improvement in resolution mandates a fourfold increase in N. Moreover, for capacity factors greater than 10 (β small compared to K_R), the term $[k'/(k' + 1)]$ tends to unity, whence resolution is governed simply by the quantity in brackets involving α, and $N^{1/2}$. For example, suppose that two solutes exhibit partition coefficients of 50 and 60, and that the Craig apparatus at hand contains 500 tubes each of $v_S = v_M$. Then, according to eq. 162b, $k'(2) = 60$ while $k'(1) = 50$. Therefore, α is $60/50 = 1.2$ and the resolution to be expected is 0.93, that is, close to $4\sigma_m$. If we had taken into account the term $[k'(2)/(k'(2) + 1)]$, we would have calculated an expected resolution of 0.92. Now, in contrast, assume that the capacity factors of the two solutes are 1.2 and 1, respectively. Therefore, α is still 1.2, with $R_s = 0.93$, when the capacity-factor term in eq. 191 is ignored. However, when $k'(2)$ is taken into account, R_s is calculated to be 0.51, that is, almost half of the approximate value.

Thus the capacity factors of the solutes play an important role in governing the resolution between solutes: separations become increasingly more difficult (require more tubes) as the values of k' decrease substantially below 10. This then belies the seemingly attractive notion put forth earlier that the choice of which solvent to use as S and which to let be M be made on the basis of minimizing K_R. In fact, we see that the criterion actually amounts to which combination will yield solute capacity factors that will provide the requisite resolution with the number of tubes comprising the system at hand.

The number of tubes required to effect a particular separation for a given alpha and k', that is, N_{req}, can be calculated directly from eq. 191.

Suppose that we desired a resolution of unity ($4\sigma_m$ separation). Then, solving for $N(=N_{req})$,

$$N_{req} = 16 \left(\frac{\alpha}{\alpha - 1}\right)^2 \left(\frac{k' + 1}{k'}\right)^2 \tag{192a}$$

Alternatively, if a $6\sigma_m$ separation were desired,

$$N_{req} = 36 \left(\frac{\alpha}{\alpha - 1}\right)^2 \left(\frac{k' + 1}{k'}\right)^2 \tag{192b}$$

The price to pay for $6\sigma_m$ resolution is therefore $36/16 = 2.25$ times the number of tubes required for a $4\sigma_m$ separation.

Equations 192 were first derived by Purnell (145) with regard to open-tubular columns in GC, but the conclusions to be drawn apply equally to the Craig apparatus as well as any form of chromatography. The point to be made is that a very large number of tubes is not sufficient by itself to ensure that a separation will be achieved. The solute capacity factors must also be taken into account. As a result, Purnell proposed that an effective number N_{eff} of tubes (plates, stages, etc.) be defined such that

$$N_{eff} = N \left(\frac{k'}{k' + 1}\right)^2 \tag{193}$$

Now, since according to eqs. 174c and 182, N is given by

$$N = 16 \left(\frac{V_R^0}{w_b}\right)^2 \tag{194}$$

it is possible to show via eq. 193 that N_{eff} is equal to

$$N_{eff} = 16 \left(\frac{V_R^0 - V_M}{w_b}\right)^2 \tag{195}$$

Subtraction of V_M from V_R^0 is a sensible adjustment to the elution volume, since the difference, which is referred to as the "net" retention volume V_N, represents the overall volume of M phase passed through the last plate minus the volume of M required just to fill all the tubes in the system. That is, to get any solute into the last tube of the system, a volume V_M must at least be put into the system to start with. Comparison of the amounts of M solvent in excess of this amount that are required to get solutes into the last plate, V_N, then offers a fair description of differences of the extent to which the solutes are retained by the system. It is also useful to note that the net retention volume is zero for solutes of $K_R = 0$, that is, for solutes that are completely insoluble in (unretained by) the stationary (S) phase.

As a corollary, we note the relation (cf. eqs. 67 and 179b, and ref. 139):

$$K_R = \frac{V_R^0 - V_M}{V_S} = \frac{V_N}{V_S} \tag{196}$$

which emphasizes that differences in net retention volumes arise strictly as a consequence of differences in solute partition coefficients.

The Separation Factor

There is in some instances a considerable degree of inconvenience in the measurement of values of R_s since, in order to do so, peak widths either at the baseline or at some related height interval (e.g., h/e) must be determined. Moreover, in order to calculate solute capacity factors, the system volume of M phase must either be known or be amenable to measurement. While this poses no difficulty in the use of a Craig apparatus or, for that matter, in gas chromatography, the column volume of M phase in liquid chromatography is very often ambiguous. As a result, Jones and Wellington (146) derived what is known as the separation factor S_f as an alternative to R_s:

$$S_f = \frac{2R_s}{N^{1/2}} = \frac{V_R^0(2) - V_R^0(1)}{V_R^0(2) + V_R^0(1)} \tag{197}$$

That is, the difference between the retentions of two solutes, divided by their sum, provides a measure of the separation achieved in terms both of resolution and the number of system tubes. Furthermore, the need to measure V_M and w_b has been entirely eliminated. Values of S_f, analogous to the relative volatility parameter α, can therefore be determined unambiguously, and will be used later in the optimization of LC mobile phase compositions.

2.5.3 Extension of Plate Theory to a Continuum

Liquid Chromatography

The keen insight of Martin and Synge (147) in 1941 was that, in contrast to countercurrent distributions wherein two liquid solvents (phases) are made to move through each other, there is no reason either from a fundamental or a practical standpoint that, in seeking to achieve separations, one of the liquids cannot be held fixed and the second made to move over or past it. As a result, they immobilized one liquid (the stationary phase) on the surface of an inert solid (paper or silica gel) and then caused another liquid (the mobile phase) to flow past it. During the course of passage through the system, solutes were then partitioned to different extents between the two phases in accordance with their partition coefficients, thus providing separations. The technique is therefore entirely analogous to the Craig apparatus except that there are no discretely defined "tubes"; moreover, the results obtained are precisely those to be derived from consideration of the latter as given in the preceding sections (134).

Consider, first, that a particular liquid-chromatographic system consists of an immobilized liquid dispersed on the surface of a quantity of diatomaceous earth, and that the material (packing) is contained in a vertically held glass tube such that a second liquid will flow through the system under the influence of gravity at some fixed volume per unit time. Then, after allowing a certain volume of the second liquid (the mobile phase) to pass through the column, hence filling all spaces (as well as support pores) within the tube not occupied by the packing, we arrange that an infinitely small amount of solute be deposited on the top of the column, and then washed through it by the mobile phase. We assume at this point that the sorption isotherm of the solute is completely linear; that, as the solute travels through the column, the molecules stay grouped as an infinitely thin band; and that the rate of solute longitudinal traversal through the system is slow compared to the time required to maintain an equilibrium distribution between the two phases. Thus, the time of emergence of solute from the system will be independent of the amount of solute injected into the system; the elution profile will be a single vertical line as opposed to a Gaussian band; and the elution volume corresponding to the emergence of the solute will be related directly to the partition coefficient.

This model is known as the linear ideal case since the sorption isotherm is taken to be linear; since the solute molecules hypothetically do not diffuse radially or longitudinally upon passage through the system; and since the attainment of equilibrium is taken to be so fast that molecules are not spread apart as a result of slow kinetics of mass transport. However, while the first of these assumptions can frequently be met in practice, the latter two are never achieved (discussion of which is deferred to Sec. 2.5.4). Even so, the model clearly represents an advance over the consideration of discrete tubes since it corresponds directly to the chromatographic process, that is, represents the continuum limit of the Craig apparatus.

We next consider the actual time, called the retention time t_R, that a solute takes to pass through the liquid-chromatographic column. The velocity, or rate of travel, of a solute is proportional to the fractional amount of time that it spends in the mobile phase as well as the linear velocity of the mobile phase. Thus, we consider that the solute does not move when in the stationary phase. The pertinent relation is

$$\text{rate of travel} = \bar{u}x_1^M = \bar{u}\left[\frac{C_1^M V_M}{C_1^S V_S + C_1^M V_M}\right] = \bar{u}\left[\frac{1}{1 + K_R/\beta}\right] \tag{198}$$

The rate of travel of a solute through the LC system (distance per unit time) can also be thought of in terms purely of a velocity, that is, the column length L (distance traversed) divided by the retention (travel) time t_R:

$$\text{rate of travel} = \frac{\text{column length, } L}{\text{retention time, } t_R} \tag{199}$$

We now set eqs. 198 and 199 equal:

$$\frac{L}{t_R} = \bar{u}\left[\frac{1}{1 + K_R/\beta}\right] \tag{200a}$$

or, upon rearrangment,

$$t_R = \frac{L}{\bar{u}}(1 + K_R/\beta) \tag{200b}$$

The quantity $L/\bar{u}$ just the time a nonsorbed ($K_R = 0$) solute takes to pass through the system, which we choose to give the symbol t_M. Thus eq. 200b becomes

$$t_R = t_M\left(1 + K_R\frac{V_S}{V_M}\right) \tag{201}$$

where β has been replaced by the ratio of the volumes of the phases. We now multiply both sides of eq. 201 by the mobile-phase flow rate F and find

$$V_R^0 = V_M\left(1 + K_R\frac{V_S}{V_M}\right) = V_M + K_R V_S \tag{202}$$

that is, exactly the result predicted by eq. 179 [which, if nothing else, provides considerable justification for the (much-maligned) exposition of chromatographic principles via the Craig apparatus]. Moreover, we can now also cast the solute capacity factor in terms of retention times:

$$k' = \frac{t_R - t_M}{t_M} \tag{203}$$

In doing so, we recognize that these relations are yet other forms of the fundamental retention equation of chromatography.

For historical reasons, and in addition to being a matter of convenience, solute retentions in thin-layer and paper chromatography are reported very frequently as "retardation factors" R_f, that is, in terms of the distance traveled by the spot d_i divided by the distance traveled by the solvent front D: $R_f = d_i/D$. However, we suspect immediately that R_f must somehow be related to t_R, t_M, k', K_R, and β. The relation is

$$R_f = \frac{d_i}{D} = \frac{t_M}{t_R} \tag{204}$$

Inversion of this definition, followed by subtraction of unity from each side, eventually leads to the expression

$$R_f = \frac{1}{1 + k'} = \frac{1}{1 + K_R/\beta} = x_1^M \tag{205}$$

That is, retardation factors amount simply to the fraction of solute to be found in the mobile phase, and it is always possible to convert from R_f to k' and vice versa. This is a sensible result also when we recognize that t_R is the sum of the time that the solute spends in the mobile phase t_M plus the time spent in the stationary phase, t_S: $t_R = t_M + t_S$. Thus

$$x_1^M = \frac{t_M}{t_M + t_S} \tag{206a}$$

$$x_1^S = \frac{t_S}{t_M + t_S} \tag{206b}$$

It is of interest to note that, since $\alpha_{2/1} = k'(2)/k'(1)$, relative retentions are calculable in thin-layer chromatography from R_f values; substitution of eqs. 203 and 204 readily provides

$$\alpha_{2/1} = \frac{R_f(1)}{R_f(2)} \left[\frac{1 - R_f(2)}{1 - R_f(1)} \right] \approx \frac{d_1}{d_2} \tag{207}$$

where the approximation holds in the limit that $R_f(i) \ll 1$.

Gas Chromatography

According to the linear ideal model of chromatography there is no band-spreading and so, there is no *fundamental* difference in the elution equations pertinent to liquid chromatography and gas chromatography except those that take into account compressibility and virial effects of the carrier gas, as discussed in Secs. 2.2.5 and 2.4. We recall, first, the James-Martin pressure-correction factor j:

$$j = \frac{3}{2} \left[\frac{(p_i/p_o)^2 - 1}{(p_i/p_o)^3 - 1} \right] \tag{25}$$

and that the average carrier velocity $\bar{u}$ is given by $u_o j$, where u_o is the linear velocity measured at the column outlet. The "raw" solute retention volume V_R and the retention volume of a nonsorbed ($K_R = 0$) solute V_A (i.e., those observed at the column outlet) are then related to the "corrected" retention volume V_R^0 and the true volume of the column mobile phase V_M by

$$jV_R = V_R^0 \tag{208a}$$

$$jV_A = V_M \tag{208b}$$

$$j(V_R - V_A) = (V_R^0 - V_M) = V_N \tag{208c}$$

in gas chromatography. Note that since in liquid chromatography $j = 1$ (the carrier is assumed to be incompressible below $\sim 10^4$ psig),

$$V_R \equiv V_R^0 \tag{209a}$$

$$V_A \equiv V_M \tag{209b}$$

$$(V_R^0 - V_M) \equiv V_N \tag{209c}$$

In gas chromatography, however, we must distinguish between those volumes that have been fully corrected for compressibility effects, and those that have not. To do so requires simply that the left-hand sides of eqs. 209 each be multiplied by j as given by eqs. 208.

There remains only one other form of retention volume of interest in gas chromatography, namely, the so-called adjusted retention volume V'_R (that has been "adjusted" for V_A but not j), that is derived from the difference between V_R and V_A:

$$V'_R = V_R - V_A = \frac{V_R^0 - V_M}{j} = \frac{V_N}{j} \tag{210}$$

It is thereby possible to show that

$$V_R = V_A\left(1 + K_R \frac{V_S}{V_M}\right) \tag{211a}$$

and that

$$t_R = t_A\left(1 + K_R \frac{V_S}{V_M}\right) \tag{211b}$$

which are precisely identical to the relations derived earlier for the Craig apparatus, eqs. 179, and for liquid chromatography, eq. 202, as is readily shown by substitution of the identities referred to in eqs. 209.

There now remains only one additional complication introduced by compressibility effects, that is, measurement of (gaseous) carrier flow rates. The volume of gas collected per unit time at the column outlet of a GC apparatus F_{fm} is normally measured with a soap-bubble flow-meter, and must therefore be corrected both for the vapor pressure of water above the soap solution as well as to the column temperature, F_c. Application of Dalton's and Boyle's relations yields the expression

$$F_c = F_{fm}\left(\frac{T_c}{T_{fm}}\right)\left(\frac{p_{fm} - p_w}{p_{fm}}\right) \tag{212}$$

where T_c is the column temperature, and where p_w is the vapor pressure of water at the flow-meter temperature T_{fm}.

The virial correction to K_R was given previously by eq. 70:

$$\ln K_R = \ln K_R^0 + \beta p_o J_3^4 \tag{70}$$

where β was defined in eq. 71, repeated here for convenience:

$$\beta = \frac{2B_{12} - \overline{V}_1^\infty}{RT} \tag{71}$$

and where the average column pressure has been replaced by $p_o J_3^4$:

$$J_3^4 = \frac{3}{4}\left[\frac{(p_i/p_o)^4 - 1}{(p_i/p_o)^3 - 1}\right] \tag{213}$$

The importance of virial effects lies in the decrease of retention times (or volumes) with increasing column pressure drop for all carriers except helium and hydrogen (Sec 2.2.5), and in that retention order can be adjusted depending upon the identity of the carrier or blend of carriers. We defer discussion of the latter since first we must correct the linear ideal model of chromatography to take into account peak bandspreading, that is, the linear nonideal model of elution.

2.5.4 Rate Theories: the Linear Nonideal Model of Gas Chromatography

The so-called "rate" theories of chromatography seek to detail factors that contribute to dispersion (spreading) of the solute band as it passes through the system. The earliest of these efforts recognizable as applicable to the chromatographic process are due to Wilson (148) and deVault (149) in the 1940's, followed shortly thereafter by Glueckauf (150), Lapidus and Amundson (151), and Klinkenberg and Sjenitzer (152). However, it was not until the resultant relations were simplified and greatly clarified by Van Deemter, Zuiderweg, and Klinkenberg (153) that any cognizance was taken of the significance of kinetic processes extant in chromatographic systems. Substantiation of the importance of these effects soon followed, the theory often being cited (with considerable justification) as reaching full maturity with the prediction, thence demonstration in practice, of the advantages of open-tubular columns by Golay (154).

In the author's view, the most lucid summary of these efforts, including

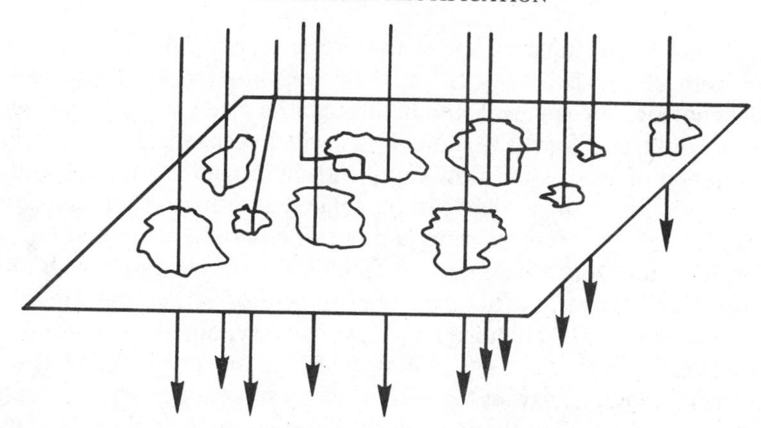

Figure 2.28. Planar cross-sectional conception of streaming of solute molecules through interparticle spaces in a packed column.

the various stochastic approaches (155–159), remains that by Purnell (ref. 45, Chs. 8, 9), where the conclusions reached in that work (applicable in fact both to GC and LC) have since been altered largely only in form. The various contributions to solute bandbroadening recounted here follow that treatment, where the conceptually simpler aspects of gas chromatography are considered first and then extended to liquid chromatography.

Streaming

Consider a planar cross-section of a packed GC column, as illustrated schematically in Fig. 2.28: the "pores" in the plane correspond to interstitial spaces between packing particles of diameter d_p. Suppose that this were 120/140-mesh (105–125 μ) diatomaceous earth, and that the column were packed such that the interstitial spaces were (optimistically) on average 1% of the particle diameter, that is, ~1–1.2 μ. Now suppose that hydrogen were the carrier gas at a pressure above the plane just slightly above atmospheric and at atmospheric pressure just below the plane. Then, if the volume of the container above the plane were 22.4 dm^3 and the temperature 0°C, the mean free path l for a molecular diameter of ~2 Å amounts roughly to 0.2 μ. That is, molecules would "stream" through the plane, as opposed to "effusing" through the pores as in Knudson flow. However, the rate of streaming (just as in Knudson flow) would nevertheless be proportional to the quantity $(P/2\pi\rho)^{1/2}$, that is, directly proportional to the square root of the pressure, but inversely so to the square root of the density of the gas. For example, assuming ideal gas-phase behavior, nitrogen will stream through the plane at a given pressure approximately one fourth as quickly as will hydrogen. The immediate consequence of this is that flow rates with hydrogen carrier gas will be roughly four times those that can be achieved with nitrogen at a given inlet/outlet column pressure ratio and packing particle diameter.

Consider next that very many molecules are streaming through interstitial spaces between particles of uneven size or of nonuniform distribution in a packed GC column (i.e., very many hypothetical planes of the type illustrated in Fig. 2.28). The molecules will be spread symmetrically along the longitudinal axis of the system as a result of differences in transit times through each (hypothetical) plane. Let the average thickness of each plane correspond to the average particle diameter. The time required for an individual molecule to pass through a plane (i.e., past a particle of diameter d_p) is then given simply by $d_p/\overline{u}$, where $\overline{u}$ is the average linear velocity of the molecules at that point in the system. We can therefore write the standard deviation τ of what is presumed to be a Gaussian distribution of the times of emergence of molecules from the plane in the form

$$\tau = (2\lambda)^{1/2}\left(\frac{d_p}{\overline{u}}\right) \tag{214}$$

where τ is in units of time and where the factor $(2\lambda)^{1/2}$ is an empirical constant. Squaring both sides, and recognizing that $\overline{u}^2\tau^2 = \sigma^2$, the localized length variance, yields

$$\overline{u}^2\tau^2 = 2\lambda d_p^2 = \sigma^2 \tag{215}$$

The number of times that molecules are redistributed is given by the number of particle diameters in the column (number of hypothetical planes to be traversed), that is, the column length divided by the diameter of an individual particle, L/d_p. We therefore conclude that the variance per unit length of molecules passing through the entire length of the system is

$$\frac{\sigma_S^2}{L} = 2\lambda d_p \tag{216}$$

where a subscript S has been included in the variance symbol to indicate that it arises as a result of inhomogeneities in the streaming process.

The derivation of this first contribution to band-broadening has been brought out in full in order to demonstrate that the process actually has nothing to do with so-called eddying, as it is commonly referred to, nor with diffusional phenomena, as many have pointed out. Part of the confusion that has arisen over the years on this point undoubtedly stems from the unfortunate choice of $(2\lambda)^{1/2}$ in eq. 214 since, as we shall see, a similar term is involved in what is in fact properly a diffusional term. However, in streaming, band dispersion arises simply as a result of longer or shorter path-lengths that molecules must travel in order to pass from one end of the column to the other. It is as if we had caused a gas at some common inlet pressure to pass through a number of bundled and completely empty capillary tubes, each of identical diameter but of different length. The distribution of the gas molecules as they emerged would then be a reflection of the diversity of the capillary lengths that had been employed, and could be described, as above, in terms of a variance per unit length of open tube.

[It was on this basis that Golay (154) proposed utilizing a (wetted-wall) capillary tube as a column, thereby eliminating altogether band-broadening due to multiple-capillary (packed-column) effects. The success of his prediction is of course now universally recognized.

In contrast, the problem associated with the use of several "bundled" (i.e., parallel-operated) capillary GC columns (for example, in order to provide higher sample capacity) is tantamount to incurring band dispersion on account of empty tubes of identical length but of varying diameter. This method can therefore be expected to be practicable only if the diameters of the tubes (in addition to the amount of stationary phase in each) are very closely matched. The limit required appears to be a standard deviation of ~1% or less (160).]

By analogy, it is also possible to say that band broadening due to stream flow maldistribution can be minimized in packed-column GC (as well as in LC) by employing packings of the smallest d_p of the highest possible uniformity commensurate with the highest pressure limit permitted by the system. Moreover, a column packed with particles ranging widely in size will give rise to band dispersion governed by those of the largest d_p, while the pressure drop will be dictated by those of the smallest, both effects of course being detrimental to column performance. In either event, it is also clear that increasing the particle size can only result in greater band spreading.

We make mention that a great many experimental evaluations yield a range of 0.5 to 1.5 for λ and that, in the routine use of gas chromatography, the optimal particle size appears to be 120/140 mesh.

Longitudinal Diffusion

The Einstein equation for the variance due to diffusional spreading from a point source during some time t was given in Sec 2.3 as $\sigma^2 = 2\overline{D}_{1M}t$, where $\overline{D}_{1M}$ is the average diffusion coefficient of the solute in the mobile phase. Recognizing that $\overline{u} = L/t$, the relation becomes

$$\frac{\sigma_L^2}{L} = \frac{2\overline{D}_{1M}}{\overline{u}} \tag{217}$$

where $\overline{u}$ is as usual the average linear velocity of the carrier, and where a subscript L has been appended to the variance symbol to distinguish it from that used for stream dispersion. Equation 217 is the form expected for the dependence of diffusional bandspreading along the longitudinal axis of a chromatographic column, if for no other reason than the higher the linear velocity the more brief will be the time that molecules have to diffuse.

Van Deemter and his colleagues (153) have suggested that a "labyrinth" factor γ be included in eq. 217 such that, for packed columns,

$$\frac{\sigma_L^2}{L} = \frac{2\gamma\overline{D}_{1M}}{\overline{u}} \tag{218}$$

where γ is said to lie between 0.5 and 1, and where it is often found in practice that such a correction term must be added in order that experimental data conform to the relation.

A point often overlooked in the utilization of eqs. 217 and 218 is that the (pressure-dependent) term $\overline{D}_{1M}$ corresponding to some value $\overline{u}$ is not in fact directly accessible from experiment. On the other hand, $\overline{D}_{1M}/\overline{u} = D_{1M}^o/u_o$, where the o's indicate the column outlet, so that eq. 201 can be written as

$$\frac{\sigma_L^2}{L} = \frac{2\gamma D_{1M}^o}{u_o} \tag{219}$$

Stationary-Phase Resistance to Mass Transfer

The kinetics of mass transport across the stationary-phase/mobile-phase interface are of finite duration such that the solute distribution between each is displaced locally from equilibrium. A hypothetical system is il-

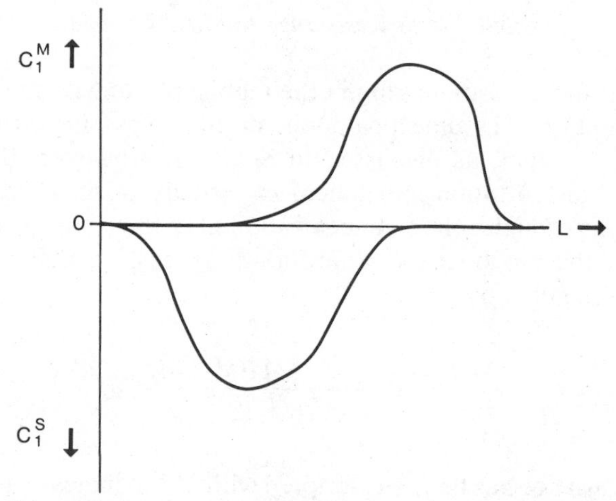

Figure 2.29. Solute bands in mobile and stationary phases shown displaced from equilibrium.

lustrated in Fig. 2.29: the maximal concentration of solute in the mobile phase has progressed further down the column than has that in the stationary phase. The resultant band dispersion per unit length of column is given by

$$\frac{\sigma_{ST}^2}{L} = \frac{2k'd_f^2}{3(1 + k')^2 D_{1S}} \bar{u} \tag{220}$$

where d_f is the stationary-phase film thickness, and where D_{1S} is the diffusion coefficient of the solute dissolved in it. The variance is given a subscript ST to indicate finite stationary-phase mass transfer.

We see, first, that when k' tends to zero, σ_{ST}^2 likewise becomes negligible, that is, if there is no solution there can be no band spreading due to solution/dissolution kinetics. We note, too, that the variance is proportional to the square of the thickness of stationary-phase film: thinner coatings of stationary phase will therefore yield columns of higher efficiency. Moreover, since the square of d_f is involved, the improvement on passing from, say, a layer of phase of 1 μ to one of 0.5 μ can be expected to be a factor of four. The variance as written is also directly proportional to the average linear velocity: the faster the flow, the greater the displacement from equilibrium. Finally, the appropriate value of the linear velocity is apparently the average (as opposed to the outlet) value, since D_{1S} is approximately independent of the column pressure drop (Sec. 2.2.3).

Mobile-Phase Resistance to Mass Transfer

The flow profile of a solute plug in the mobile phase will, in all instances save turbulent flow, assume a parabolic profile on passing through a chromatographic system, as discussed in Sec. 2.4. However, the resultant dispersion (concentration imbalance) is partially offset by diffusion between regions of high solute concentration near the walls of the tube and the head of the parabola. An appropriate expression that relates these phenomena is given by

$$\frac{\sigma_{MT}^2}{L} = \frac{1 + 6k' + 11(k')^2}{24(1 + k')^2} \frac{d_p^2}{D_{1M}^o} u_o \tag{221}$$

where the variance has been subscripted with *MT* to indicate mobile-phase mass transfer.

As one would intuit, eq. 221 predicts that the flow-profile variance will be relaxed by particles of reduced diameter, and by higher values of the gas-phase solute diffusion coefficient. Small values of the solute capacity factor are also favored (result in a smaller variance), as are low linear carrier velocities.

Summation Expression for the Column Variance

We now make the assumption that the variances are additive, and that the sum gives the overall variance exhibited by a solute on passing through a packed chromatographic column

$$\frac{\sigma_{\text{col}}^2}{L} = \frac{1}{L}(\sigma_S^2 + \sigma_L^2 + \sigma_{ST}^2 + \sigma_{MT}^2)$$

$$= \frac{1}{L}\Bigg[\underset{\text{(eq. 216)}}{2\lambda d_p} + \underset{\text{(eq. 219)}}{\frac{2\gamma D_{1M}^o}{u_o}} + \underset{\text{(eq. 220)}}{\frac{2k'}{3(1 + k')^2}\frac{d_f^2}{D_{1S}}\bar{u}}$$

$$+ \underset{\text{(eq. 221)}}{\frac{1 + 6k' + 11(k')^2}{24(1 + k')^2}\frac{d_p^2}{D_{1M}^o} u_o}\Bigg] \tag{222}$$

In the instance of open-tubular columns, we need apparently only drop the streaming term and replace the packing particle diameter by the tube radius r (161):

$$\frac{\sigma_{col}^2}{L} = \frac{1}{L}(\sigma_L^2 + \sigma_{ST}^2 + \sigma_{MT}^2)$$

$$= \frac{1}{L}\left[\frac{2\gamma D_{1M}^o}{u_o} + \frac{2k'}{3(1+k')^2}\frac{r^2}{D_{1S}}\overline{u} + \frac{1+6k'+11(k')^2}{24(1+k')^2}\frac{r^2}{D_{1M}^o}u_o\right] \tag{223}$$

The variance per unit length of column clearly must somehow or other be related to the number of theoretical plates N. It can be shown that

$$\frac{\sigma}{L} = \frac{w_b/4}{t_R} \tag{224}$$

so that, according to eq. 187,

$$N = (L/\sigma)^2 \tag{225}$$

Dividing this result into the column length then provides

$$\frac{L}{(L/\sigma)^2} = \frac{L}{N} = \frac{\sigma^2}{L} \tag{226}$$

where it is understood that the variance is that arising from the sum of all bandspreading contributions incurred during passage of the solute through the chromatographic column, and where the subscript col has been dropped for convenience.

The length corresponding to a single plate in the chromatographic system, L/N, is given the symbol H and is called the height equivalent to a theoretical plate (abbreviated at times as HETP), or simply the plate height. [Use of the word "height" rather than "length" or "distance" stems from the initial derivation of eqs. 224–226 in terms of plant-scale distillation towers (132, 133).] The importance of H derives from eq. 226, that is, H is equal to the variance of the solute band per unit length of column. The square root of the product (HL) is then the standard deviation of the solute band as it emerges from the system.

Equations 222 and 223 are often represented in simplified form by the expressions

$$H = A + B/u_o + C_S\overline{u} + C_M u_o \tag{227}$$

$$H = B/u_o + C_S\overline{u} + C_M u_o \tag{228}$$

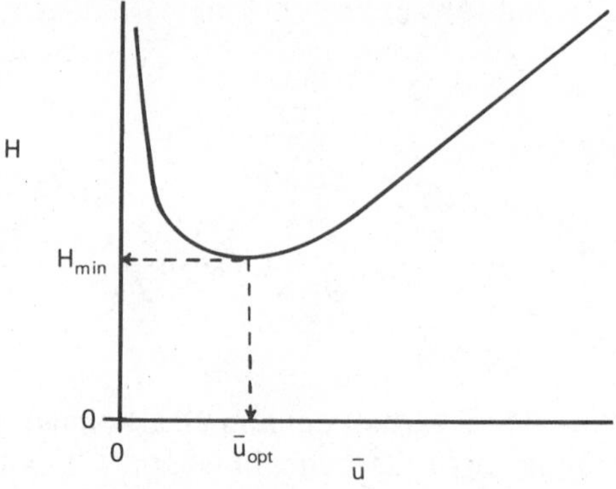

Figure 2.30. Typical form of van Deemter plots of plate height against linear mobile-phase velocity.

We note in addition, and by analogy with eqs. 193 and 195, that an "effective" plate height H_{eff} can be defined such that

$$H_{\text{eff}} = \frac{L}{N_{\text{eff}}} \tag{229}$$

Disputation of the Additivity of Variances

An elegant argument has been made by Giddings (158) that, in fact, variances arising from the A and C_M terms are not independent, but instead are said to be mutually dependent, or "coupled". The extended rate equation is accordingly written as

$$H = \frac{B}{u_o} + C_S\bar{u} + \sum \left(\frac{1}{A} + \frac{1}{C_M u_o}\right)^{-1} \tag{230}$$

where the similarity to (and disparity with) eq. 227 is immediately apparent.

In comparative tests of the two relations with packed columns, it is found that there are indeed deviations from the hyperbolic shape (Fig. 2.30; see below) of plots of H against $\bar{u}$ at high $\bar{u}$: that predicted by eq. 227 is a straight line extending to infinity, whereas eq. 230 indicates a leveling-off effect at high linear mobile-phase velocity. However, in practice, difficulties are encountered in attempts at establishing the validity of one model over another due to wall effects at high flow rates. Moreover, second and third virial effects come into play at pressures utilized, for example, by Giddings, yet were not accounted for in his data reduction procedures. Even so, the matter rarely if ever arises in practical analytical gas chromatography, since the system flow rate generally is adjusted to no more than 10–20% beyond optimum, that is, well within the region described by eqs. 227 and 228.

Experimental Evaluation of Contributions to the Plate Height

As mentioned above, it is found experimentally that plots of H against $\bar{u}$ are invariably of the form of a hyperbola, Fig. 2.30. However, eqs. 227 and 228 contain terms both in u_o and $\bar{u}$. It is therefore not possible to derive values of A, B, C_S, or C_M directly from such plots, a point that is frequently overlooked. Nor is the difference necessarily trivial, since in GC, $\bar{u} = u_o j$, and since j can assume values of from zero to unity. Schupp (162) has summarized four techniques suitable for independent evaluation of each of the van Deemter terms relevant to gas chromatography.

The leftmost curved portion of plots such as that in Fig. 2.30 is due to longitudinal diffusion, while the right-hand curved branch arises as a result of slow stationary- and mobile-phase mass transport. According to eqs. 222 and 223, the factors that govern the magnitudes of each are the particle diameter (or tube radius); the solute gas-phase and liquid-phase diffusion coefficients; and the solute capacity factor (hence the partition coefficient and phase ratio). A comprehensive understanding of the effects of each of these parameters is obviously desirable in order, if for no other reason, that columns of the highest possible efficiency may be fabricated reproducibly (thence operated) in accordance with strictly defined guidelines. However, while much work in the area has been carried out over the years, it is fair to say even today that the complexity of the chromatographic process is such that the manner and magnitude of interplay of the various band-spreading processes (and even at times their existence or absence) largely remain open to question. Even so, the practical consequences, that is, observable changes of H arising from alteration of the gas-chromatographic system parameters (e.g., flow rate, temperature, and so forth) are for the most part well documented, and are considered in turn in what follows.

In the earliest work on mass-transfer nonequilibrium, it was commonplace to measure (in error) what amounts to a composite C term, that is, $C = C_M + C_S$. It is therefore difficult to interpret the results in terms of extended relations such as eq. 227, although some qualitative guidelines can be derived. In general, it has been found that plots of C against solute capacity factor k' (i.e., liquid loading) rise initially from the origin in a geometric fashion, but then give way eventually to a straight line. The transition from one curve form to another is usually found at quite low amounts of stationary phase. Several interpretations have been placed on the phenomenon, including purported pore-filling (as opposed to thin-film spreading); droplet formation and distribution; and gas–liquid and gas–solid (i.e., bare-packing or bare column-wall) interfacial adsorption. More recently, it has been argued that when liquids are spread very thinly onto virtually any surface, there is no longer random molecular ordering, as in bulk stationary phase, but rather semicrystalline mesomorphism, presumably induced by the solid surface. Solute diffusion coefficients as well as retentions in such liquids must then be considerably different from their counterparts in thick-film systems, as has been verified experimentally by Serpinet (163).

In the absence of anomalous thin-film behavior and gas/bare-packing or wall adsorption, it must evidently be true that as the stationary-phase loading is decreased, the overall plate height will asymptotically approach the B term at low $\bar{u}$ and, in the instances of packed columns, the $A + C_M$ terms (the former being ideally zero in open-tubular columns) at high $\bar{u}$. Al-Thamir, Laub, and Purnell (164) have shown that this is in fact the case with hydrocarbon solutes eluted from gas-solid systems, where "inert dilution" of adsorbent packings with nonsorbent materials (e.g., glass beads or Chromosorb) resulted in a substantial improvement in column efficiency. This can also be expected to be the case for capillary systems wherein alumina, silica, or some other adsorbent is made to adhere to the column walls: blending the stationary-phase suspension with crushed HMDS-treated diatomaceous earth prior to coating should markedly reduce the overall plate height as well as solute capacity factors and hence elution times. However, the capacity of such systems will be reduced accordingly, as is also the case with thin-film liquid loadings in gas-liquid chromatography. Overall, therefore, and in the absence of further comprehensive experimental study, it must be said at this time that the gains in speed of analysis and efficiency may well be outweighed by sample input requirements (particularly in trace analysis); the merits of reduced liquid loadings in GLC as well as inert-dilution in GSC consequently must be left at present to the subjective judgment of the analyst.

In accordance with eq. 164, variation of the phase ratio V_M/V_S in gas-liquid chromatography has the immediate effect of altering the solute capacity factor:

$$k' = K_R \frac{V_S}{V_M} \tag{164}$$

which in turn affects the stationary- and mobile-phase mass-transfer terms

Figure 2.31. Reservoir for fabricating packed GC columns (165).

(cf. eqs. 220 and 221). At low liquid loadings the former contribution to the peak variance is negligible, while at high liquid loadings both terms come into play in determining the overall H. However, the C_S term will in general dominate, and so better column efficiency (as well as fastest analysis time) will be found with the lowest amount of stationary phase that is compatible with the limitations of sample input and sensitivity of detection.

It is found experimentally that the number of theoretical plates exhibited by packed-column GC systems increases monotonously as the column length for lengths up to 50 ft while, for open-tubular column systems, the additivity of N with L is obeyed over at least 100 m. Moreover, there can be no doubt that long lengths of eighth- or quarter-inch i.d. packed columns are found at times to be more useful than wide-bore open-tubular columns in terms, for example, of sample capacity. Nor is the fabrication of such systems as difficult as might be supposed: for example, Laub and Purnell (165) made use of a simple reservoir, Fig. 2.31, from which coated support was displaced by gas pressure into eighth- or quarter-inch metal or glass columns, which in turn ensured uniform packing as well as minimum dead space with little attendant damage to the diatomaceous earth. Fig. 2.32 presents a chromatogram of 40 solutes obtained with a 50-ft $\frac{1}{8}$-in. stainless-steel column containing squalane phase deposited on Chromosorb G (120/140-mesh; AW-DMCS-treated; 120-psig helium inlet pres-

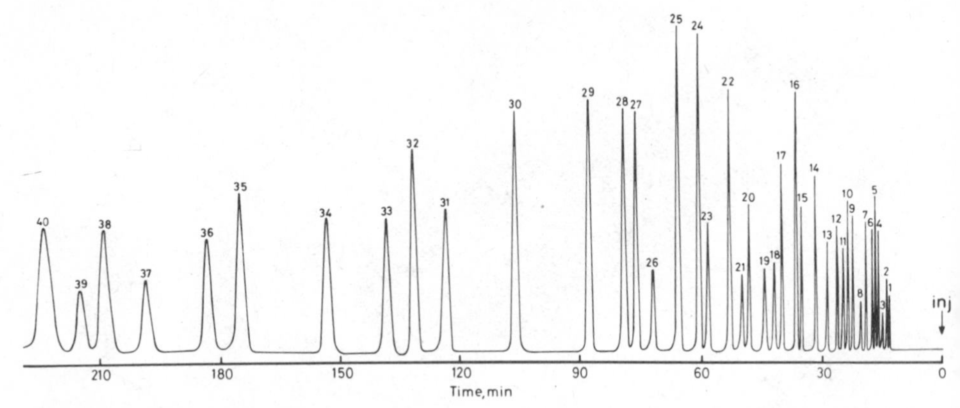

Figure 2.32. Packed-column chromatogram of 40 solutes at 100°C. Reprinted with permission from R.J. Laub, J.H. Purnell, and P.S. Williams, *J. Chromatogr.*, **134,** 249 (1977).

sure) operated isothermally at 100°C, that yielded on average a thousand theoretical plates per foot. Fig. 2.33 illustrates a 50-ft $\frac{1}{4}$-in. i.d. packed glass column, where the radiator design of the coils (assembled in banks) gives a very compact system that will fit into virtually all modern gc ovens.

The number of theoretical plates obtained per unit length of packed column hardly varies as a function of the column diameter for systems of up to ~5 mm i.d. Thereafter, deviations of the carrier flow profile at the column walls from that toward the center become appreciable, which of course degrades accordingly the peak variance. Therefore, it is common practice to include column baffles of one kind or another in preparative-scale systems, the design and placement of which must apparently be carried out empirically. Alternatively, sections of the column may themselves comprise the baffling: rather than construct, say, a 10-ft column of 1-in. diameter from a single tube, better performance is achieved if ten 1-ft sections are connected in series via short lengths of $\frac{1}{16}$-in. tube. Solutes passing from one column into the next are thereby redistributed at the head of each successive section, the result being a substantial improvement in overall efficiency.

Equation 222 contains two terms in the particle diameter d_p, where the first is almost certainly negligible in comparison with the second in well-packed columns. As a result, it is found that packed-column efficiency can be improved rather dramatically with a seemingly small reduction in the packing particle size. However, the column pressure drop increases unmanageably beyond packings of ~120/140-mesh (105–125 μ), and so this range is commonly viewed as the optimum practical lower limit to d_p.

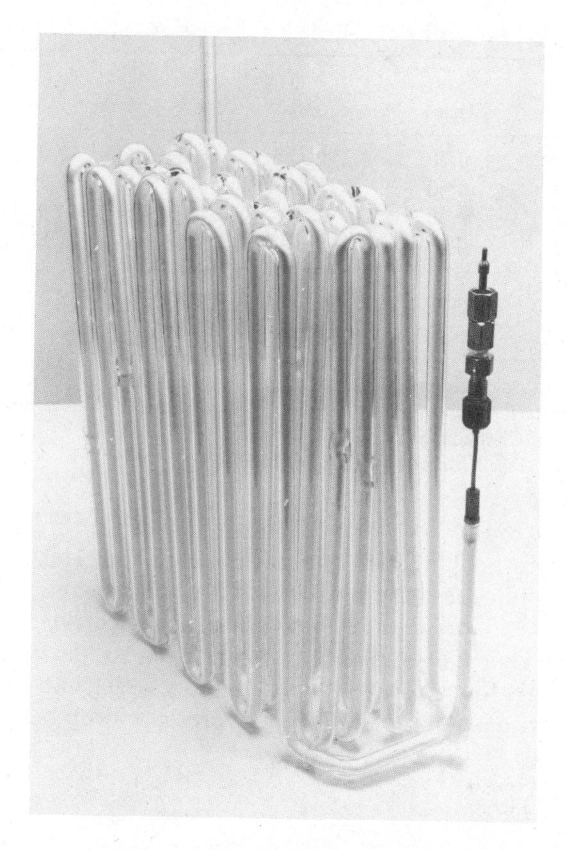

Figure 2.33. Fifty-ft. quarter-in. packed glass GC column (167).

Equation 223 indicates that column efficiency varies as the square of the tube radius with open-tubular systems, which is largely borne out in practice so long as there is an according adjustment made to keep the phase ratio constant. In contrast, if the stationary-phase film thickness is held constant while the radius is reduced, the solute capacity factor is increased which in turn limits the decrease in H approximately to a linear function of r. Table 2.13 presents some typical data that illustrate the variation of the phase ratio, solute capacity factor, optimum linear (hydrogen) carrier velocity, and efficiency per unit length with open-tubular systems of various radii (168). The efficiency data are shown plotted in Fig. 2.34, while the price that must be paid in terms of the inlet pressure necessary to achieve the optimum linear carrier velocity is shown in Fig. 2.35. A further aspect of some interest is that there exists a maximum in

Table 2.13 Typical Properties of Capillary Columns of Indicated Dimensions

Col. No.	ID (mm)	d_f (μm)	V_M/V_S	k'^a	$\bar{u}_{opt}$ (cm/sec)	N (m)
1	0.278	0.30	231.5	6.83	41.6	3770
2	0.238	0.30	197.5	8.50	40.5	4190
3	0.191	0.30	158.1	10.18	44.6	4850
4	0.149	0.30	123.0	13.12	51.5	6430
5	0.107	0.30	88.42	18.48	43.1	8590
6	0.066	0.30	54.25	29.47	35.7	12,750
7	0.0345	0.10	85.50	17.24	31.4	21,000

Reprinted with permission from B.J. Lambert, R.J. Laub, W.L. Roberts, and C.A. Smith, in *Ultra-High Resolution Chromatography*, ACS Symposium Series, S. Ahuja (Ed.). Copyright 1984 American Chemical Society.

[a] Decane solute at 65°C; hydrogen carrier; SE-30 stationary phase.

plots of $\bar{u}_{opt}$ against column i.d., as shown in Fig. 2.36. This portends that for a fixed film thickness there exists for a given separation an optimal column radius that will require the highest linear carrier velocity, and hence yield the fastest overall analysis time. Even so, the optimum practical tube diameter for capillary systems, broadly taking into consideration sample capacity, ease of fabrication, and pressure drop per unit length,

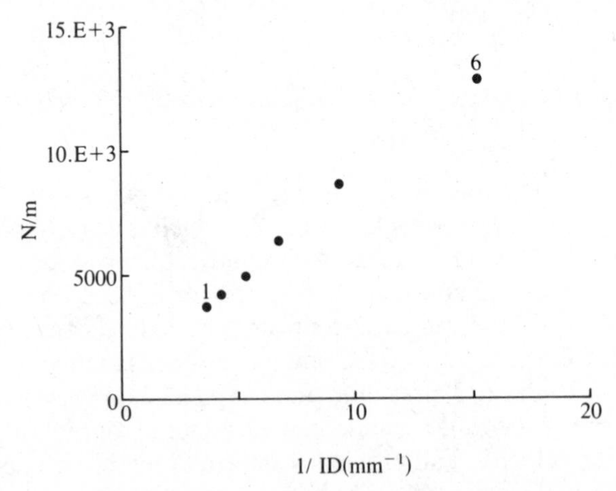

Figure 2.34. Plot of efficiency against internal diameter for open-tubular columns of Table 2.13. Reprinted with permission from B.J. Lambert, R.J. Laub, W.L. Roberts, and C.A. Smith, in *Ultra-High Resolution Chromatography*, ACS Symposium Series, S. Ahuja (Ed.). Copyright 1984 American Chemical Society.

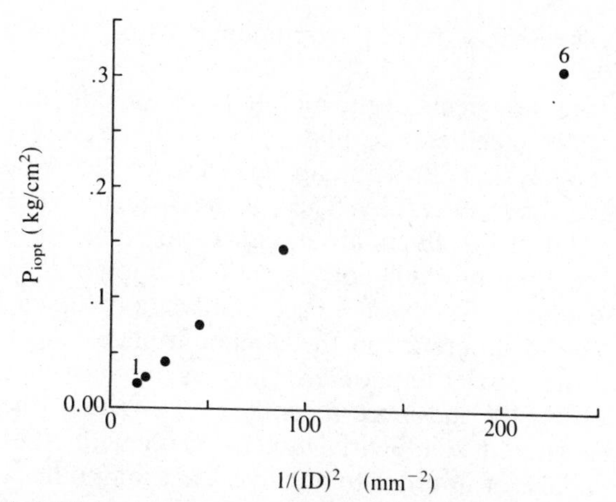

Figure 2.35. Plot of inlet pressure against internal diameter for open-tubular columns of Table 2.13. Reprinted with permission from B.J. Lambert, R.J. Laub, W.L. Roberts, and C.A. Smith, in *Ultra-High Resolution Chromatography*, ACS Symposium Series, S. Ahuja (Ed.). Copyright 1984 American Chemical Society.

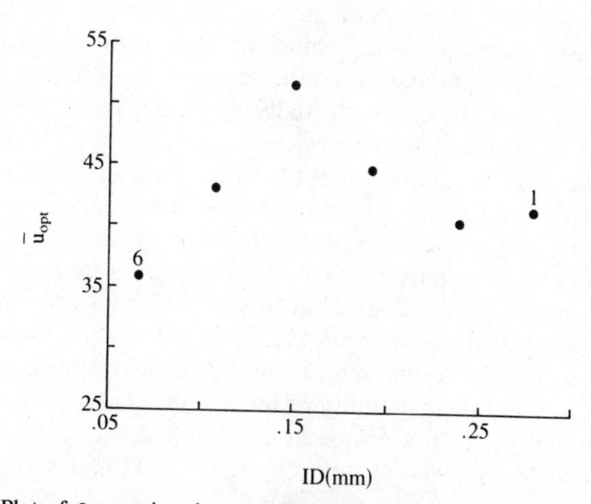

Figure 2.36. Plot of $\bar{u}_{opt}$ against internal diameter for columns of Table 2.13. Reprinted with permission from B.J. Lambert, R.J. Laub, W.L. Roberts, and C.A. Smith, in *Ultra-High Resolution Chromatography*, ACS Symposium Series, S. Ahuja (Ed.). Copyright 1984 American Chemical Society.

coincidentally is on the order of the optimum particle diameter for packed columns, ~100 μ.

In considering the effects of the mobile phase on column efficiency, it is clear that, as deduced earlier, hydrogen should be used as the carrier gas for fastest analysis both with capillary- and packed-column systems in analytical gas chromatography. There is the further advantage with this mobile phase that virial effects are minimal such that, for a fixed linear carrier velocity, retentions will not vary as a function of the column pressure drop. However, it is also the case that when columns are operated at $\bar{u}_{opt}$, there is no difference in the minimum plate height that can be achieved from one carrier to another. Thus, as demonstrated experimentally by Pretorius (10) and more recently by Laub (54), there may well arise instances wherein solute virial effects arising with other carriers can be taken advantage of in order to achieve what may otherwise prove to be difficult separations. For example, it is not uncommon that various amounts of mobile-phase additives such as steam or Freons are used to enhance resolution and, as indicated by the data of Table 2.2, the effects may indeed be substantial. The use of carriers other than hydrogen should therefore not be overlooked as a means of enhancing analytical separations although, as indicated above, any such improvement can, apparently be achieved only at the expense of analysis time if the system is operated at $\bar{u}_{opt}$.

The effect of temperature on column efficiency is reflected largely in the change of the solute partition coefficient, hence k'. With packed columns and moderate liquid loadings, any increase in k' will result in a decrease in the stationary-phase mass-transfer term C_S that dominates the overall C term, thus resulting in a decrease in H as the system temperature is lowered. However, partially offsetting this is the decrease in D_{1S}, which will result in an increase in C_S. Even so, the former effect easily wins out, and so operation at the lowest possible temperature commensurate with some user-defined requirement of analysis time generally is favored. In contrast, with lightly loaded packed columns and capillary columns, the C_M term may prove to be more important, in which case higher efficiency will be achieved at low k' and high D_{1M}, hence higher operating temperature. However, the resultant decrease in relative capacity factors, that is, alpha values, may well raise the theoretical-plate requirement beyond the capability of the system, in which case there will exist an optimal operating temperature that amounts to a compromise between speed of analysis, system efficiency, and resolution.

2.5.5 Extension of Rate Theories to Liquid Chromatography

Historically, even though the original work of Martin and Synge (147) in 1941 was with LC systems, the development of "high-performance" liquid chromatography was in fact preceded by gas chromatography by at least 15 years. This was due in part to the advent only relatively recently of small sorbent stationary-phase particles of uniform size; of suitable high-pressure liquid pumps; and of appropriate injection and detection devices capable of handling microliter-amounts of samples at pressures of several hundred atm. That is, development of instrumentation appropriate for modern column LC required a substantially different effort than was the case in GC.

In a similar vein, and perhaps partly as a result of the divergence of instrumentation, there were at one time considerable doubts about the applicability of GC rate theories to LC. For example, the prevailing view prior to 1980 was that longitudinal diffusion is negligible in LC, and that plots of H against u must therefore differ substantially in shape from those in GC (169,170). Experimental data sets purportedly supporting this supposition have in fact only recently been shown to be comprised, at least in substantial measure, of peak variance arising from extracolumn factors, that is, large instrumental (as opposed to column) dead volumes, slow detector time constants, and so forth (171). In apparatus designed carefully to eliminate extracolumn effects, curves of precisely the same shape as that shown in Fig. 2.30 are in fact obtained (172). As a result, one is led to conclude that, while of considerable and stimulating benefit at one time, other formulations [such as those involving semiempirical exponentials (173)] can no longer be assumed to represent the kinetics of mass transport in LC any more faithfully than does eq. 227. Moreover, the (gas-chromatographic) relations from which the latter equation was derived must therefore also apply.

Variation of Plate Height with Carrier Velocity

Fig. 2.37 provides the van Deemter plots (best-fits of eq. 227) of the data of Reese and Scott (172) for toluene ($k' = 0$) and benzyl acetate (upper curve: $k' = 1.94$; lower curve: $k' = 2.04$) solutes eluted with 5% v/v ethyl acetate/n-heptane mobile phase from two columns 25 cm in length by 9 mm i.d. (upper curve) and 4.6 mm i.d. (lower curve) containing microparticulate silica. The best-fit values of the van Deemter equation are presented in Table 2.14, where the magnitudes of the A, B, and C terms immediately provide measures of the contributions of streaming, longitudinal diffusion, and stationary- and mobile-phase mass transfer to the overall plate height H. For example, we see that for both solutes with each column, the latter contribution predominates, and hence governs very largely the overall observed peak variance. In contrast, values of

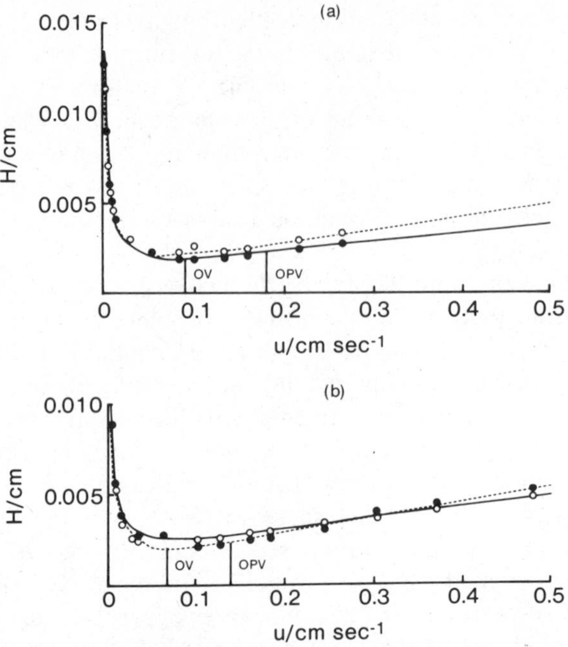

Figure 2.37. Van Deemter curves for LC systems of Table 2.14. Reproduced from the *Journal of Chromatographic Science* by permission of Preston Publications, Inc.

the longitudinal-spreading term are at least two orders of magnitude smaller, while those for the streaming term fall somewhere in between.

The data presented in Fig. 2.37 and Table 2.14 confirm that eq. 227 does in fact apply to band-spreading processes in liquid chromatography. Moreover, the resultant van Deemter constants appear to substantiate what might have been supposed in advance from a consideration in general of the magnitude of liquid-phase diffusion coefficients, namely that a major source of denigration of column efficiency in LC is mass-transfer nonequilibrium.

Consequences of Slow Mass Transfer

Reference to eqs. 220–222 shows that both mobile- and stationary-phase diffusional terms enter into the C term. Since diffusion of a solute along an adsorbent surface (i.e., D_{1S}) is negligible, while solute diffusion in liquid mobile phases (D_{1M}) is very slow compared to that extant in gases, relaxation of the nonequilibrium distribution of solute between the mobile and stationary phases in LC must accordingly be very much slower than

Table 2.14 Best-Fit van Deemter Parameters for LC Columns Containing Microparticulate Silica

Column diameter:	4.6 mm		9.0 mm	
Column length:	24. cm		25. cm	
Mean particle diameter:	7.8 μ		7.8 μ	
Dead volume:	2.9 cm^3		12.5 cm^3	
Total column volume:	4.15 cm^3		17.4 cm^3	
Solutes:	Toluene	Benzyl acetate	Toluene	Benzyl acetate
k'	0	2.03	0	2.03
Constant				
A (cm):	9.3×10^{-4}	1.56×10^{-3}	1.04×10^{-3}	8.9×10^{-4}
B (cm^2/sec):	4.21×10^{-5}	4.09×10^{-5}	3.97×10^{-5}	4.64×10^{-5}
C (sec):	9.0×10^{-3}	6.81×10^{-3}	7.73×10^{-3}	5.68×10^{-3}
H_{min} (cm):	2.16×10^{-3}	2.62×10^{-3}	2.15×10^{-3}	1.92×10^{-3}
$\bar{u}_{opt}$ (practical) (cm/sec):	6.84×10^{-2}	7.74×10^{-2}	7.17×10^{-2}	9.04×10^{-2}

From Reese and Scott (172). Reproduced from the *Journal of Chromatographic Science* by permission of Preston Publications, Inc.

is the case in GC. Thus, if nothing else, the linear carrier velocity at the van Deemter optimum will be considerably slower, usually by at least an order of magnitude. In addition, the particle size of packings must be much smaller than those employed in gas chromatography if one is to achieve comparable plate heights.

Mass-transfer nonequilibrium is also the source of the often-voiced (pejorative) observation that separations in LC are achieved only in the first few centimeters along the column length and that there is therefore rarely if ever justification for use of columns of length greater than ~15 cm. Indeed, it is sometimes said that, at best, longer columns serve only to preserve what separation was achieved at the head of the column and, at worst, to degrade the separation at hand. In point of fact, separation occurs along the entire length of LC columns; however, nonequilibrium effects also accumulate such that band broadening cancels to some extent what resolution is achieved. The effect, although operative both in GC and in LC, is much more pronounced in the latter because diffusional relaxation of the solute nonequilibrium distribution is approximately three orders of magnitude less efficient. Thus, while the number of theoretical plates is additive in gas chromatography for packed-column lengths of up to at least 50 ft, the additivity of N with length of conventional (2–10 mm i.d.) columns in liquid chromatography holds over no more than a meter or so.

Particle Diameter and Column Length

A quantitative assessment of the variation of H with L and d_p is carried out as follows (172). We recognize, first, that the linear carrier velocity corresponding to the minimum of van Deemter plots, for a column and packing of fixed dimensions, represents a compromise between the A, B, and C terms of eq. 227 only for an individual solute. In order to take into account the column length L and particle diameter d_p in terms of separations of pairs of solutes, we set R_s equal to unity, and then solve the resolution relation (eq. 191) for the number of theoretical plates N:

$$N = \frac{16(\alpha)^2(1 + k')^2}{(k')^2(\alpha - 1)^2} \tag{231}$$

The ratio α/k' is approximately equal to the inverse median of the solute capacity factors $1/k'$ for difficult separations (i.e., $\alpha \sim 1.05$), so that eq. 231 becomes

$$N = \frac{16(1 + k')^2}{(k')^2(\alpha - 1)^2} \tag{232}$$

If we now let $k' = 1$, as is often the case in liquid chromatography, eq. 232 reduces to

$$N = \frac{64}{(\alpha - 1)^2} \tag{233}$$

The column length is related to the plate height and number of theoretical plates by the expression $L = NH$, so that substituting L/H for N in eq. 233 and then solving the result for L yields

$$L = \frac{64H}{(\alpha - 1)^2} \tag{234}$$

A "reduced" plate height h next is defined, such that $h = H/d_p$:

$$L = \frac{64hd_p}{(\alpha - 1)^2} \tag{235}$$

Equation 235 thereby provides the means of assessing the column length required for a given separation as a function of the packing particle diameter, so long as the value of h is known or can be measured. For example, the average of the reduced plate heights found for the systems in Table 2.14 was 2.8 particle diameters at the average optimum linear carrier velocity, 0.077 cm/sec. However, since the right-hand branches of curves such as those in Fig. 2.37 are comparatively flat, one would in practice operate the system at a mobile-phase velocity approximately double this, 0.15 cm/sec [the optimum practical velocity (174)]. Analysis times would then be very nearly halved while, according to Fig. 2.37, the plate height would be increased only by ~15%. H would thereby be increased to $3.2d_p$ and, upon substitution of this value into eq. 235, we find for k' and R_s set to unity,

$$L = \frac{205d_p}{(\alpha - 1)^2} \tag{236}$$

Equation 236 thus provides for calculation of the column length required to achieve 4σ separation of a pair of solutes of some specified α value. However, we can proceed further than this by recalling the definitions

$$t_A = \frac{t_R}{(1 + k')} = \frac{L}{\bar{u}} \tag{237}$$

Substitution of 0.15 cm/sec for $\bar{u}$ and unity for k', followed by rearrangements, then yields the relation

$$t_R/\text{min} = \frac{45.6d_p}{(\alpha - 1)^2} \tag{238}$$

where the particle diameter is in centimeters. We can therefore calculate the analysis time required for given separations with whatever columns are to hand. For example, the particle diameter utilized by Reese and Scott was 7.8 μ; hence, according to eq. 238, a value of α of 1.05 will require an analysis time of 14 min for two solutes of average k' of unity, with resultant R_s also of unity. In contrast, for an α of 1.1, the separation can be achieved in 3.6 min while, for $\alpha = 1.01$, 6 hr is required. The respective column lengths required (eq. 237) are 64, 14, and 1600 cm. The second example, that of $\alpha = 1.05$, corresponds very approximately to what can be regarded as an average degree of difficulty of many LC separations, which accounts for the current popularity of columns of length of 15 cm.

Column Diameter

In order to explore the effects of the column diameter, we consider the worst case, namely, the variance of band-shapes of solutes that are not retained by the stationary phase. Squaring both sides of the exact form of the standard-deviation relation, eq. 183, provides (175)

$$\sigma^2 = \frac{V_M^2}{N} = \frac{V_M^2 h d_p}{L} \tag{239}$$

The column void volume is next cast in terms of the fraction θ of the total column volume not occupied by the packing (172).

$$V_M = \theta \pi r^2 L \tag{240}$$

Substitution of eq. 240 into 239, followed by incorporation of the expression derived previously (eq. 235) for the column length, with R_s and k' set to unity, yields after some rearrangement

$$r^2 = \frac{\sigma(\alpha - 1)}{8\pi\theta h d_p} \tag{241}$$

If we now combine the numerical constants with a value of h of $3.2d_p$ and a particle diameter of 7.8×10^{-4} cm, eq. 241 becomes

$$r^2 = 23\sigma(\alpha - 1) \tag{242}$$

which relates the column radius required to effect unit resolution to the solute-pair alpha value and the average standard deviation of the peaks. Thus, for fixed alpha, if peak band-spreading is increased (as, for example, in poorly packed columns or in systems with large extracolumn dead volume), a larger column radius is mandated in order to maintain the separation. Alternatively, as the column radius is reduced the peak standard deviation is reduced, but only as the square root. Even so, eq. 242 clearly establishes the potential utility of columns of ~0.1–1 mm i.d. (i.e., "microbore" columns, as opposed to conventional systems of 2–8 mm i.d.), where the attendant advantages also include low solvent consumption, increased mass sensitivity, appreciably reduced radial mass-transfer nonequilibrium with a concomitant increase in column efficiency, and additivity of N with L up to at least 10 m (N = 1,000,000 has thus far been achieved). However, there must be an accompanying reduction in the extracolumn (i.e., instrument) variance if the gains in microbore-column efficiency are to be realized in practice. Studies to these ends are well under way at the present time, as evidenced by several recent symposia on microbore columns; indeed, the subject holds promise of capturing the same intensity of interest as that accompanying the introduction of open-tubular columns in gas chromatography.

Temperature

In those instances where only one retention mechanism is operative (which may in fact be rare; cf. Sec. 2.6), an immediate improvement in LC column efficiency can be realized by increasing the column temperature, thereby increasing D_{1M}, which in turn will result in a decrease in H. Furthermore, the (van Deemter) optimal linear carrier velocity will be higher at elevated temperatures, which portends faster analysis time. Higher operating temperature will also result in a decrease in the mobile-phase viscosity, that is, a decrease in the column inlet pressure required to maintain a given flow rate. Even so, the effects by and large have not been examined in any great quantitative detail, the redress of which awaits further experiment.

Indeed, while it is fair to say at this time that, in general, the extension of eq. 227 to liquid chromatography has gone far in advancing our understanding both of fundamental aspects of solute elution as well as practical considerations of column and instrument design, considerable experimental clarification is nevertheless required of virtually all aspects of rate theories pertinent to LC. Moreover, it must in all fairness be emphasized that the formulations applied thus far in this work (and elsewhere) to GC as well as LC at best represent crude models that most likely only approximate the actual mechanisms of the dynamical mass transport of solutes through chromatographic systems. The situation is therefore comparable to that described by Flory, Orwoll, and Vrij (176) as recalled by Martire (177) with regard to the prediction of activity coefficients: "While the level of refinement may (yet) leave much to be desired, the relationships derived (to date) are (at least) manageable, and correlation(s) with experiment (are) not altogether disappointing."

2.6 OPTIMIZATION OF GAS- AND LIQUID-CHROMATOGRAPHIC SEPARATIONS

Broadly speaking, four factors come into play in formulating an optimization strategy for selection of the most favorable system parameters for particular chromatographic separations. These are: resolution, speed of analysis, sample capacity, and, in some circumstances, economics. More often than not, each of these is also directly dependent on (is some or other function of) the other three. Moreover, each may be subject to several additional constraints inherent in the nature of the system; the solutes; and/or the reagents utilized in the analysis.

For example, consider that the efficiency of a particular liquid-chromatographic system can be improved by operation at elevated temperature; however, if the compounds to be separated are thermally labile, the system must in fact be used at some reduced temperature, with concomitant longer analysis time and greater solvent consumption. In contrast, suppose that a petroleum company seeks to separate only the aliphatic fraction (to be quantitated as "total aliphatics") from the aromatic portion (as "total aromatics") of a gasoline sample and, furthermore, to do so in as rapid an analysis time as can be achieved, since many such samples must be run in a given day. On the other hand, assume that a hypothetical research team has recently synthesized a new organometallic compound and requires a gram-sample of the substance for further analysis and testing; whatever effort is needed, time of analysis notwithstanding, will therefore be expended in order to isolate and purify this amount of the material.

The analytical requirements of the above examples clearly are substantially different: temperature stability is a major factor in the first; whereas only partial separation is required in the second, where the time of analysis is also important. Complete separation is demanded in the third example, where system efficiency as well as sample capacity far outweigh the importance of analysis time. That is, the relative importance given each of the four factors comprising a separation must inevitably be weighted at least in part in accordance with subjective judgments rendered by the analyst. In doing so, a hierarchy of priorities (i.e., requirements) pertaining to the analysis must be formulated.

Even so, and irrespective of the mixture to be analyzed, there can be little doubt also that the *primary* goal of analytical chromatography invariably involves at least *some* degree of *separation* of the sample at hand. According to eq. 191, the resolution relation, there are two means of achieving this end. In the first, the system efficiency simply is increased. This approach frequently proves to be the quickest and most economical way of accomplishing a separation; in fact, in the limit of infinite system efficiency, all solutes will be separated. However, there are practical upper limits to the number of theoretical plates N that can be obtained from conventional column systems, these corresponding at the present time to approximately 500,000 for 50-m lengths of open-tubular columns in GC, and ~100,000 plates for four 25-cm tandem-connected columns in LC. The upper practical limits to alpha, assuming unit resolution and k' > 10, are then 1.006 and 1.013, respectively. These levels are nevertheless difficult to achieve routinely, and system efficiencies corresponding to alpha values of ~1.04–1.05 (~5,000–10,000 plates) are by far the more common.

For systems of more or less fixed efficiency, the second method of achieving some desired level of resolution thereby comes into play, namely, increasing the system selectivity (alpha). The advantage in doing so is that less sophisticated (i.e, less efficient) column systems can then be employed. For example, while a capillary column containing a "boiling-point" stationary phase may be required for the gas-chromatographic separation of a pair of solutes, a packed column containing a phase that is selective on some basis other than volatility (e.g., a liquid crystal) may also provide the desired resolution. Furthermore, if the analytes are trace constituents in a complex matrix, the packed column may well be preferred overall as a result of higher attendant sample capacity.

The former method of optimizing separations, that is, increasing the system efficiency, is straightforward and would be carried out following eqs. 227 and 228 as described in Sec. 2.5. The latter strategy for enhancing resolution, that is, altering the system selectivity, requires a more detailed knowledge of the factors that govern solute elution behavior, and is therefore more difficult to carry out in practice. Furthermore, in order to effect separations in this manner, we require functional relationships that involve the chromatographic system variables (e.g., temperature), as well as procedures with which to relate absolute retentions to separations as these variables are altered. To this end, we set about in what follows in detailing, first, a generalized plenary (i.e., complete) optimization strategy that permits the a priori forecasting, via graphical analysis, of the global set of values of system parameters that will yield maximal separation. Examples are then provided that illustrate the compelling utility of the methodology in achieving the desired level of system selectivity.

2.6.1 The Plenary Window-Diagram Optimization Strategy

The three elements common to all chromatographic techniques are the stationary phase, the mobile phase, and the system temperature. Changing any one of these will usually alter absolute retentions and hence, potentially, separations. Unfortunately, it is impossible at the present time to predict values of absolute retentions with this or that pure stationary phase, while the situation with regard to pure mobile phases is little better. Retentions usually vary in a predictable manner with the column temperature, but this, too, is subject to qualification, particularly in LC, but also in GC. As a result, blended stationary and/or mobile phases have heretofore been used only infrequently in gas chromatography, while the choice of mobile-phase compositions in liquid chromatography is little better than subjective at this time. Moreover, coherent guidelines for isothermal or temperature-programmed operation in GC as well as LC are virtually nonexistent. Nevertheless, controlled variation of each of these parameters constitutes the most powerful method yet devised for achieving (or enhancing) chromatographic selectivity.

Intelligent-Search Routines for Maximizing Relative Retentions: Localized Minima

A number of mathematical techniques have been developed over the years that address so-called multivariate problems, that is, where one or more solutions are sought for sets of multivariable (and potentially interrelated) functions. The task has been made much easier in recent years with the advent of small yet powerful micro and minicomputers, so much so that problem-solving today via what amounts essentially to successive approximation frequently turns out to be the most expeditious approach. An entirely new subdivision of analytical chemistry, dubbed "chemometrics", has simultaneously emerged, where the topics of major concern include parameter estimation, pattern recognition, optimization, multivariate statistics, and so forth (178). However, in all of these, the problem of entrapment in local minima remains a major difficulty. For example, consider successive guesses at solutions to the quadratic equation $X^2 + 4X - 9 = 0$. An initial guess of $X = 1.0$ provides a remainder of -4.0, while 2.0 yields 3.0. Since there is a reversal in the sign of the remainder, the answer lies between 1.0 and 2.0. The next guess, 1.5, gives -0.75, while a fourth guess of 1.6 results in -0.040. To two significant figures, $X = 1.6$ is thereby arrived at as one of the roots. The path taken as a result of each guess is shown plotted in Fig. 2.38, where the remainders are seen to draw successively closer to ("ring" about) $Y = 0$ (179). However, suppose that we employ the same strategy in an effort to find the other root. An initial guess of -1.0 gives a remainder of -12, while a second guess of -2.0 provides -13. Since the second remainder is farther away from zero than the first, the value of the second guess appears to be too negative. A sensible third guess might then be $+1.0$, which would place us on the path indicated in Fig. 2.38 and which would again lead to the solution of 1.6. In contrast, if the first and second guesses were -1.0 and -10, the remainders would be -12 and 51. Since a reversal

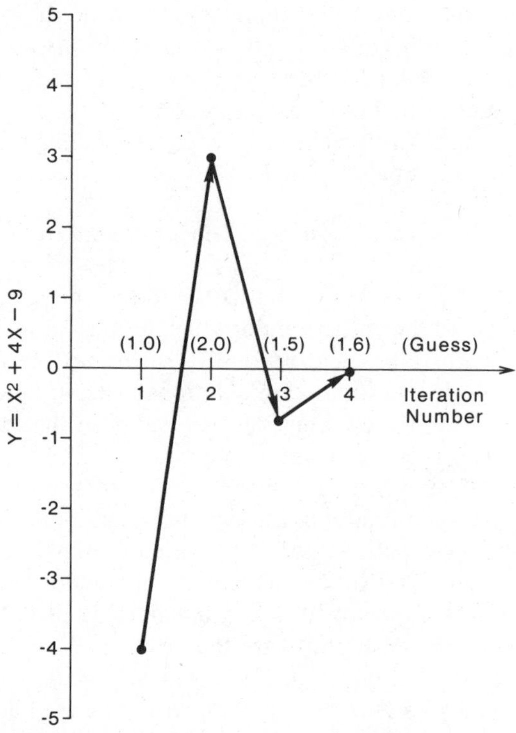

Figure 2.38. Plot of the remainder of the indicated quadratic with successive guesses of $x = 1.0, 2.0, 1.5$, and 1.6.

of signs occurs within these limits, the second root lies somewhere between -1 and -10, and is in fact -5.6.

The example illustrates that stringent requirements must often be placed on guessed-at solutions even in the case of quadratic functions, and that failing to do so may well result in one or more solutions to a given problem being overlooked. Moreover, these and other difficulties are often severely compounded in problems of higher complexity, for which mathematical means of reaching exact solutions frequently are not available. The strategy generally employed in such instances amounts to defining, then passing across, the contour lines of a topological "map" of the problem, the goal being to reach the points of highest "elevation" (the solutions). This would be straightforward if the entire "map" were available for prior inspection; however, in techniques such as SIMPLEX, it is in fact defined simultaneously as the solutions are searched for. As a result, when the process is carried out with a computer, it is sometimes the case that a small "elevation summit" (false solution) is reached. And, since further exploratory attempts fail to identify *immediately adjacent* contours of higher elevation, the algorithm terminates. This problem is referred to at times as that of false optima, or, in the instance of searching for solutions comprised of valleys in a planar (or higher-dimensioned) surface, localized minima.

Window Diagrams: A Global Alternative

It would seem that a reasonable alternative to the above approach would be definition, first, of the entire topology of solutions (global set of optima) for a given problem and then, secondly, selecting this or that overall "best" or most desirable solution or solutions in accordance with criteria established by the end-user. For example, suppose that a particular optimization problem reduces to the cubic equation: $X^3 - 9X^2 + 23X - 15 = 0$, for which only the real and positive roots are required. There are a variety of ways of solving the equation mathematically as well as graphically. The most expeditious of the latter is illustrated in Fig. 2.39, where the absolute values of the calculated remainders Y are shown plotted against positive values of X taken at intervals of 0.2 (solid curve). The three real and positive roots are thereby found *by inspection* to be 1, 3, and 5, that is, the values of X for which $Y = 0$. The dashed lines were constructed by solving the equation with integer values of X, which requires far fewer calculations yet which, in this instance, nevertheless still yields the correct answers. (If the roots had not been integers, the effort saved by reducing the number of calculations would of course be offset by a loss in accuracy.) In any event, either version of Fig. 2.39 represents a "map" of *solutions*, that is, values of Y for assumed values of X. Furthermore, the plots were generated without regard initially for what the solutions might be, other than employing the criterion that the roots (i.e., values of X for which $Y = 0$) must be real and positive (hence, the abscissa was defined beginning at an origin of $X = 0$ and extending to X positive). In addition, the roots were determined simply by inspecting the final graph.

We now apply the graphical method of identifying roots of equations to the optimization of chromatographic variables. To begin, consider that the gas-liquid chromatographic partition coefficients of five solutes vary as linear functions of some or other continuous system variable V. Plots of K_R° for each hypothetical compound, taken at integer V, might then resemble those shown in Fig. 2.40, whereas the actual chromatograms (with the peaks drawn as sticks) would appear as indicated in Fig. 2.41.

We note, first, that if it were known with certainty that K_R° were related

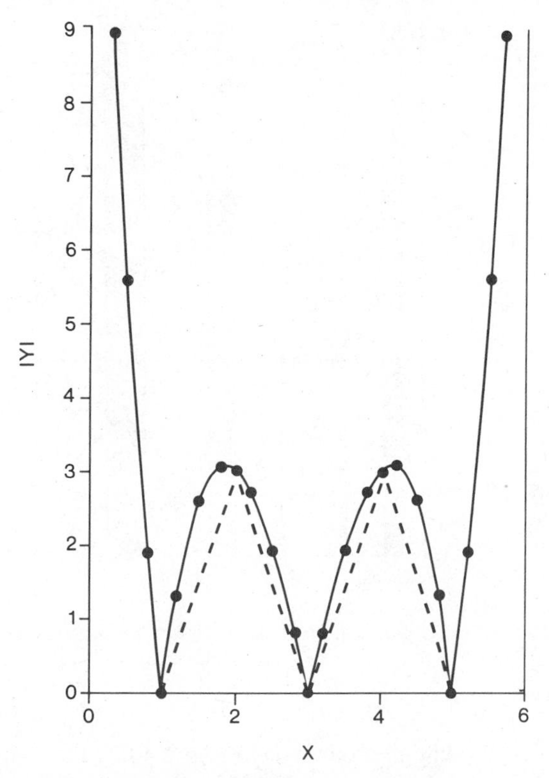

Figure 2.39. Graphical means of determining the three real and positive roots of the equation: $x^3 - 9x^2 + 23x - 15 = 0$. Solid curve: $| y |$ calculated at intervals of x of 0.2; dashed lines: $| y |$ calculated at integer-intervals of x.

linearly to V for all solutes, only the end-point data for each compound would have to be measured experimentally. Straight lines could then be constructed immediately between the respective limits, here, $V = 0$ and $V = 6$ as shown. However, if the data exhibited curvature, it would also be necessary to measure retentions at intermediate V in order to define the (nonlinear) relations that represent the elution behavior of each solute. Nevertheless, and even though supplemental experimental measurements would be required, no fundamental difficulties are presented, since subsequent optimization is independent of the form of the fitted function, and since even hand-held calculators are today capable of nth-order polynomial curve-fitting.

Secondly, we see that separation of the five solutes is not a trivial matter, since nos. 3 and 4 overlap (have identical K_R°) when V is set to zero, whereas 2 and 3 are unresolved at $V = 6$. However, there do appear

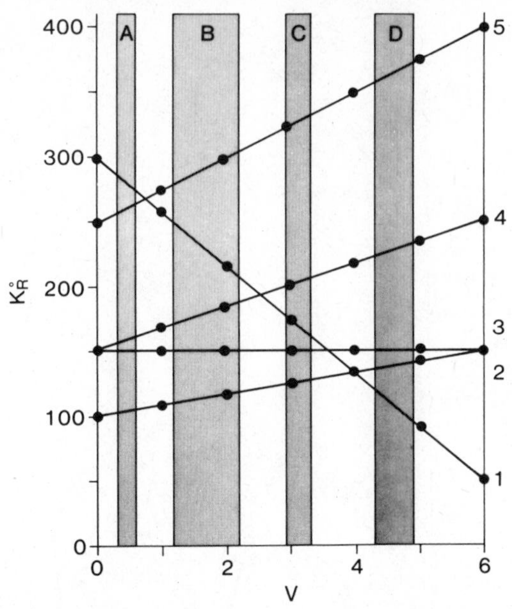

Figure 2.40. Plots of GLC partition coefficients of five hypothetical solutes against the system variable V.

to be regions of V (shaded areas A-D in Fig. 2.40) within which resolution of all compounds might be possible, that is, that do not contain intersections of lines. For example, with $V = 0.5$, the expected elution order is 2, 3, 4, 5, 1; whereas with $V = 4.7$, the order would be 1, 2, 3, 4, and then 5. Thus the solutes appear to be separable at least to some extent if the system were to be operated within region A ($V = 0.3$–0.6), region B (1.2–2.2), region C (2.9–3.3), or region D (4.3–4.9).

Many workers have used this and similar means of depicting graphically the variation of retentions with one or another system parameters; however, Laub and Purnell (180) were the first to devise a complementary method of pictorial data representation and analysis with which *precise* values of *all* optima can be identified. Recognizing that the primary goal of any separations task is maximizing alpha, we rewrite eq. 69 for a pair of solutes i and j of Figs. 2.40 and 2.41 as follows:

$$\alpha_{i/j} = \frac{K^\circ_{R(V)i}}{K^\circ_{R(V)j}} \tag{69}$$

where the subscripts V refer to some value of the system variable. Since

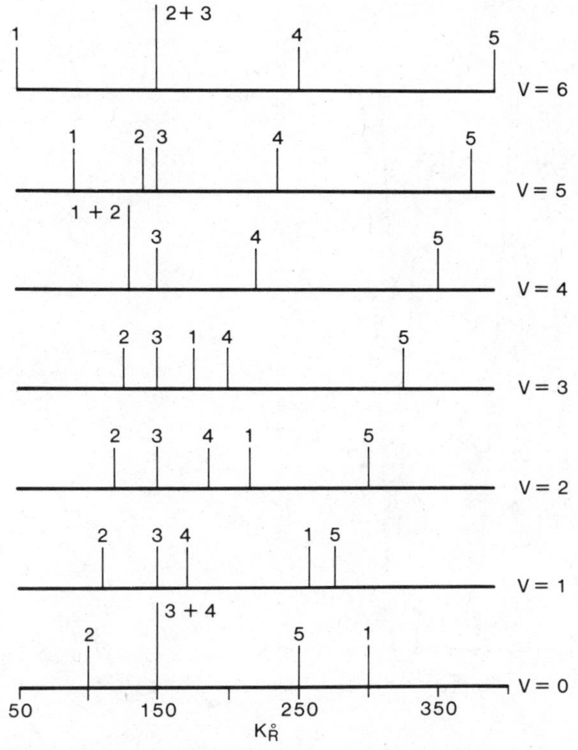

Figure 2.41. Skeletal chromatograms of solutes of Fig. 2.40.

the partition coefficients of all of the solutes regress linearly with V, we can also write a general equation that describes the straight lines in terms of the slopes and end-points:

$$K^\circ_{R(V)} = \frac{V}{6} [K^\circ_{R(6)} - K^\circ_{R(0)}] + K^\circ_{R(0)} \tag{243}$$

where, again, the subscripts refer to partition coefficients at the specified values of V, that is, $V = 0$ (left-hand ordinate of Fig. 2.40), $V = 6$ (right-hand ordinate), or some intermediate value between these limits.

We now substitute eq. 243 into 69:

$$\alpha_{i/j} = \frac{[(K^\circ_{R(6)} - K^\circ_{R(0)})V + 6K^\circ_{R(0)}]_{\text{solute } i}}{[(K^\circ_{R(6)} - K^\circ_{R(0)})V + 6K^\circ_{R(0)}]_{\text{solute } j}} \tag{244}$$

Equation 244 is an important result: values of alpha (i.e., the separation)

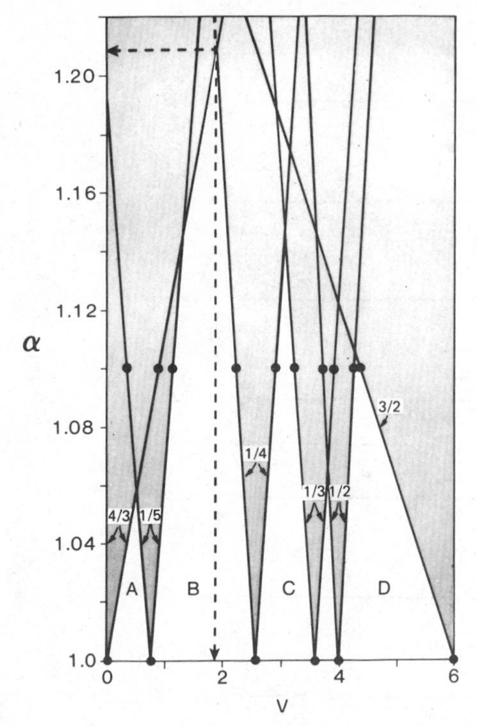

Figure 2.42. Window diagram for solutes of Fig. 2.39.

for the solute pair i/j can be predicted at any intermediate value of V solely from the end-point partition coefficients, that is, those observed with $V = 0$ and at $V = 6$.

We next graph α against V for all $[5!/2!(5 - 2)! = 10]$ pairs of solutes, as shown in Fig. 2.42. When alpha is maintained greater than or equal to unity (by reversal of the identities of solutes i and j), the plots resemble inverted and partially overlapped triangles. Regions in which there are no overlapped lines appear as windows, and the graph is therefore called a window diagram. The tallest four are labeled A through D, and correspond to the shaded regions of Fig. 2.40.

Window diagrams such as that shown in Fig. 2.42 provide an extraordinary amount of *quantitative* (i.e., *precisely-defined*) data. First, we note that only the lower boundary of the plots need be emphasized, since if at a particular window-value of V the most difficult solute pairs can be separated, all others will also be resolved. Secondly, we deduce that resolution of the solutes will be easiest with the value of V given by a perpendicular dropped from the apex of the tallest window to the abscissa, since the respective most-difficult alpha (horizontal line drawn from the

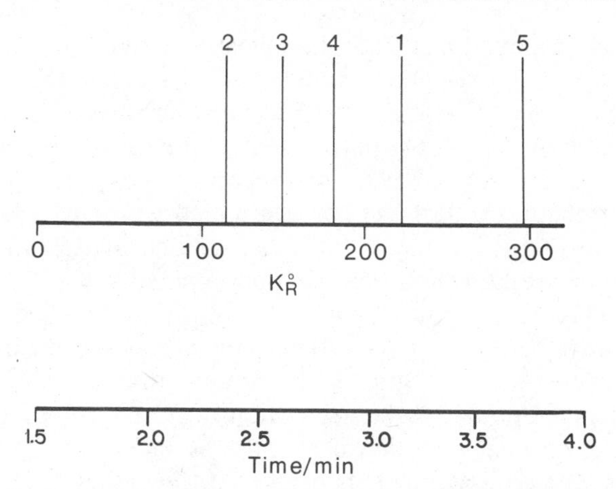

Figure 2.43. Predicted skeletal chromatogram of solutes of Fig. 2.40 with $V = 1.85$.

apex to the left-hand ordinate) for this window comprises the *overall-easiest* of the most-difficult window separations. In the present instance, the most-difficult alphas given by windows *A* through *D* are 1.056, 1.208, 1.148, and 1.106, while the respective optimal values of *V* are 0.6, 1.85, 3.05, and 4.25. The *overall-easiest* of these is clearly *B*, since, *by inspection,* the most-difficult alphas given by the other three (shorter) windows are smaller, that is, they represent more difficult separations. Thus the overall optimal value of *V* is 1.85. Further, the most-difficult solute pairs (4/3 and 1/4) that comprise window *B* are immediately identified. Equation 192 can then be used to calculate the number of theoretical plates required for their separation (hence that of all other pairs as well); here, for $k' > 10$, $N = 210$ plates required to effect 6σ resolution. In addition, reference back to the "straight-line" diagram, Fig. 2.40, provides the order of elution; for window *B* we find it to be 2, 3, 4, 1, and then 5. Finally, having identified the overall optimal *V*, the partition coefficients of the solutes can be calculated from eq. 243. Then, for a specified phase ratio and system void time, the chromatogram can be drawn, as shown in Fig. 2.43 for the case at hand with the hypothetical values $\beta = 200$ and $t_A = 1.5$ min (very roughly typical of open-tubular columns).

Window diagrams thereby provide, first, the *global set* of optimal values of the independent system variable (here, *V*) in terms of maximal separation (alpha) of the solutes at hand. Secondly, selection of the overall-best of these is carried out simply by visual inspection, whereupon the most-difficult solute pairs are identified that comprise the window boundary. The overall-easiest of the window alphas is then determined by drawing a horizontal line from the tallest window to the left-hand ordinate, while the overall-optimal value of the system variable V is found by dropping a perpendicular from the window apex to the abscissa. The number of theoretical plates required to effect the separation of all solutes is next calculated from eq. 192, whereupon the absolute retentions (as well as the retention order) can be determined with eq. 243. The chromatogram can then be constructed if the column phase ratio and void volume or time are known or can be approximated.

It is important to note that the data derived from the window-diagram procedure *must* be correct, since the window diagram itself is nothing more than a new method of graphical representation of retentions. Thus, if at a particular optimal V the separation does not correspond to that predicted, the original retention data must be incorrect. Conversely, if the retention data are accurate to begin with, the window-diagram strategy cannot fail. The situation could hardly be more favorable, particularly in view of the simplicity of the window-diagram method of graphical data presentation and interpretation.

Generalized Retention Equation for Window Diagrams

Three reviews of the window-diagram strategy have been published by Laub (181), these including discussions of the use of three- (or higher-) dimensional Cartesian or triangular coordinate systems as well as appropriate dependent-ordinate parameters other than alpha (e.g., R_s, S_f, etc.). Also described and discussed are various situations in which retentions may or may not vary linearly with one or another system variables. For example, eq. 92 indicates that liquid-gas partition coefficients will regress nonlinearly as a function of the volume-fraction based composition of binary mixed liquid stationary phases, while eq. 99 provides for linear regression in the absence of solvent-solvent or solute-solvent association (79). Similarly, plots of log (retention) against inverse absolute temperature (van't Hoff plots) must be linear in GC and LC in the absence of multiple retention mechanisms (182), since the enthalpy of solution or adsorption hardly varies over the temperature span appropriate to either technique. However, since multiple retention mechanisms appear to be inevitable in liquid chromatography, van't Hoff plots of LC data are found more often than not to be curved (183).

Variation of the mobile-phase pressure and/or composition in GC produces retentions that vary in a well-defined (logarithmic) manner (eqs. 10 et seq.; eq. 70). In contrast, plots of retention against mobile-phase composition in LC may be linear, concave, convex, or exhibit one or more inflection points, a result, certainly, of the complex interplay of solute/

carrier adsorption-displacement mechanisms extant both with (hydrophilic) adsorptive as well as (hydrophobic) derivatized stationary packings commonly utilized today in all modes of liquid chromatography (87, 88). Madden, McCann, Purnell, and Wellington (184) were the first to address this problem from the standpoint of Langmuir sorption of mobile-phase constituents on the surface of LC packings, then to recast accordingly the retention relation first derived by Scott and Kucera (111,185). For a binary mobile phase comprised of solvents B and C,

$$\frac{1}{t_{R(M)}} = \phi_C \left[\frac{1}{t_{R(C)}} + \frac{b\phi_B}{1 + b'\phi_B} \right] + \frac{\phi_B}{t_{R(B)}} \tag{245}$$

where ϕ_i represents a volume fraction, and where $t_{R(i)}$ is the raw solute retention time observed with pure mobile phases B or C, or with some mixture of the two ($M = B + C$). The two constants b and b' are empirical fitting parameters that must at present be derived from an analysis of the experimental data.

Equations 245 and 114–115 (the latter due to Langmuir) are formally nearly identical, and each derives from considering that the equilibrium distribution of a solute between a solid (or sorptive liquid) surface and the fluid above it is proportional to the ratio of forward and reverse rate constants pertaining, respectively, to sorption onto, and desorption from, the surface. However, the relation proposed originally by Langmuir conforms only to concave-shaped curves and so, Purnell and his coworkers included the parameter b' (instead of b) in the denominator of eq. 245. The result is an equation of remarkable generality that fits linear as well as convex or concave curves, and that therefore appears to be applicable to optimization of retentions in most situations encountered both in gas and in liquid chromatography. Indeed, as demonstrated recently by Hsu, Laub, and Madden (186), there have yet to be identified systems (excluding those that exhibit inflection points; see below) to which the relation does not apply.

For example, when b and b' are set to zero, eq. 245 reduces to the LC analog of the Laub-Purnell GC relation, eq. 99 (linear either in ϕ_B or ϕ_C, since $\phi_C = 1 - \phi_B$):

$$\frac{1}{t_{R(M)}} = \frac{\phi_B}{t_{R(B)}} + \frac{\phi_C}{t_{R(C)}} \tag{246}$$

Alternatively, in the instance of linear variation of liquid-gas partition coefficients with stationary-phase volume fraction, $K^{\circ}_{R(i)}$ ($i = B$, C, or B

$+ C$) would be substituted for $1/t_{R(i)}$ with b and b' set to zero. In contrast, suppose that retentions with a particular GC system varied nonlinearly with inverse absolute temperature. The behavior might well be described by a relation such as

$$\log K_R^\circ = \frac{A}{T} + \frac{B}{T^2} + \cdots$$

that is, a kind of virial expansion in reciprocal T. In order that the data be made to conform to eq. 245, log K_R° would first be plotted against a lower abscissa of T^{-1}, and would exhibit curvature. Next, an upper hypothetical abscissa of "ϕ_C" of from zero to unity (spanning the range of the lower abscissa) would be drawn such that (x,y) data pairs of ("ϕ_C", log K_R°) would be generated. Equation 245 could then be applied immediately, with $1/t_{R(i)}$ set equal to log K_R°. Were the data to exhibit one or more inflection points, the curves could be broken down into segments that conform to the relation. Optimization would then be carried out over individual regions of T^{-1}, rather than over the entire range (see also ref. 186).

*Computer Generation of Window-Boundary Data: Program WINDOW**

Equation 245 appears to be of sufficiently general form that it can be used to represent retentions as functions (whether linear or not) of a variety of system parameters. As a result, Laub (187) devised a simple algorithm that makes use of it in generating window-boundary data. In the program and description that follow, it is assumed that the window boundary is comprised of separation-factor data (eq. 197) arising from solute retentions that vary with the composition of a liquid-chromatographic mobile phase in accordance with eq. 245. For the sake of clarity, the program statements have not in many instances been concatenated where it would otherwise be possible (and even beneficial) to do so and, for the same reason, potential savings in execution time are sacrificed in favor of presentation of the logic in expanded form.

Data Input (Statements 1000–1200)

```
1000 REM DATA INPUT—INPUT THE SOLVENT AND SOLUTE NAMES,
     AND THE RESPECTIVE RETENTIONS. THEN DISPLAY THESE VAL-
     UES.
1010 HOME : PR#0 : DIM N$(51), A(51), S(51), B1(51), B2(51), X(500),
     Y(500), M$(500)
```

* Computer program and description from R. J. Laub, *J. Liq. Chromatogr.*, **7**, 647 (1984).

```
1020  PRINT : PRINT : PRINT : PRINT : PRINT : PRINT
1030  PRINT "SOLVENT 'A' IS: ";
1040  INPUT A$
1050  PRINT : PRINT
1060  PRINT "SOLVENT 'S' IS: ";
1070  INPUT S$
1080  PRINT : PRINT
1090  PRINT "THE NUMBER OF SOLUTES (MAXIMUM OF 50) IS: ";
1100  INPUT N
1110  HOME
1120  PRINT : PRINT
1130  PRINT "ENTER THE RESPECTIVE SOLUTE NAMES AND RETEN-
      TIONS WITH SOLVENTS 'A' AND 'S' "
1140  PRINT : PRINT
1150  PRINT "SOLUTE NAME, TR(A), TR(S), B1, AND B2" : PRINT
1160  FOR I = 0 TO N - 1 : INPUT N$(I), A(I), S(I), B1(I), B2(I): NEXT
      I
1165  HOME : PRINT : PRINT : PRINT : PRINT
1170  PRINT "THE LOWER MOBILE-PHASE COMPOSITION PERCENT
      TO BE CONSIDERED IS (WHOLE NUMBER) ";
1175  INPUT DL
1180  PRINT : PRINT "THE UPPER MOBILE-PHASE COMPOSITION PER-
      CENT TO BE CONSIDERED IS (WHOLE NUMBER) ";
1185  INPUT DU
1190  PRINT : PRINT "THE MOBILE-PHASE COMPOSITION PERCENT
      INTERVAL TO BE CONSIDERED IS (WHOLE NUMBER; SMALLEST
      PERMISSIBLE IS 1%) ";
1200  INPUT D
```

These statements first clear the screen (1010), dimension the variables, and then query the user for the names of the solvents and the number of solutes. The program then clears the screen again (1110) and asks for the names of the solutes, the respective retentions with solvents A and S, and the fitted values of b (B1) and b' (B2) (1130 ff.). The data entry format is as shown, namely, SOLUTE NAME (comma), TR(A) (comma), TR(S) (comma), B1 (comma), B2, then ⟨RETURN⟩. The program then asks for the mobile-phase composition range and interval (e.g., every 1%, every 5%, etc.) to be considered (1170-1200); note that the lowest permitted interval, for reasons of memory conservation, is 1%.

Data Verification (Statements 1210–1360)

```
1210  PR#1
1220  PRINT : PRINT
1230  PRINT TAB(26), "*****RETENTION DATA*****"
1240  PRINT : PRINT
1250  PRINT TAB(5); "SOLVENT 'A' IS "; A$
1260  PRINT TAB(5); "SOLVENT 'S' IS "; S$
1270  PRINT : PRINT
1280  PRINT TAB(5); "SOLUTE"; TAB(20); "TR(A)"; TAB(35); "TR(S)";
      TAB(52); "B1"; TAB(27); "B2"
1290  PRINT
1300  FOR I = 0 TO N - 1
1310  PRINT TAB(5); LEFT$ (N$(I),10); TAB(20); A(I); TAB(35); S(I);
      TAB(50); B1(I); TAB(55); B2(I): NEXT I
1330  PRINT : PRINT : HOME
1340  PRINT "MIXTURES OF 'A' WITH 'S' WILL BE CONSIDERED AT
      EVERY "; D; "% FROM 'A' = "; DL; " TO "; DU; "%."
1350  PRINT : PR#0: PRINT : PRINT : PRINT : PRINT : PRINT
1360  PRINT "FIRST, HOWEVER, THE RELEVANT PAIRS OF SOLUTES
      FOR CALCULATION OF THE WINDOW DIAGRAM WILL BE DE-
      TERMINED."
```

The solute and solvent data are printed out on the hard-copy device PR#1. The program uses a simple loop (1300,1310) to do so after the title (1230) and column headings (1280) are printed. Note that the solute names are contained as strings in the array N$(I), and that the retentions with solvents A and S (names A$ and S$) are in the arrays A(I) and S(I), respectively.

Determination of Relevant Pairs of Solutes (Statements 1500–1980)

```
1500  REM THIS SECTION OF THE PROGRAM WILL DETERMINE THE
      RELEVANT PAIRS OF SOLUTES FOR CALCULATION OF THE WIN-
      DOW-DIAGRAM ARRAY.
1510  PRINT : PRINT : PRINT
1520  PRINT "ENTER THE UPPER LIMIT OF SEPARATION FACTOR (>0)
      TO BE CONSIDERED: ";
1530  INPUT MAX
1540  Z = 0
1550  Z1 = 0
1560  FOR J = 0 TO N - 2
1580  HOME : PRINT : PRINT : PRINT : PRINT : PRINT "THE NUMBER
      OF RELEVANT PAIRS": PRINT : PRINT : PRINT "FOUND SO FAR
      IS :";Z1
```

```
1620  FOR I = J + 1 TO N - 1
1670  LP = (A(I) - A(J))/(A(I) + A(J))
1680  IF (ABS(LP)) < MAX THEN GOTO 1730
1690  LQ = (S(I) - S(J))/(S(I) + S(J))
1710  IF (ABS(LQ)) > MAX THEN IF (LP/LQ) > 0 THEN GOTO 4000

      4000  FOR P = DL TO DU STEP D
      4020  COMP = P * 0.01
      4030  L1 = COMP * ((1/A(I)) + (B1(I) * (1 - COMP)/(1 + B2(I) *
            (1 - COMP)))) + (1 - COMP)/S(I)
      4040  L2 = COMP * ((1/A(J)) + (B1(J) * (1 - COMP)/(1 + B2(J) *
            (1 - COMP))) + (1 - COMP)/S(J)
      4050  SF = (L1 - L2)/(L1 + L2)
      4060  IF (ABS(SF)) > MAX THEN GOTO 4080
      4070  GOTO 1730
      4080  NEXT P
      4090  GOTO 1850

1730  Z1 = Z1 + 1
1740  HOME : PRINT : PRINT : PRINT : PRINT : PRINT "THE NUMBER
      OF RELEVANT PAIRS": PRINT : PRINT : PRINT "FOUND SO FAR
      IS :"; Z1: FOR PAUSE = 1 TO 100:NEXT PAUSE
1750  K = J
1760  FOR Z = Z TO (Z + 1)
1770  X(Z) = A(K)
1775  Y(Z) = S(K)
1780  M1(Z) = B1(K)
1785  M2(Z) = B2(K)
1790  M$(Z) = N$(K)
1795  K = I
1800  NEXT Z
1850  NEXT I
1900  NEXT J
1905  IF Z1 = 0 THEN GOTO 3300
1910  HOME : PR#1 : PRINT : PRINT
1915  PRINT TAB(26); "************************"
1920  PRINT : PRINT : PRINT TAB(5); "THE NUMBER OF RELEVANT
      PAIRS OF SOLUTES IS "; Z1; "."
1930  PRINT : PRINT
1940  PRINT TAB(5); "THE RELEVANT PAIRS ARE:" : PRINT
1960  FOR Z = 0 TO (Z1 * 2 - 1) STEP 2
1970  PRINT TAB(15); (LEFT$ (M$(Z),10)); "/"; (LEFT$ (M$(Z + 1),10))
1980  NEXT Z
```

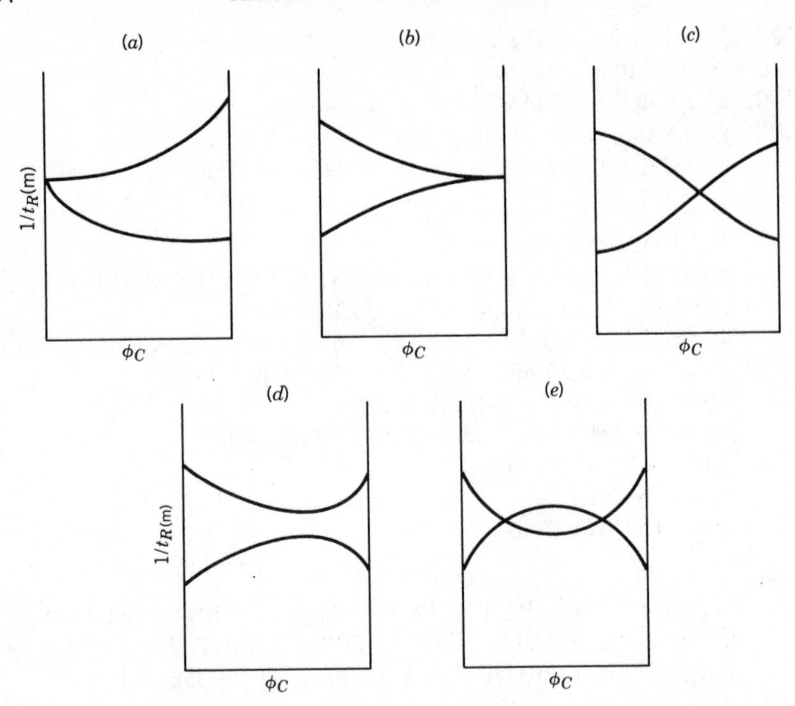

Rather than calculating the separation factors for all pairs of solutes at all compositions, the program first determines the number and identity of pairs of solutes that have values of S_f less than the user-defined limit MAX at some point within the specified composition range of DL to DU% of A in $(A + S)$. The task is straightforward when the variation of solute retentions is known as a function of column composition. Five situations arise generally, as seen in the accompanying figure. In situations (*a*) and (*b*), full overlap of the solutes occurs at one or the other of the ordinates. S_f is therefore 0 at each of these points. In the third case, (*c*), the order of elution of the solutes is reversed on passing from one extremum to the other. Hence, while S_f is greater (or less) than 0 at one ordinate, it will be less (or greater) than 0 at the other. Finally, situations (*d*) and (*e*) encompass those instances where the curves do no intersect at any or at more than one composition. These can be identified only by examination of the solute retentions at intermediate mobile-phase compositions.

In order to test for each of the above possibilities (and hence identify the relevant pairs), the separation factors for each solute pair are calculated at each of the ordinates (1540–1900) and, where necessary, at intermediate compositions (subroutine 4000–4090). First, however, and following a displayed message so indicating, the user is prompted to enter the upper limit of S_f which will be used to define what constitutes a relevant pair. Judicious choice of the limiting separation factor can lead to an enormous savings in the time of calculation of the window boundary, since whatever pairs are eliminated at this point will not be considered again. (An S_f of 0.02828 corresponds to a column of 5000 plates and minimum resolution R_s of unity.) If no relevant pairs are found, the program branches at 1905 to statement 3300 and displays a message so informing the user:

```
3300 HOME : PRINT : PRINT : PRINT : PRINT : PRINT : PRINT "NO
     PAIRS FOUND—ALL COMPOSITIONS WILL PROVIDE GOOD
     RESOLUTION. WANT TO TRY A HIGHER VALUE OF SF (Y/N)?":
     INPUT ANS$
3310 IF ANS$ = "N" THEN GOTO 3270
3320 PRINT : PRINT : PRINT : GOTO 1520
```

The final task of this section of the program (1910–1980) gives a hardcopy printout of the number and identity of the relevant pairs of solutes.

Calculation of the Window Boundary Array (Statements 3000–3200)

```
3000 REM THIS SECTION OF THE PROGRAM CALCULATES THE WIN-
     DOW DIAGRAM ARRAY, HERE, SF AS A FUNCTION OF MOBILE-
     PHASE COMPOSITION FOR LIQUID CHROMATOGRAPHY.
3010 HOME : PR#0
3020 DIM Q$(101), R$(101), SFP(101)
3040 BSFP = 0
3045 FOR P = DL TO DU STEP D
3050 HOME : PRINT : PRINT : PRINT : PRINT : PRINT : PRINT : PRINT
     : PRINT
3060 PRINT "THE COLUMN COMPOSITION CURRENTLY BEING":
     PRINT : PRINT "CONSIDERED IS "; P; "%"
3070 SFP(P) = MAX
3075 COMP = P * 0.01
3080 Q$(P) = "(NONE)"
3085 R$(P) = "(NONE)"
3100 FOR Z = 0 TO (Z1 * 2 - 1) STEP 2
3110 L1 = COMP * ((1/X(Z)) + (M1(Z) * (1 - COMP)/(1 + M2(Z) * (1
     - COMP)))) + (1 - COMP)/Y(Z)
3115 L2 = COMP * ((1/X(Z + 1)) + (M1(Z + 1) * (1 - COMP)/(1 +
     M2(Z + 1) * (1 - COMP)))) + (1 - COMP)/Y(Z + 1)
```

```
3120  SF = (L1 - L2)/(L1 + L2)
3125  IF (ABS(SF)) > SFP(P) THEN GOTO 3170
3130  SFP(P) = ABS(SF)
3140  Q$(P) = M$(Z): R$(P) = M$(Z + 1)
3170  NEXT Z
3175  IF SFP(P) < BSFP THEN GOTO 3200
3180  BSFP = SFP(P)
3185  BA$ = Q$(P)
3190  BS$ = R$(P)
3195  OPT = P
3200  NEXT P
```

Once the relevant pairs of solutes have been identified, separation factors for each are calculated in turn at each column composition and the lowest (most difficult) is saved in the array subscripted as P. Thus, SFP(P) (3130) is the most-difficult (window-boundary) value of S_f at the column composition corresponding to P, while solutes Q$(P) and R$(P) (3140) are the names of the solutes. The *overall best* value of SFP(P), BSFP (3180), is updated on each pass through the outer loop, as are the names of the corresponding most-difficult solutes, BA$ (3185) and BS$ (3190). The overall best (optimum) column composition is also stored (3195) as OPT.

This section of the program is by far the slowest, the rate-limiting statements being 3110 and 3115. To indicate that the computer is still working (and to time the program if desired), the composition currently being considered is displayed.

SFP(P), Q$(P), and R$(P) default (3070,3080,3085) to the value of MAX and the string "(NONE)" if, at a given column composition, the separation factors of all relevant pairs of solutes exceed that of MAX (see later).

Data Output (Statements 3205–3290)

```
3205  PR#1
3210  HOME : PRINT : PRINT : PRINT TAB(5); "THE WINDOW-BOUND-
      ARY DATA ARE:"
3215  PRINT : PRINT
3220  PRINT TAB(11); "SOLUTE"; TAB(36); "COL."; TAB(57); "SEPN."
3225  PRINT TAB(12); "PAIR"; TAB(36); "COMP."; TAB(16); "FACTOR"
3230  PRINT : PRINT
3235  FOR P = DL TO DU STEP D
3240  PRINT TAB(5); LEFT$ (Q$(P),10); "/"; LEFT$ (R$(P),10); TAB(37);
      P; TAB(54); (INT(10 ^ 5 * (SFP(P)) + 0.02))/10 ^ 5
3245  NEXT P
3250  HOME : PRINT : PRINT
```

```
3255 PRINT "THE BEST COLUMN COMPOSITION IS: "; OPT; "%."
3260 PRINT : PRINT "THE MOST-DIFFICULT SEPARATION FACTOR AT
     THIS COMPOSITION IS: "; BSFP; "."
3265 PRINT : PRINT "THE MOST DIFFICULT SOLUTES TO SEPARATE
     AT THIS COMPOSITION ARE: "; BA$ " FROM "; BS$; "."
3270 PR#0
3275 PRINT : PRINT : PRINT : PRINT : PRINT : PRINT : PRINT : PRINT
     PRINT TAB(10); "*****THAT'S ALL, FOLKS*****"
3290 END
```

A hard-copy printout of the window-boundary array is accomplished by the loop, 3235–3245. For easier reading, the separation-factor data are truncated (3240) to five places. If at a given column composition the separation factors of all relevant pairs exceed the value of MAX, the solute-pair printout is (NONE)/(NONE) and the separation factor printed is MAX. (A plot of the data in this composition region thus would show a flat top.) Also printed out (3255–3265) are the overall best column composition, the most difficult S_f at this composition, and the associated (most difficult) solute pair.

Commands Indigenous to APPLE™ BASIC

The only three commands used here which may not be compatible with other versions of BASIC are PR#1, PR#0, and HOME. The first two of these specify the hardcopy printer and the display unit, respectively, while the third command causes the display to clear and the cursor to be positioned in the upper left-hand corner of the screen. These commands appear in the following statements:

Command	Statement Nos.
PR#1	1210, 1910, 3205
PR#0	1010, 1350, 3010, 3270
HOME	1010, 1110, 1165, 1910, 3010, 3050, 3210, 3250, 3300

There may also be difficulty with multiple TAB statements depending upon the printer employed (here, an Epson MX-70). Substitution of POKE (36,*nn*) for TAB (*nn*) solves this problem.

2.6.2 Application of the Window-Diagram Strategy to Samples of Initially Unknown Content and Complexity

Rarely does an analyst know the precise number of components in a sample nor, frequently, their identities or amounts. The window-diagram strategy was extended successfully by Laub and Purnell (188) to mixtures of initially unknown content and complexity, and the resultant methodology has since proved to be a very powerful tool in analytical chromatography.

Consider that a solute mixture is to be separated by packed-column gas-liquid chromatography. The traditional method of determining an appropriate stationary phase and operating temperature amounts to one of trial-and-error: the analyst first chooses a column temperature so as to yield reasonable elution times, followed by attempting the separation with several stationary phases, these being chosen primarily on the basis of experience and/or literature reports (i.e., selected virtually at random). The results with any given solvent may or may not be satisfactory and, consequently, the search for appropriate phases and conditions may be somewhat protracted.

In contrast, suppose that the mixture were chromatographed with four columns, each of which contained different amounts (say, 0%, 33%, 67%, and 100% by volume) of two stationary phases B and C, fabricated by mixing mechanically appropriate amounts of pure-phase packings (for which instances eq. 99 must apply exactly). The resultant chromatograms might then appear as shown in Fig. 2.44, where the peaks are represented by lines of equal length and are drawn according to increasing partition coefficient. [Capacity factors or even retention indices could alternatively be employed if the columns were of equal dead volume and weight-percent liquid loading (189).]

Since columns 1 and 2 yield four peaks, while columns 3 and 4 show five solutes, there must be *at least* five solutes in the mixture, with some overlaps occurring with stationary-phase compositions 1 and 2. (There would of course be a change in heights and areas to indicate which peaks are comprised of overlapped solutes but, for simplicity, these are ignored in this example.) That is, a *floor* has been established for the number of sample constituents.

The data are shown plotted in Fig. 2.45 as K_R° against % C (v/v). Since eq. 99 is known to apply, each datum must lie on a straight line connecting the left- and right-hand ordinates, where each line corresponds to each solute. However, it is impossible to determine at this point where the lines should be drawn. Therefore, *all possible* lines are drawn through *all*

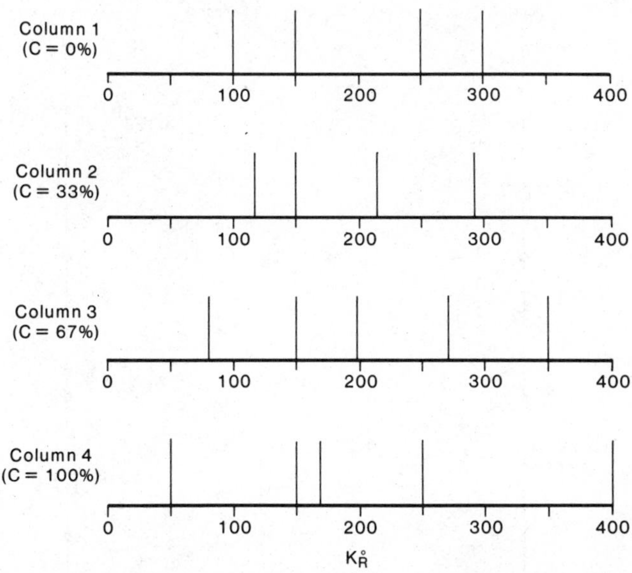

Figure 2.44. Skeletal chromatograms obtained for mixture of initially unknown content and complexity with columns of 0, 33, 67, and 100% stationary phase C in admixture with B.

possible sets of points, such that each line has four (but only four) points on it (corresponding to the four compositions of phases B and C).

The result is shown in Fig. 2.46 where the eight possible lines have been constructed. These comprise the *ceiling* number of solutes (labeled 1–8) that can be present in the mixture. Note that in drawing the lines, some points are used more than once: these correspond to overlapped solute bands. In contrast, we cannot construct the dashed line along the bottom of the figure since, if the line were real, a point (peak) of $K_R^{\circ} \approx$ 80 should have been seen with column 2. Since no peaks were detected at $(x,y) = (33\%,80)$ (empty parentheses), the line is fictitious. On the other hand, the line indicated as no. 8 must be real, since the points at $(x,y) =$ (67,350) and (100,400) arise as a result of actual peaks seen in the chromatograms. Thus we have established at this point that there is a minimum of five, and a maximum of eight, solutes present in the mixture and, moreover, that their partition coefficients regress against the stationary-phase composition as shown.

It is important to recognize that in deducing the number of regression lines, it is entirely unnecessary to know the identities of the solutes. In addition, only four chromatographic runs have been carried out at this point, that is, one run with each of the four columns.

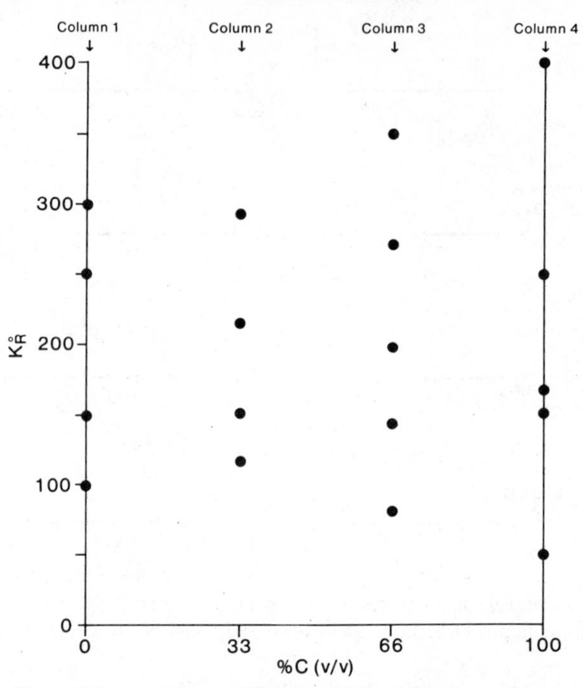

Figure 2.45. Plots of the partition coefficients obtained for the peaks seen in Fig. 2.44.

Three of the lines shown in Fig. 2.46 may potentially be fictitious, that is, a set of four points may coincidentally define a straight line as in solute-lines 2, 4, 5, and 6 (recall that the floor was five solutes). In order to eliminate the fictitious solutes, one could in fact construct a fifth column corresponding to some or other stationary-phase blend such that all real solutes would be at least partially resolved. For example, a column of 6% *C* (leftmost vertical dashed line in Fig. 2.46) could be fabricated, with which eight partially resolved solutes (if eight were in fact present) should be seen. Another such composition occurs at 45% *C* (central vertical dashed line). However, rather than construct such a column, we proceed instead to calculate the window diagram for all eight solutes, which is shown in Fig. 2.47. The overall best column composition occurs at 45% *C*, with which all solutes will be resolved with a column yielding 3135 plates, irrespective of whether the mixture is comprised of 5, 6, 7, or 8 components. If all eight are present, eight baseline-resolved peaks will be found; on the other hand, if there are only five solutes, these, too, will be baseline-resolved.

The chromatogram of our hypothetical solute mixture with $C = 45\%$ is shown in Fig. 2.48; the solid vertical lines represent actual solutes,

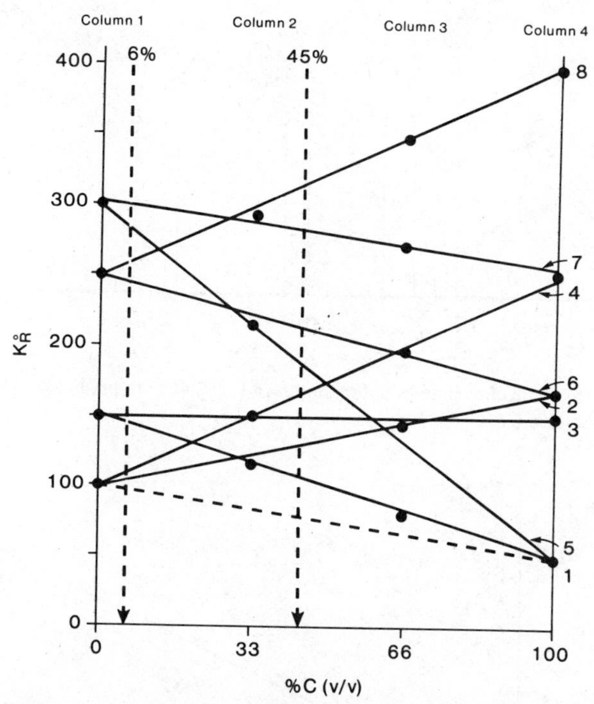

Figure 2.46. Eight possible lines drawn through sets of points of Fig. 2.45. Dashed line must be fictitious since no point is found for it with column containing 33% *C*.

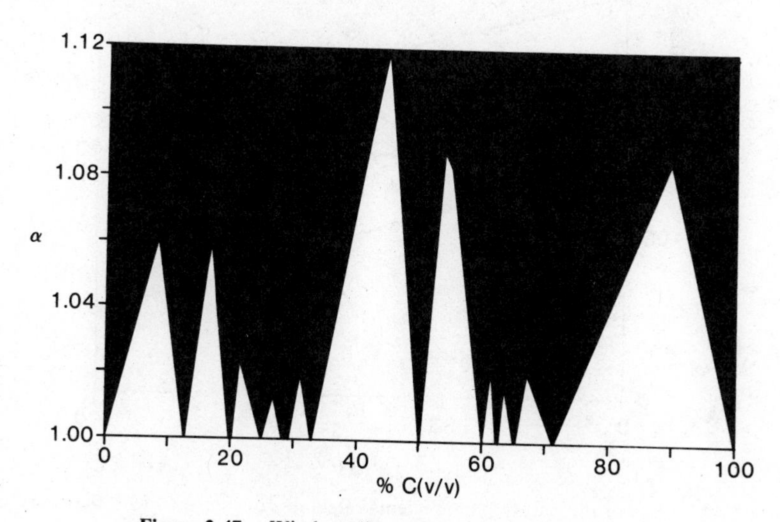

Figure 2.47. Window diagram for solutes of Fig. 2.46.

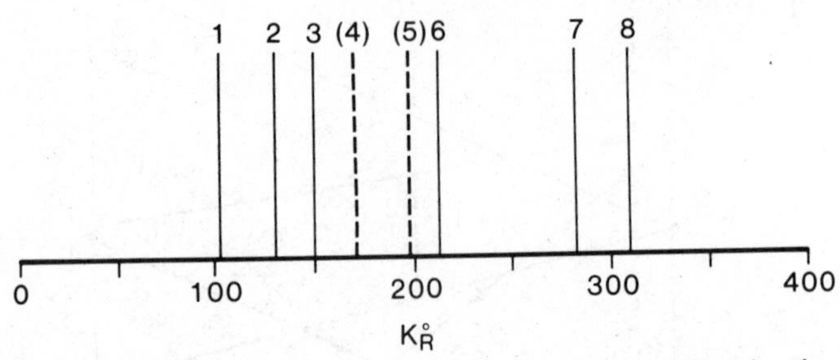

Figure 2.48. Skeletal chromatogram for solutes of Fig. 2.46 with predicted optimum column composition of 45% C.

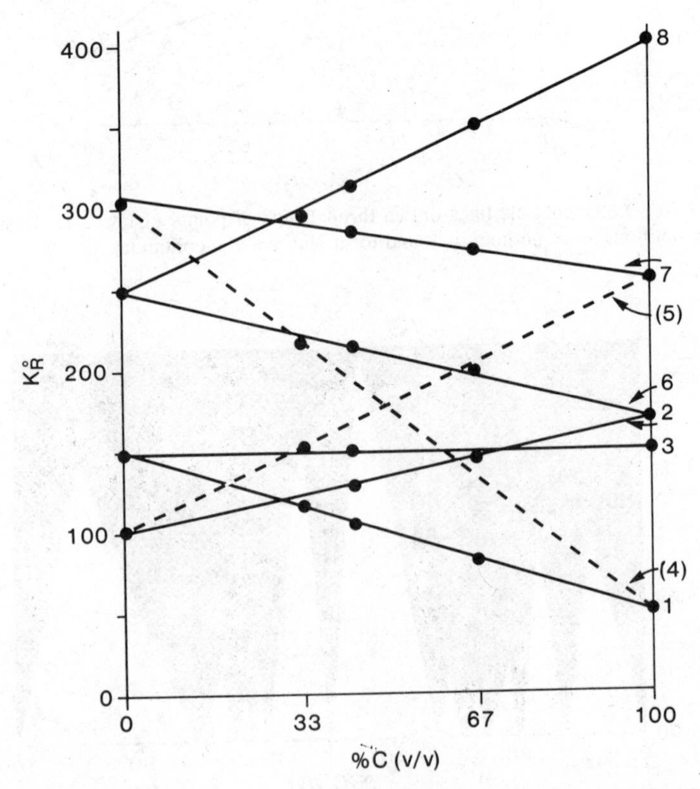

Figure 2.49. Plots of the partition coefficients of solutes of Fig. 2.46 with fictitious lines shown dashed.

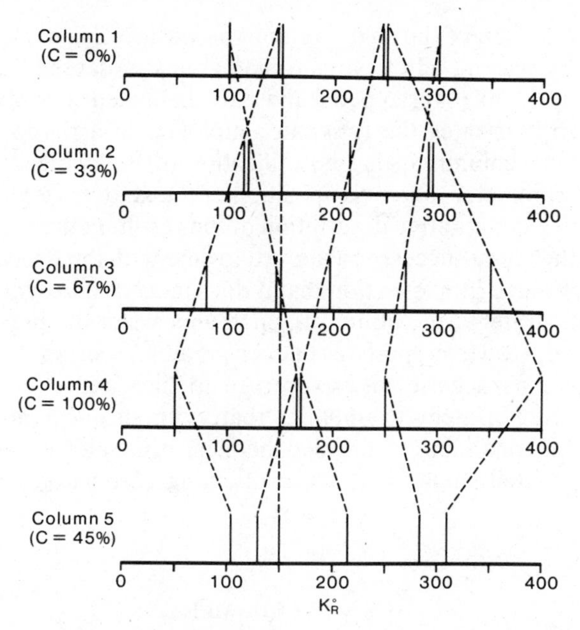

Figure 2.50. Summary skeletal chromatograms of the solutes, showing the variation of partition coefficients with column composition.

while the positions of the dashed lines indicate where the fictitious peaks would have appeared. Thus there are in fact six solutes present in the sample, all of which have been fully resolved after only five chromatographic runs.

The partition coefficients of the actual solutes are shown in Fig. 2.49, where the fictitious lines are now given as dashed. (The empty parentheses indicate where the data points for these solutes would have appeared.) We see immediately that the separation could also be carried out at 15% *C*, but with very little overall gain in analysis time. This might not be true generally, however, and so, once having identified the actual number of solutes present in a mixture, it might well be worthwhile to do a second window diagram in order to identify stationary-phase compositions more suitable from the standpoints of, for example, time of analysis; efficiency with one or the other of the stationary phases; the demands of a particular elution order in the instance of a trace constituent eluting near a compound present in much larger amounts; or other local criteria defined by the analyst. Even so, and regardless of these, the primary goal, that is, separation of the solute mixture, has in fact at this point been accomplished.

Finally, variation of the solute retentions with % *C* is depicted in summary form by placing all of the chromatograms in descending order, as shown in Fig. 2.50 (overlapped bands are indicated as doublets). The example also illustrates the rationale employed in arriving at the final separation with column 5, that is, utilization of the controlled variation of a system parameter (here, composition of the stationary phase) in conjunction with a quantitative description of the resultant data (eqs. 69, 99).

Overall, the many successes achieved to date with the window-diagram methodology leave little question that it does indeed represent a universal optimization strategy, the utility of which appears at the present time to be unbounded. However, perhaps of even greater importance (and despite nescient opinions occasionally expressed to the contrary), the plenary window-diagram strategy exemplifies that, without question, even a rudimentary understanding of physicochemical principles cannot but lead to very substantial gains in efforts at effecting chromatographic separations.

ACKNOWLEDGMENT

Support provided for this work in part by the Alcoa Foundation, the Department of Energy (LC), and the National Science Foundation (GC) is gratefully acknowledged.

REFERENCES

1. R.J. Laub and R. L. Pecsok, *Physicochemical Applications of Gas Chromatography*, Wiley-Interscience, New York, 1978, p. 3.
2. M.L. McGlashan and D.J.B. Potter, *Proc. Roy. Soc. Ser. A.*, **267,** 478 (1962).
3. E.A. Guggenheim and M.L. McGlashan, *Proc. Roy. Soc. Ser. A.*, **206,** 448 (1951).
4. R.R. Dreisbach, *Physical Properties of Chemical Compounds*, American Chemical Society, Washington, D.C., Vol. 1, 1955; Vol. 2, 1959; Vol. 3, 1961.
5. J.E. Lennard-Jones and W.R. Cook, *Proc. Roy Soc. Ser. A.*, **115,** 334 (1927).
6. G.H. Hudson and J.C. McCoubrey, *Trans. Faraday Soc.*, **56,** 761 (1960). See also: R.J. Munn, *Trans. Faraday Soc.*, **57,** 187 (1961).
7. F.M. Zado and R.S. Juvet Jr., in *Aspects in Gas Chromatography*, H.G. Struppe (Ed.), Akadamie Verlag, Berlin, 1971, p. 206.

8. D.H. Everett, B.W. Gainey, and C.L. Young, *Trans. Faraday Soc.*, **64,** 2667 (1968).
9. A. Goldup, G.R. Luckhurst, and W.T. Swanton, *Nature,* **193,** 333 (1962).
10. V. Pretorius, *J. High Resolut. Chromatogr. Chromatogr. Commun.*, **1,** 199 (1978).
11. Mention must also be made of the concept of acentricity, introduced by Pitzer and his colleagues, where distortion of the intermolecular potential function relative to that of simple spherical molecules is taken into account. The goodness of fit of B_{12} data to systems of interest here appears to be somewhat limited, but the method requires only critical and vapor-pressure data and is therefore more convenient (if more empirical) than eq. 14. The interested reader is referred to Lewis and Randall, *Thermodynamics*, Second Revised Ed., Pitzer and Brewer, McGraw-Hill, New York, 1961, Appendix 1; K.S. Pitzer and R.F. Curl, Jr., *J. Am. Chem. Soc.*, **79,** 2369 (1957).
12. J.P. Kuenen, *Commun. Phys. Lab. Univ. Leiden,* **4,** (1892). See also: J.S. Rowlinson and F.L. Swinton, *Liquids and Liquid Mixtures*, Third Ed., Butterworths, London, 1982, Ch. 6.
13. D. Bartmann and G. M. Schneider, *J. Chromatogr.*, **83,** 135 (1973).
14. G.M. Schneider, *Chem. Eng. Progr. Symp. Ser.*, **64** (88), 9 (1968).
15. J.C. Giddings, M.N. Myers, L. McLaren, and R.A. Keller, *Science,* **162,** 67 (1968).
16. M. Novotny, W. Bertsch, and A. Zlatkis, *J. Chromatogr.*, **61,** 17 (1971). See also: R. Board, D. McManigill, H. Weaver, and D. Gere, *CHEMSA*, (June) 12, (1983).
17. R.E. Jentoft and T.H. Gouw, *J. Chromatogr. Sci.*, **8,** 138 (1970).
18. R.E. Jentoft and T.H. Gouw, *J. Polym. Sci. B,* **7,** 821 (1969).
19. S.T. Sie, J.P.A. Bleumer, and G.W.A. Rijnders, *Gas Chromatography 1968,* C.L.A. Harbourn (Ed.), Butterworths, London, 1969, p. 235.
20. R.E. Jentoft and T.H. Gouw, *Anal. Chem.*, **44,** 681 (1972).
21. E. Klesper, A.H. Corwin, and D.A. Turner, *J. Org. Chem.*, **27,** 700 (1962).
22. S.T. Sie, W. van Beersum, and G.W.A. Rijnders, *Separ. Sci.*, **1,** 459 (1966).
23. S.T. Sie and G.W.A. Rijnders, *Separ. Sci.*, **2,** 699, 729, 755 (1967).
24. T.H. Gouw and R.E. Jentoft, *J. Chromatogr.*, **68,** 303 (1972); *Adv. Chromatogr.*, **13,** J.C. Giddings, E. Grushka, R.A. Keller, and J. Cazes (Eds.), Marcel Dekker, New York, 1975, Ch. 1.
25. W. Heintz, *Jahresber.*, **2,** 342 (1849).
26. Alternatively, suppose that an organic reaction in methanol solvent is known to proceed farther toward completion under pressure: a thermostated reaction vessel would commonly be filled partially with reactants and solvent and then pressurized with inert gas to a value that is taken to be both judicious and prudent. This represents a hazardous situation, insofar as the vessel contains a gas and may therefore fail explosively. A simpler and safer procedure would seem to entail filling the vessel com-

pletely with (presumed liquid or solid) reactants and solvent at some temperature below that desired. Reference to Table IV then shows that the system pressure can be increased by 10 atm per degree, and so the desired pressure would be achieved simply by adjusting the system thermostat. Furthermore, the pressure is entirely hydraulic, that is, is released immediately if the vessel fails, and the system is therefore considerably safer than one containing pressurized gas.

27. The mean free path and collison cross-sectional area of gaseous molecules can in fact be extracted from viscosity data.
28. R.C. Hardy and R.L. Cottingham, *J. Res. Nat. Bur. Stand.*, **42,** 573 (1949).
29. Plots of log η_l against T^{-1} would seem to offer an alternative (although comparatively speaking an experimentally difficult) method of detection of weak first-order phase transitions inter alia in liquid crystals.
30. Strictly, a lowering of the system free energy.
31. P. Lecomte du Noüy, *J. Gen. Physiol.*, **1,** 521 (1918); **7,** 403 (1925).
32. J.W. McBain and A.M. Bakr, *J. Am. Chem. Soc.*, **48,** 690 (1926).
33. Accurate work requires calculation of an effective height h_{eff} of the liquid column, which in essence takes into account the weight of liquid in the meniscus. For convex menisci,

$$h_{\text{eff}} = h + r/3 - 0.1288\, r^2/h + 0.1312\, r^3/h^2$$

The signs are reversed for concave menisci.
34. N.K. Adam and G. Jessop, *J. Chem. Soc.*, **127,** 1863 (1925); N.K. Adam, *The Physics and Chemistry of Surfaces*, Third Ed., Oxford University Press, Oxford, England, 1941.
35. M.G. Voronkov, L.P. Ignat'eva, E.V. Kukharskaya, and V.M. Makaraskaya, *Zh. Prikl. Khim.* (*Leningrad*), **54,** 1392 (1981).
36. Contact-angle and surface-tension measurements offer in addition yet another method of assessment of solutions phenomena, such as phase transitions of liquid crystals. Mercury would seem to be an ideal surface for such studies, but in practice it is found that the surface cannot be kept clean enough for accurate work [Tronstad and Feachem, *Proc. Roy. Soc. Ser. A.*, **145,** 115 (1934)]. Gallium and indium, both of which wet glass, may be useful alternatives.
37. H. McLeod, *Trans. Faraday Soc.*, **19,** 38 (1932).
38. F.M. Perel'man, *Phase Diagrams of Multicomponent Systems: Geometric Methods*, D.A. Paterson (Trans.), Consultants Bureau, New York, 1966.
39. A.C. de Kock, *Z. Phys. Chem.*, **48,** 129 (1904).
40. H.T. Peterson and D.E. Martire, *Mol. Cryst. Liq. Cryst.*, **25,** 89 (1974).
41. T. Boublík, V. Fried, and E. Hála, *The Vapour Pressures of Pure Substances*, Elsevier, Amsterdam, 1973.
42. E.g., ref. 4.

43. G.P. Baxter and M.R. Grose, *J. Am. Chem. Soc.,* **37,** 1061 (1915).
44. E.A. Moelwyn-Hughes, *Physical Chemistry*, Pergamon Press, London, 1957.
45. J.H. Purnell, *Gas Chromatography*, Wiley, London, 1962, Ch. 3.
46. R.J. Laub, J.H. Purnell, P.S. Williams, M.W.P. Harbison, and D.E. Martire, *J. Chromatogr.,* **155,** 233 (1978). See also Ref. 1, Chs. 2,3.
47. W. Thomson, *Phil. Mag.,* **42,** 448 (1871).
48. J.R. Conder and C.L. Young, *Physical Measurement by Gas Chromatography*, Wiley-Interscience, Chichester, England, 1979, pp. 509–512.
49. J.S. Rowlinson, in *Handbuch der Physik*, Vol. 12, S. Flugge, Ed., Springer, Berlin, 1958.
50. C.M. Knobler, in *Chemical Thermodynamics,* Vol. 2, M.L. McGlashan (Ed.), Specialist Periodical Reports, The Chemical Society, London, 1978, Ch. 7.
51. J. von Zawidski, *Z. Phys. Chem.,* **35,** 129 (1900).
52. D.H. Everett, *Trans. Faraday Soc.,* **61,** 1637 (1965).
53. A.J.B. Cruickshank, M.L. Windsor, and C.L. Young, *Proc. Roy. Soc. Ser. A.,* **295,** 259 (1966).
54. R.J. Laub, *Anal. Chem.,* **56,** 2110, 2115 (1984).
55. A.J.B. Cruickshank, B.W. Gainey, C.P. Hicks, T.M. Letcher, R.W. Moody, and C.L. Young, *Trans. Faraday Soc.,* **65,** 1014 (1969).
56. F.M. Zado and R.S. Juvet Jr., in *Gas Chromatography 1966,* A.B. Littlewood (Ed.), Institute of Petroleum, London, 1967, p. 283.
57. C.-F. Chien, M.M. Kopečni, and R.J. Laub, *J. High Resolut. Chromatogr. Chromatogr. Commun.,* **4,** 539 (1981).
58. R.J. Laub, D.E. Martire, and J.H. Purnell, *J. Chem. Soc. Faraday Trans. I,* **73,** 1685 (1977); *J. Chem. Soc. Faraday Trans. II,* **74,** 213 (1978).
59. A.B. Littlewood, *Gas Chromatography*, Academic Press, New York, 1970, pp. 66–70.
60. M. Barbe and D. Patterson, *J. Solut. Chem.,* **9,** 753 (1980).
61. D.D. Deshpande, D. Patterson, H.P. Schreiber, and C.S. Su, *Macromolecules,* **7,** 530 (1974).
62. G.N. Lewis and M. Randall, *Thermodynamics*, Second Revised Ed., K.S. Pitzer and L. Brewer, McGraw-Hill, New York, 1961, Ch. 29.
63. R.L. Martin, *Anal. Chem.,* **35,** 116 (1963).
64. D.E. Martire, R.L. Pecsok, and J.H. Purnell, *Nature,* **203,** 1279 (1964); *Trans. Faraday Soc.,* **61,** 2496 (1965). See also: R.L. Pecsok, A. de Yllana, and A. Abdul-Karim, *Anal. Chem.,* **36,** 452 (1964); R.L. Pecsok and B.H. Gump, *J. Phys. Chem.,* **71,** 2202 (1967); and the review: D.E. Martire, in *Progress in Gas Chromatography*, J.H. Purnell (Ed.), Wiley-Interscience, New York, 1968, p. 93.
65. R.H. Perrett and J.H. Purnell, *J. Chromatogr.* **7,** 455 (1962).

66. G.M. Janini and D.E. Martire, *J. Chem. Soc. Faraday Trans. II*, **70,** 837 (1974).
67. M.W.P. Harbison, R.J. Laub, D.E. Martire, J.H. Purnell, and P.S. Williams, *J. Phys. Chem.*, **83,** 1262 (1979).
68. J.H. Purnell and J.M. Vargas de Andrade, *J. Am. Chem. Soc.*, **97,** 3585, 3590 (1975).
69. R.J. Laub and J.H. Purnell, *J. Am. Chem. Soc.*, **98,** 30, 35 (1976).
70. It is important to note that no approximations of any kind have been made in the derivation of eq. 98 and 99 other than that the solvents exhibit either complete immiscibility or complete ideality upon comixing.
71. R.J. Laub and C.A. Wellington, in *Molecular Association*, Vol. 2, R. Foster (Ed.), Academic Press, London, 1979, Ch. 3.
72. P.F. Tiley, *J. Chromatogr.*, **179,** 247 (1979).
73. R.J. Laub, J.H. Purnell, and D.M. Summers, *J. Chem. Soc. Faraday Trans. I*, **76,** 362 (1980).
74. J. Klein and H. Widdecke, *J. Chromatogr.*, **147,** 384 (1978).
75. C.-F. Chien, M.M. Kopečni, and R.J. Laub, *J. Chromatogr. Sci.*, **22,** 1 (1984).
76. C.-F. Chien, M.M. Kopečni, and R.J. Laub, *Anal. Chem.*, **52,** 1402 (1980).
77. R.J. Laub and J.H. Purnell, *Anal. Chem.*, **48,** 799 (1976).
78. The summation should therefore apply also to solvents of a highly specialized nature, such as liquid ion-exchange media and the remarkable series of compounds known as liquid clathrates developed by Atwood and his coworkers [U.S. Patent 4,024,170, May 1977; see the review: J.L. Atwood, in *Recent Developments in Separations Science*, Vol. 3 (Pt. B), N.L. Li, Ed., CRC Press, West Palm Beach, Florida, 1977, p. 195)] which are of the form: $M[Al_2R_6X]$, where M is an alkali metal or pseudometal (e.g., NH_4^+), R is an alkyl group, and X is a halogen or pseudohalogen (e.g., NO_3^-, RCO_2^-, etc.).
79. R.J. Laub, *Glas. Hem. Drus. Beograd*, **48,** 377 (1983).
80. W.D. Harkins, *The Physical Chemistry of Surface Films*. Van Nostrand-Reinhold Co., New York, 1952.
81. It is indeed found that injection volumes much smaller than is usually the norm must be used with liquid-crystalline stationary phases when these are in one or another mesomorphic forms. Pinkevich, Reshetnyak, and Sugakov appear to be the most recent to offer comments on ordering in the neighborhood of impurities in liquid crystals: *Krystallografiya*, **28,** 400 (1983).
82. D. Patterson, Y.B. Tewari, H.P. Schreiber, and J.E. Guillet, *Macromolecules*, **4,** 356 (1971).
83. R.N. Lichtenthaler, J.M. Prausnitz, C.S. Su, H.P. Schreiber, and D. Patterson, *Macromolecules*, **7,** 136 (1974).

84. A.J. Ashworth, C.-F. Chien, D.L. Furio, D.M. Hooker, M.M. Kopečni, R.J. Laub, and G.J. Price, *Macromolecules,* **17,** 1090 (1984).

85. It must be said, of course, that a very large contributory factor is the advanced state of surface analysis techniques that permit the direct and facile characterization of surfaces and adsorbed species, in contrast to (experimentally difficult) methods such as low-angle X-ray or neutron analysis of liquid solutions.

86. A.V. Kiselev, *Gas-Adsorption Chromatography,* Plenum, New York, 1969.

87. D.H. Everett (Ed.), *Colloid Science*, Specialist Periodical Reports, The Chemical Society, London, Vol. 1 (1973); Vol. 2 (1975); Vol. 3 (1979).

88. A.W. Adamson, *Physical Chemistry of Surfaces*, Third Ed., Wiley-Interscience, New York, 1976.

89. S.J. Gregg and K.S.W. Sing, *Adsorption, Surface Area, and Porosity*, Second Ed., Academic Press, London, 1982, and many references therein.

90. V. Bosáček, *Coll. Czech. Chem. Commun.,* **29,** 1797 (1964). See also: R.M. Barrer and A.B. Robins, *Trans. Faraday Soc.,* **49,** 807 (1953).

91. I. Langmuir, *J. Am. Chem. Soc.,* **38,** 2221 (1916); **40,** 1361 (1918).

92. K.S.W. Sing, ref. 87*a*, Ch. 1.

93. J.P. Hobson, *J. Phys. Chem.,* **73,** 2720 (1969).

94. S. Brunauer, P.H. Emmett, and E. Teller, *J. Am. Chem. Soc.,* **60,** 309 (1938).

95. S. Brunauer, L.S. Deming, W.S. Deming, and E. Teller, *J. Am. Chem. Soc.,* **62,** 1723 (1940). See also: A.W. Adamson, *J. Colloid Interfac. Sci.,* **27,** 180 (1968); M.E. Tadros, P. Hu, and A.W. Adamson, *J. Colloid Interfac. Sci.,* **49,** 184 (1974) for examples of isotherm types labeled VI and VII.

96. It has been argued on occasion that absolutely symmetric peaks are found only with a very shallow Type III isotherm since solute molecules at the back edge of an elution band spend a longer time in the column, and hence have the opportunity to spread (deviate from the average) more than do those that elute more quickly. However, this is not commonly observed experimentally except at very high carrier flow rates.

97. W. Drost-Hansen, in *Chemistry and Physics of Interfaces*, Vol. 2, S. Ross (Ed.), American Chemical Society, Washington, D.C., 1971, p. 203.

98. G.E. Van Gils, *J. Colloid Interfac. Sci.,* **30,** 272 (1969).

99. D.H. Everett, *Trans. Faraday Soc.,* **60,** 1803 (1964); **61,** 2478 (1965); ref. 87*a*, Ch. 2; 87*b*, Ch. 2; 87*c*, Ch. 2; in *Adsorption at the Gas–Solid and Liquid–Solid Interface*, J. Rouquerol and K.S.W. Sing (Eds.), Elsevier, Amsterdam, 1982, p. 1.

100. S.G. Ash, D.H. Everett, and G.H. Findenegg, *Trans. Faraday Soc.,* **64,** 2639 (1968); **66,** 708 (1970).

101. F. Köster and G.H. Findenegg, *Chromatographia,* **15,** 743 (1982).

102. D.E. Martire and D.C. Locke, *Anal. Chem.,* **43,** 68 (1971).

103. D.C. Locke, *J. Chromatogr. Sci.*, **12,** 433 (1974).
104. R.E. Boehm and D.E. Martire, *J. Phys. Chem.*, **84,** 3620 (1980).
105. D.E. Martire and R.E. Boehm, *J. Liq. Chromatogr.*, **3,** 753 (1980); *J. Phys. Chem.*, **87,** 1045 (1983).
106. R.E. Boehm, D.E. Martire, D.W. Armstrong, and K.H. Bui, *Macromolecules*, **16,** 466 (1983).
107. J. Stranahan and S.N. Deming, *Anal. Chem.*, **54,** 2251 (1982).
108. L.H. Ngoc, J. Ungvarai, and E. sz. Kovats, *Anal. Chem.*, **54,** 2410 (1982).
109. P.E. Hare and E. Gil-Av, *Science*, **204,** 1226 (1979).
110. R.P.W. Scott, in *Chromatography, Equilibria, and Kinetics*, D.A. Young (Ed.), The Royal Society of Chemistry, London, 1980, p. 49; R.P.W. Scott and C.F. Simpson, *Chromatography, Equilibria, and Kinetics*, D.A. Young (Ed.), Royal Society of Chemistry, London, 1980, p. 69.
111. R.P.W. Scott and P. Kucera, *J. Chromatogr.*, **112,** 425 (1975); **149,** 93 (1978).

112. The relation represents the form generally for the transport of matter, energy, pressure, charge, and so forth:

$$J_x = -\kappa_T \left(\frac{dT}{dx}\right)$$

$$J_x = -\kappa \left(\frac{dV}{dx}\right)$$

$$J_x \propto -\frac{1}{\eta}\left(\frac{dp}{dx}\right)$$

where (dT/dx), (dV/dx), and (dp/dx) represent thermal, potential, and pressure gradients, and where κ_T, κ, and η are the coefficients of thermal conductivity, electrical conductivity, and viscosity. The equations are known, respectively, as Fourier's law, Ohm's law, and Poiseuille's law.

113. The formal proof of this is as follows. Let the chemical potential of a solute be u_A at position A, and u_B at position B within a system. If $u_A \neq u_B$, chemical equilibrium has not been established, that is, the overall free energy of the system is not zero. An amount dN of material will therefore pass from A to B such that the change in the free energy at A is $-u_A dN$, while that at B is $+u_B dN$:

$$dG = (-\ u_A dN) + (u_B dN) = (u_B - u_A)\ dN$$

If $u_A > u_B$ (as here), dG must be negative and the transport of mass from A to B must therefore be spontaneous. At equilibrium, no further mass is transported, and so $u_A = u_B$, that is, $dG = 0$, the thermodynamic condition of equilibrium.

114. Broadly speaking, the predictions hold in practice approximately to the same degree to which the ideal gas law is obeyed.
115. M. Knudsen, *Ann. Phys.*, **28,** 75 (1909).
116. It has therefore been argued on occasion that long columns packed with coarse particles can, at least in principle, match or even exceed the chromatographic "efficiency" (that is, minimum band-spreading) exhibited by (even longer capillary) open-tubular columns. In practice, open-tubular columns are overwhelmingly favored generally in gas chromatography because of the limitation of solute thermal lability. However, the argument has merit in liquid chromatography since this limitation no longer applies. It is therefore not inconceivable that LC columns of ca. 0.5 mm i.d. by 10–50 m in length, packed with particles of uniform size lying between 40–100 μ, may well outperform more conventional systems of 2–4 mm i.d. by 5–25 cm in length and containing 3- or 5-μ packings. It would seem, in fact, that the only limitation to the length (hence, efficiency) of the former (coarse-packing) type of column would be the pressure limit of the mobile-phase pump.
117. H. Darcy, *Les Fontaines Publiques de la Ville de Dijon*, Victor Dalmont, Paris, 1856.
118. B. Alder, *J. Chem. Phys.*, **23,** 263 (1955).
119. P.C. Carman, *Flow of Gases Through Porous Media*, Butterworths, London, 1956. There is general agreement on this point irrespective of the type of packing: D.H.M. Bowen, *Flow Through Porous Media*, American Chemical Society, Washington, D.C., 1970.
120. J. Bohemen and J.H. Purnell, in *Gas Chromatography 1958*, D.H. Desty (Ed.), Butterworths, London, 1958, p. 6.
121. J. Kozeny, *Akad. Wiss. Wien, IIA,* **136,** 271 (1927).
122. P.C. Carman, *Trans. Inst. Chem. Eng.* (*London*), **23,** 150 (1937).
123. F.C. Blake, *Trans. Am. Inst. Chem. Eng.*, **14,** 415 (1922).
124. A.J.P. Martin and A.T. James, *Biochem. J.*, **50,** 679 (1952).
125. D.H. Everett, *Trans. Faraday Soc.*, **61,** 1637 (1965).
126. J.C. Giddings, S.L. Seager, L.R. Stucki, and G.H. Stewart, *Anal. Chem.*, **32,** 867 (1960).
127. J.C. Giddings, *Anal. Chem.*, **35,** 353 (1963); **36,** 741 (1964).
128. J.C. Sternberg and R.E. Poulson, *Anal. Chem.*, **36,** 58 (1964).
129. J.C. Sternberg, *Anal. Chem.*, **36,** 921 (1964).
130. Parenthetically, while the distinction is now only rarely drawn, the term "rectification" seems more appropriate than "separation" in the sense that chromatography seeks to "purify" or refine (thence quantitate) mixture components, rather than "disengage" them or cause their dissolution. However, "dynamical" is an appropriate adjective for application of either term to chromatography, since solutes are in fact transported (moved) through such systems by virtue of (flowing) mobile phase.
131. Ref. 47, pp. 1077–1078.
132. D.F. Othmer, *Ind. Eng. Chem.*, **20,** 743 (1928).
133. N.R. Fenske, *Ind. Eng. Chem.*, **24,** 482 (1932).

134. The analogy has been roundly criticized on occasion on the basis that, among other things, there are no distinguishable stages in a chromatographic column [for example, see: E. Glueckauf, *Trans. Faraday Soc.*, **51,** 34 (1955)]. While it may be true that no analogy, by definition, can be an exact representation of the situation it is meant to depict, plate theory does in fact resemble very closely the chromatographic partitioning process if the average behavior of solute molecules is considered, that is, as if the average could be represented as an infinitely thin slice of the solute band (molecules) as it (they) proceed down the column.
135. Eq. 165 is a very commonly misstated relation in many texts purporting to relate multiple extractions with liquid–liquid chromatography. It is derived and presented here in the correct manner, that is, in terms of the amount of material left in the stationary phase after sweeping with mobile phase and not (as is common, for example, in separatory-funnel extractions) the amount extracted out of S into phase M. Moreover, the derivation as given is precisely analogous to that of the Fenske equation (133), the pertinent relation for vapor–liquid distillation, which has relevance to gas–liquid chromatography. Thus, eq. 165, in the form given, offers a description of the chromatographic process that is both applicable to LLC and coherent with that which would be derived for GLC.
136. L.C. Craig, *J. Biol. Chem.,* **155,** 159 (1944); *Anal. Chem.*, **22,** 1346 (1950); L.C. Craig and O. Post, *Anal. Chem.*, **21,** 500 (1949); L.C. Craig and D. Craig, in *Techniques of Organic Chemistry*, Vol. 3, Part 1, A. Weissberger (Ed.), Interscience, New York, 1956, p. 248.
137. In the special case that $x_1^M = \frac{1}{2}$, eq. 170a also applies to a small number of transfers. For example, For $R = 4$ in Fig. 25, N_{max} is correctly predicted to be 2.
138. For an elegant proof of eq. 170c, see: H.D. Young, *Statistical Treatment of Experimental Data*, McGraw-Hill, New York, 1962, Appendix *B*.
139. Eq. 170 each contain x_1^M, and are therefore quite sensitive to values of K_R and β. For example, suppose that a 101-tube apparatus of unit β were to hand, and that a particular solute had a partition coefficient of 100. x_1^M is therefore $1/101 \approx 0.01$, and R_{last}^{max} is 10^4. In contrast, let $K_R = 0.1$. x_1^M is now 1/1.1 and $R_{last}^{max} = 110$ transfers, a very considerable difference indeed.
140. The relations can also be expressed in terms of the solute capacity factor defined in eq. 163 and 164:

$$k' = \frac{1 - x_1^M}{x_1^M} = \frac{K_R}{\beta}$$

whence eq. 178a, for example, becomes:

$$R_{last}^{max} = N_{last}\,(1 + k') = N_{last}\,(1 + K_R/\beta)$$

141. We note in passing that eq. 179c can be rearranged to:

$$K_R = \frac{V_R^o - V_M}{V_S}$$

which provides the means of measuring solute partition coefficients from the datum V_R^o derived purely from experiment, and the system parameters V_M and V_S.

142. This point is often overlooked; for a particularly clear discussion of the matter see D.C. Baird, *Experimentation: An Introduction to Measurement Theory and Experimental Design*, Prentice-Hall, Englewood Cliffs, New Jersey, 1962, pp. 30–38.

143. As a check, we note that 100(2300/25,500) = 9% of the system that is occupied by two standard deviations of the peak when the maximum is in the last tube. If an abscissa of serial tube number instead of $v_M R$ volume units had been employed in graphing the peak elution profile, half of it would have been found to be spread over $0.09 \times 500 = 45$ tubes.

144. This is precisely the situation in zone refining (or for that matter in fractional crystallization), where solute solubility in the (liquid) mobile phase generally is much higher than in the (solid) stationary state. Thus, since K_R is defined as C_1^S/C_1^M, the partition coefficients inherent in these techniques usually are quite small.

145. J.H. Purnell, *J. Chem. Soc.*, 1268 (1960).

146. P. Jones and C.A. Wellington, *J. Chromatogr.*, **213,** 357 (1981).

147. A.J.P. Martin and R.L.M. Synge, *Biochem. J.*, **35,** 1358 (1941).

148. J.N. Wilson, *J. Am. Chem. Soc.*, **62,** 1583 (1940).

149. D. deVault, *J. Am. Chem. Soc.*, **65,** 532 (1943).

150. E. Glueckauf, *Disc. Faraday Soc.*, **7,** 12, 202 (1949); *Analyst*, **77,** 903 (1952); *Trans. Faraday Soc.*, **51,** 1540 (1955).

151. L. Lapidus and N.R. Amundson, *J. Phys. Chem.*, **56,** 984 (1952).

152. A. Klinkenberg and F. Sjenitzer, *Chem. Eng. Sci.*, **5,** 258 (1956).

153. J.J. van Deemter, F.J. Zuiderweg, and A. Klinkenberg, *Chem. Eng. Sci.*, **5,** 271 (1956).

154. M.J.E. Golay, in *Gas Chromatography*, V.J. Coates, H.J. Noebels, and I.S. Fagerson (Eds.), Academic Press, New York, 1958, p. 1; in *Gas Chromatography 1958*, D.H. Desty (Ed.), Butterworths, London, 1958, p. 36.

155. J.I. Coates and E. Glueckauf, *J. Chem. Soc.*, 1308 (1947).

156. E. Glueckauf and J.I. Coates, *J. Chem. Soc.*, 1315 (1947).

157. J.C. Giddings and H. Eyring, *J. Phys. Chem.*, **59,** 416 (1955).

158. J.C. Giddings, *J. Chem. Phys.*, **26,** 169, 1755 (1957); **31,** 1462 (1959); *J. Chem. Educ.*, **35,** 588 (1968); *Dynamics of Chromatography*, Marcel Dekker, New York, 1965.

159. J.H. Beynon, S. Clough, D.A. Crooks, and G.R. Lester, *Trans. Faraday Soc.*, **54,** 705 (1958).

160. **GC**: R.C.M. De Nijs, W.J.M. Houtermans, and M.L. Bal, in *Proceedings of the Fourth International Symposium on Capillary Chromatography*, R.E. Kaiser (Ed.), Huthig Verlag, Heidelberg, 1981, p. 600; **LC**: R.F. Meyer, P.B. Champlin, and R.A. Hartwick, *J. Chromatogr. Sci.*, **21,** 433 (1983).

161. However, see R.H. Perrett and J.H. Purnell, *Anal. Chem.*, **35,** 430 (1963).

162. O.E. Schupp III, *Technique of Organic Chemistry*, E.S. Perry and A. Weissberger (Eds.), Vol. 13, *Gas Chromatography*, Interscience, New York, 1968, pp. 57–62.

163. J. Serpinet, *Anal. Chem.*, **48,** 2264 (1976).

164. W.K. Al-Thamir, J.H. Purnell, and R.J. Laub, *J. Chromatogr.*, **176,** 232 (1979); **188,** 79 (1980).

165. R.J. Laub and J.H. Purnell, *J. High Resolut. Chromatogr. Chromatogr. Commun.*, **3,** 195 (1980).

166. R.J. Laub, J.H. Purnell, and P.S. Williams, *J. Chromatogr.*, **134,** 249 (1977).

167. R.J. Laub and J.H. Purnell, unpublished work.

168. B.J. Lambert, R.J. Laub, W.L. Roberts, and C.A. Smith, in *Ultra-High Resolution Chromatography*, ACS Symposium Series No. 250, S. Ahuja (Ed.), American Chemical Society, Washington, D.C., 1984, p. 49.

169. B.L. Karger, in *Modern Practice of Liquid Chromatography*, J.J. Kirkland (Ed.), Wiley-Interscience, New York, 1971, Ch. 1, Fig. 1.6.

170. R.J. Laub, *Res./Devel.* **25**(7), 24 (1974).

171. R.P.W. Scott and P. Kucera, *J. Chromatogr.*, **169,** 51 (1979).

172. C.E. Reese and R.P.W. Scott, *J. Chromatogr. Sci.*, **18,** 479 (1980); see also: E.D. Katz, K.L. Ogan, and R.P.W. Scott, *J. Chromatogr.*, **270,** 51 (1983); E.D. Katz and R.P.W. Scott, *J. Chromatogr.*, **270,** 29 (1983); J.H. Knox and H.P. Scott, *J. Chromatogr.*, **282,** 297 (1983).

173. For example, see J.H. Knox and J.F. Parcher, *Anal. Chem.*, **41,** 1599 (1969); J.H. Knox, *J. Chromatogr. Sci.*, **15,** 352 (1977); and references therein.

174. R.P.W. Scott and G.S.F. Hazeldean, in *Gas Chromatography 1960,* R.P.W. Scott (Ed.), Butterworths, London, 1960, p. 144.

175. The expression $\sigma^2 = V_M^2 (1 + k')^2/N$ should be used when the solutes of interest are retained.

176. P.J. Flory, R.A. Orwoll, and A. Vrij, *J. Am. Chem. Soc.*, **86,** 3507 (1964).

177. D.E. Martire, in *Gas Chromatography 1966,* A.B. Littlewood (Ed.), Institute of Petroleum, London, 1967, p. 21.

178. D.L. Massart, A. Dijkstra, and L. Kaufman, *The Valuation and Optimization of Laboratory Methods and Analytical Procedures*, Elsevier, Am-

sterdam, 1978; W.E. Biles and J.J. Slain, *Optimization and Industrial Experimentation*, Wiley, New York, 1980; I.E. Frank and B.R. Kowalski, *Anal. Chem.*, **54**, 232R (1982); S.A. Borman, *Anal. Chem.*, **54**, 1379A (1982). Surveys of the "chemometrics" literature are published at irregular intervals by the Chemometrics Society, G. Kateman, Secretary, Laboratory of Analytical Chemistry, University of Nijmegen, Toernooiveld, 6525 ED Nijmegen, The Netherlands.

179. A third nearly correct guess can be derived from the intersection of the first-guess/second-guess path with the X-axis ($Y = 0$), which is called the "method of linear approximation", or the "rule of false position". Alternative techniques include Newton's "method of tangents", which will in fact converge much more rapidly than the linear approximation method, provided that the initial guess is reasonably close: H.G. Bray, Department of Mathematical Sciences, San Diego State University, San Diego, California, informal discussions.

180. R.J. Laub and J.H. Purnell, *J. Chromatogr.*, **112,** 71 (1975).

181. R.J. Laub, *Am. Lab.*, **13** (3), 47 (1981); *Trends Anal. Chem.*, **1,** 74 (1981); in *Physical Methods in Modern Chemical Analysis*, T. Kuwana (Ed.), Academic Press, New York, 1983, Ch. 4. See also ref. 71.

182. R.J. Laub and J.H. Purnell, *J. Chromatogr.*, **161,** 49 (1978).

183. R.J. Maggs and T.E. Young, in *Gas Chromatography 1968,* C.L.A. Harbourn, (Ed.), Institute of Petroleum, London, 1969, p. 217; R.J. Maggs, *J. Chromatogr. Sci.*, **7,** 145 (1969); R.P.W. Scott and J.G. Lawrence, *J. Chromatogr. Sci.*, **7,** 65 (1969).

184. M. McCann, J.H. Purnell, and C.A. Wellington, in *Chromatography, Equilibria, and Kinetics*, Faraday Society Symposium No. 15, D.A. Young (Ed.), The Royal Society of Chemistry, London, 1980, p. 82; S.J. Madden, M. McCann, J.H. Purnell, and C.A. Wellington, paper presented at the 184th National Meeting of the American Chemical Society, Kansas City, Missouri, 1982; S.J. Madden, Ph.D. Thesis, University College of Swansea, Swansea, Wales, 1983; M. McCann, S.J. Madden, J.H. Purnell, and C.A. Wellington, *J. Chromatogr.*, **294,** 349 (1984).

185. R.P.W. Scott and P. Kucera, *Anal. Chem.*, **45,** 749 (1973); *J. Chromatogr. Sci.*, **12,** 473 (1974); **13,** 337 (1975); *J. Chromatogr.*, **122,** 35 (1976); **171,** 37 (1979).

186. A.-J. Hsu, R.J. Laub, and S.J. Madden, *J. Liq. Chromatogr.*, **7,** 599, 615 (1984). *Note added in proof.* Extension of eq. 245 to the form:

$$\frac{1}{t_{R(M)}} = \phi_c \left[\frac{1}{t_{R(C)}} + \sum_{i=1}^{n} \left(\frac{x_i \phi_B}{1 + x_i' \phi_B} \right) \right] + \frac{\phi_B}{t_{R(B)}}$$

has recently been shown by Laub and Madden [*J. Liq. Chromatogr.*, **8,** 155 (1985)] to enable fitting, with $n = 2$, of all known isotherm types heretofore encountered in LC, including those that are fully parabolic as well as those that exhibit inflection points.

187. R.J. Laub, *J. Liq. Chromatogr.,* **7,** 647 (1984).

188. R.J. Laub and J.H. Purnell, *Anal. Chem.,* **48,** 1720 (1976); *J. Chromatogr.,* **161,** 59 (1978).

189. D.M. Summers, R.J. Laub, J.H. Purnell, and P.S. Williams, *J. Chromatogr.,* **155,** 1 (1978); R.J. Laub, *Anal. Chem.,* **52,** 1219 (1980).

CHAPTER

3

INSTRUMENTATION FOR GAS CHROMATOGRAPHY

HAROLD M. McNAIR

Department of Chemistry
Virginia Polytechnic Institute and State University
Blacksburg, Virginia

3.1 INTRODUCTION

Since gas chromatography (GC) is a mature field and has successfully been used for inorganic (as well as organic) analysis for almost 30 years, we cannot hope to discuss all of the instrumentation contributions. We shall discuss a gas chromatographic system in terms of its basic components, state the purpose of each component and describe the most useful designs in current usage. Several textbooks have good discussions on instrumentation (1–3) and one entire text is devoted to instrumentation in gas chromatography (4).

Figure 3.1 shows schematically a gas chromatographic system. The basic components include (1) carrier gas; (2) flow control; (3) sample inlet; (4) column thermostat; (5) column; (6) detector; and (7) recorder or data-handling system.

An inert carrier gas (like nitrogen) flows continuously from a high-pressure cylinder through the injection port, the column and the detector. The flow rate of the carrier gas is carefully controlled to insure reproducible retention times and to minimize detector drift and noise. The sample is injected (usually with a microsyringe) into the heated injection port where it is vaporized and carried onto the column. More recently, and particularly for thermolabile compounds, the sample is placed directly onto the column (''on-column injection'').

There are two main types of columns: packed columns and open tubular or capillary columns. Packed columns are usually $\frac{1}{4}$- or $\frac{1}{8}$-inch outside diameter tubes of stainless steel, nickel or glass, which are tightly packed with small particles (the solid support). A thin film (usually 3–10% by weight of total packing material) of a high boiling liquid (the stationary phase) is coated uniformly over the solid support. Open tubular or capillary columns are long (10–100 meters) of narrow-bore fused silica or

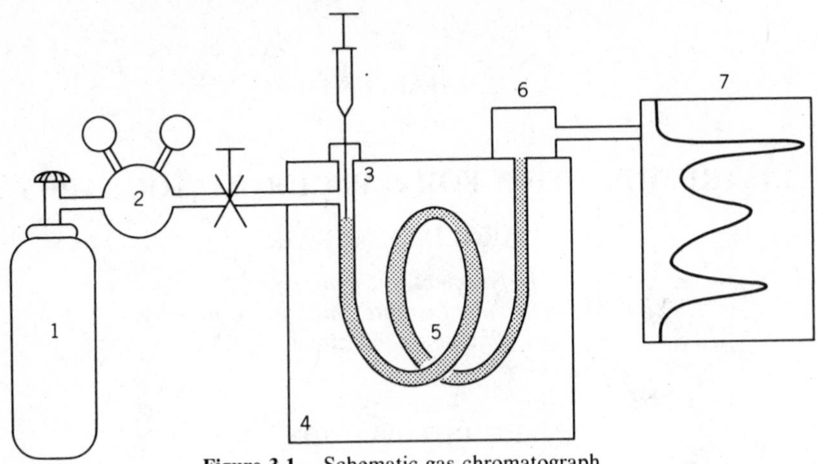

Figure 3.1. Schematic gas chromatograph.

glass with a hole in the middle. Thin films of stationary phase (0.1–1.0 μm) are deposited on the wall. In both columns, the sample partitions between the carrier gas and stationary phase and is separated into individual components.

After the column, the carrier gas and sample pass through a detector. This device generates an electrical signal that passes to a recorder and generates a chromatogram (the written record of the analysis). In many cases, a data-handling device automatically integrates the peak area, measures the retention time, performs calculations and prints out a final report. Each of these items will be discussed in the order presented here.

3.2 CARRIER GAS

3.2.1 Purpose

The purpose of the carrier gas is to carry the sample through the column. The gas should be inert and not react with either the sample or the stationary phase. The choice of carrier gas can affect both the separation and the speed of analyses.

A secondary but still essential purpose of the carrier gas is to provide a suitable matrix for the detector to measure the sample components. In the case of a thermal conductivity detector (TCD), light gases with high thermal conductivities are required to provide a good response (see section 3.2.3 for more details).

3.2.2 Purity

It is important that the carrier gas be of high purity. Impurities can chemically degrade some liquid phases. Polyester, polyglycol and polyamide columns are prone to degradation by oxygen and water. Trace amounts of water can also desorb other contaminants in the column and produce a high detector background or even "ghost peaks." Trace hydrocarbons in the carrier gas cause a high background with the flame ionization detector and decrease the detection limits.

Water and trace hydrocarbons can be easily removed by installing a molecular sieve filter between the gas cylinder and the instrument. These drying tubes are commercially available. The molecular sieve should be regenerated after each gas cylinder by heating to 300°C for 3 hours with a slow flow of nitrogen.

Oxygen is more difficult to remove and requires a special filter, such as BTS catalyst from BASF, Ludwigshaven am Rhein; Oxisorb from Supelco; or Dow Gas Purifier from Applied Science.

3.2.3 Selection Rules For Carrier Gas

1. *Thermal Conductivity Detector* (*TCD*). Use hydrogen or helium except when hydrogen or helium are to be measured; then use argon or nitrogen. Hydrogen should not be used with thermistor beads since it reacts with the rare earth oxides present.
2. *Flame Ionization Detector* (*FID*). Use helium or nitrogen. Nitrogen produces about two times the sensitivity of helium with the F.I.D. Hydrogen is sometimes used as carrier gas with capillary columns to provide very fast analyses.
3. *Electron Capture Detector*. Use very dry nitrogen or argon and 5% methane. As described earlier, special scrubber filters are necessary to remove traces of water and oxygen.

3.3 FLOW CONTROL

3.3.1 Purpose

Close control of carrier gas flow is essential both for high column efficiency and for qualitative and quantitative analysis.

For qualitative analysis it is essential to have a constant and reproducible flow rate so that retention times can be reproduced. Comparison of retention times of unknowns and standards is the quickest and easiest method for compound identification. It should be noted, however, that two or more compounds may have the same retention time. Confirmation of peak identity is not possible by GC methods; it requires the use of an auxiliary instrument such as MS, NMR, or IR.

3.3.2 Control

The first control in the flow system is a two-stage regulator connected to the carrier gas cylinder. The first stage indicates the pressure in the gas cylinder. The regulator is used to control the pressure (registered on the second stage) delivered to the gas chromatograph (usually 40 to 60 psi). This regulator does not work well at low pressures, and it is recommended that a minimum of 20 psi be used on the second stage.

In temperature programing, even when the inlet pressure is constant the flow rate will decrease as the column temperature increases. This is due to the increased viscosity of the carrier gas. An an example, with a constant inlet pressure of 24 psi and a flow rate of 22 ml/min (helium) at 50°C, the flow rate through a packed column decreases to 10 ml/min at 200°C. In all temperature-programed units and in the better isothermal units, a differential flow controller is used to assure a constant mass flow rate.

3.4 SAMPLE INLETS

3.4.1 Purpose

The separation begins with the introduction of gaseous or vaporized liquid or solid sample into the carrier gas stream. Because of the great variety of chromatographic conditions, such as column length, column diameter, weight % of stationary liquid, gas velocity, and the wide range of component concentrations, a variety of sampling techniques have to be used.

There is no optimum sample size, but some general guidelines are available. For the best peak shape and maximum resolution, the smallest possible sample size still enabling adequate detection should be used. A variety of sample sizes must be accommodated, from nanogram quantities for open tubular columns up to gram quantities for preparative columns (Table 3.1).

3.4.2 Gas Sample Valves

An alternative method for introducing gas samples is a sample valve (Fig. 3.2). In position a, the sample gas is forced continuously through the

Table 3.1 Sample Volumes for Various Column Types

Column Type	Sample Volume	
	Gas	Liquid
Preparative: 20 mm i.d., 20% liquid	0.5–1.0 L	0.02–1.0 ml
Regular Analytical: 4.0 mm i.d., 10% liquid	0.5–10 ml	0.2–10 μl
High Efficiency: 2 mm i.d., 3% liquid	0.1–1.0 ml	0.01–1.0 μl
Capillary: 0.25–0.50 mm i.d., 0.1–1.0-μm film	10–100 μl	0.001–1.0 μl

sample loop until the loop contains only the sample. The volume of the sample loop is controlled by the length and diameter of the tubing. Sample loops from 1 μl up to 100-ml capacity are available. In position b, the sample valve is rotated and the carrier gas pushes the sample through the sample loop and into the column. A sampling valve gives better reproducibility, requires less skill and can be more easily automated.

3.4.3 Liquid Sampling

Most of the samples in gas chromatography are liquids which are vaporized in an inlet system before entering the column. Syringes are used almost universally for liquid sampling, the most commonly used sizes are 1 and 10 μl. Liquid sample valves are available and are used routinely in process gas chromatographs and high pressure liquid chromatographs.

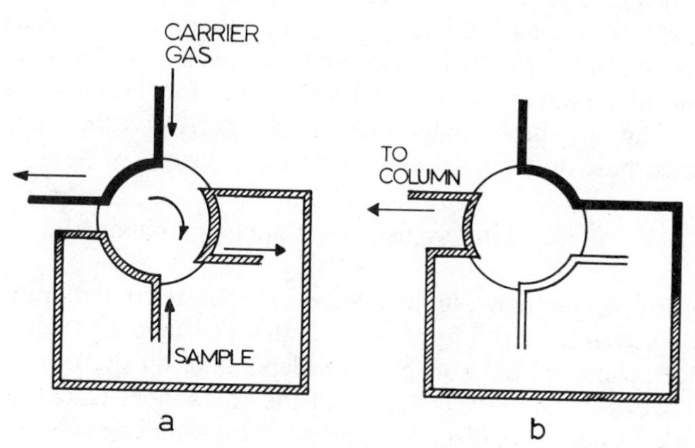

Figure 3.2. Gas sample valve.

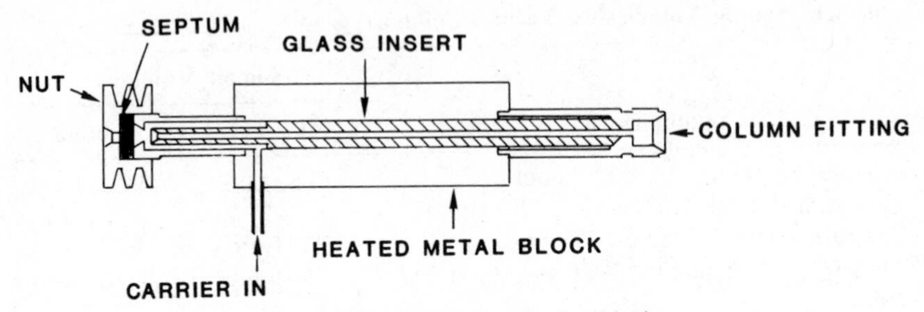

Figure 3.3. Schematic injection block.

The major limitation is that temperatures above 150°C shorten the life and reproducibility of the sample valve and column efficiencies are usually lower with sample valve injection. Various automatic sampling systems are available, some using syringes, most using sealed ampoules or capsules.

3.4.4 Solid Sampling

Solids are best handled by dissolving them in an appropriate solvent and using syringes to inject the solution. Special syringes for solid samples are available, but they have not proven convenient to use.

3.4.5 Inlet Systems for Packed Columns

A schematic of an injection block for packed columns is shown in Fig. 3.3. The heated inlet block insures a constant temperature in the vaporization tube. The dead volume is kept to a minimum to insure rapid introduction of the vaporized sample into the column. The use of glass inserts or on-column injection is highly desirable for certain unstable sample types. Among these compounds are the derivatives of amino acids, steroids, carbohydrates, pesticides, and many drugs.

3.4.6 Inlet Systems for Capillary Columns

Small-bore open tubular columns, however, operated under optimum conditions, are characterized by low flow rates (~1 ml/min). Here different sampling techniques have to be used depending on the type and characteristics of the capillary column and the component concentrations in the sample. A complete discussion on different sampling systems for capillary chromatography has been presented by Schomburg (5). Here we will limit ourselves to the characteristics of the most commonly used injection techniques.

1. *Split Injection*. As can be seen from Table 3.1, the sample capacity of capillary columns is one or two orders of magnitude less than that of conventional packed columns. An indirect sampling procedure is generally utilized. A relatively large (0.1–1 μl) liquid sample is injected into the sample inlet. The vaporized sample mixed with the carrier gas is split unequally; the smaller flow goes onto the column, while the larger flow is discarded. The ratio of these two flows is called the split ratio. As given by the split ratio, only a small fraction of the originally injected volume enters the column. In order to be effective, the stream splitter must be nondiscriminatory, that is, all sample components independent of their molecular weight and concentrations must be divided in the same ratio.

Typically, split ratios used for wall-coated open-tubular (WCOT) columns are 1:50 down to 1:1000. Sample splitting is not compatible with trace analysis because such a small amount of sample is introduced into the column.

2. *Splitless Injection*. In trace analysis it is oftentimes necessary to introduce the entire sample into the column. One way to accomplish this is the splitless sampling technique, originally described by Grob (6,7). In the splitless mode, sample is first diluted 1:10 with a volatile solvent such as hexane. A relatively large amount of dilute sample (1–5 μl) is introduced, vaporized, and carried onto the column. A regular split injector is used, but the split flow path is closed off. The amount of sample should be less than 50 ng to prevent column overloading. The large amount of volatile solvent will produce a long solvent tail, obscuring fast-eluting peaks. To minimize this solvent tail, the injection port is back flushed (the split valve open) 30–60 seconds after injection. In this way, the solvent tail is limited without loss of the components of interest. Splitless injection works only if the sample components that are broadened in the injection part are reconcentrated into narrow bands at the head of the cool column. After injection the column is temperature programmed.

The first method proposed by Grob utilizes a "solvent effect" for reconcentrating. The most widely used solvents include methylene chloride and hexane. The initial column temperature should be 10–30°C below the boiling point of the solvent. Another method of reconcentrating the components in the front of the column is to have the initial column temperature low enough to condense these solutes ("cold trapping"). A general guideline is that the initial column temperature must be 150°C lower than the boiling points of the components. The splitless technique is well-suited for trace analysis and compatible with WCOT columns.

3. *On-Column Injection*. The introduction of a liquid sample directly into the column was first described by Schomburg (5). He described two versions (macro and micro) of this technique, when the sample never encounters temperatures higher than the column temperature.

This method proved to be superior to all other sampling techniques for both the separation of compounds of low volatility and for samples of a wide boiling range. Not only is the precision of the analyses better, but the discrimination of compounds is less as well.

Grob and Grob (6,7) and Galli, Trestianu, and Grob (8,9) described on-column systems where the sample is introduced by means of a special syringe into even narrow-bore capillary columns. Galli and Trestianu include in their sampling system an additional cooling of the column inlet to overcome problems caused by the sudden vaporization of volatiles in the syringe needle.

3.5 COLUMN TEMPERATURE

3.5.1 Purpose

The column is thermostated so that the separation will occur at a reproducible temperature. In addition, it is necessary to maintain the column at a wide variety of temperatures, from −180 up to 400°C. The control of column temperature is one of the easiest and most effective ways to influence the separation, since partition coefficients are very temperature-dependant.

Since the column is fixed between the heated injection port and a heated detector, it seems appropriate to discuss the reasons why these coponents are also heated.

1. *Injection-Port Temperature*. The injection port should be hot enough to vaporize the sample so rapidly that no loss in efficiency results from the injection technique. On the other hand, the injection port temperature must be low enough so that thermal decomposition or rearrangement of the sample is avoided.

A general rule is to have the injection temperature about 50°C hotter than the boiling point of the sample. On-column injection systems use temperatures only slightly warmer than the column temperature and are recommended for thermolabile compounds.

2. *Detector Temperature*. The influence of temperature on the detector depends upon the type of detector employed. As a general rule, it can be said that the detector and the connections from the column to

detector must be hot enough so that the condensation of the sample and/or "column bleed" does not occur. Peak broadening and loss of component peaks are characteristic of condensation in these connections. The stability and resultant usable sensitivity of a thermal conductivity detector depends upon the stability of the detector temperature control. It should be ± 0.1°C or better. For flame ionization detectors, temperature must be maintained high enough to avoid not only condensation of the samples, but also of the water or by-products formed in the combustion process. A reasonable minimum temperature for a flame ionization detector is 125°C.

3. *Column Temperature*. The column temperature controls the sample distribution between the carrier gas and the stationary phase. The column temperature should be high enough that the analysis is accomplished in a reasonable length of time, and low enough that the desired separation is obtained by the selective solubility in the stationary phase. According to a simple approximation made by Giddings (10) the retention time doubles for every 30°C decrease in column temperature. For most samples, the lower the temperature the better the separation.

3.5.2 Isothermal vs. Programmed Temperature

Isothermal means a chromatographic analysis at constant column temperature. Programmed temperature (PTGC) means a linear increase of column temperature with time. Temperature programming is very useful for the analysis of wide boiling sample mixtures.

As shown in Fig. 3.4, isothermal operation limits gas chromatographic analysis to a narrow boiling sample. At constant temperature the early peaks, representing low boiling components, emerge so rapidly that sharp overlapping peaks result while higher boiling materials emerge as flat, immeasurable peaks. In some cases, high boiling components are not eluted, and may appear in a later analysis as baseline noise or "ghost" peaks that cannot be explained.

With temperature programming, a lower initial temperature is used and the early peaks are well resolved. As the temperature increases, each higher boiling component is "pushed" out by the rising temperature. High boiling compounds emerge as sharp peaks, similar in shape to the early peaks. Trace components emerge as sharp peaks that can be more easily distinguished from the baseline. Total analysis time is shorter for complex samples, although one must cool the column after temperature programming before another run can be initiated.

The decision to use PTGC is based on consideration of the boiling points of the sample components. Generally, if the range of boiling points is 100°C or more, temperature programming is advisable.

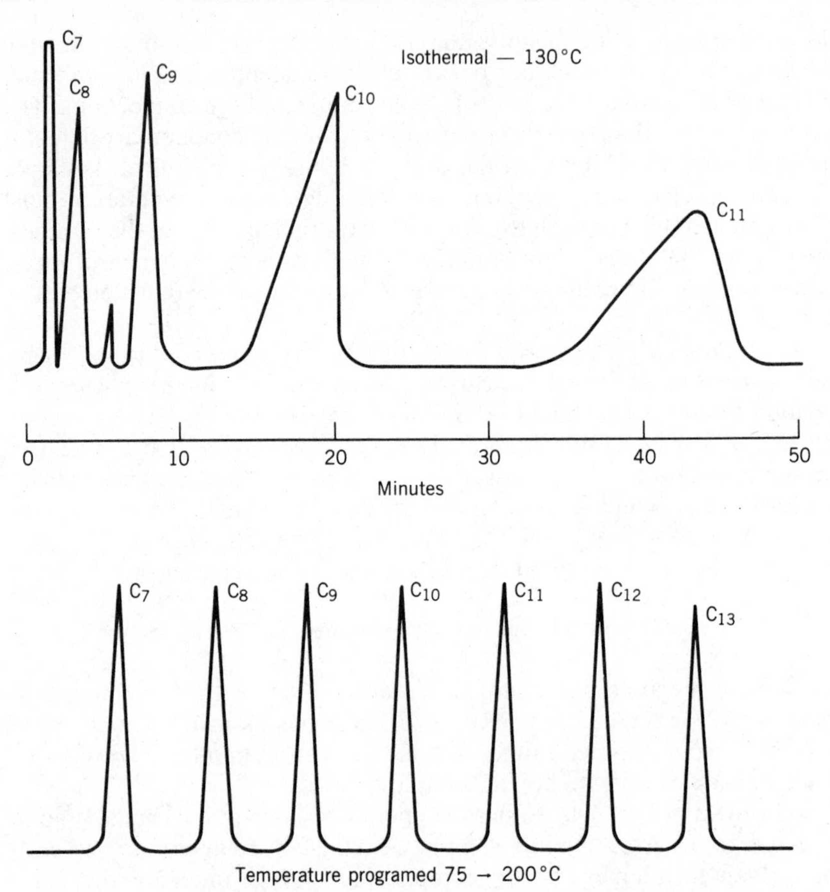

Figure 3.4. Comparison of isothermal and temperature programmed chromatograms.

3.5.3 Temperature Control

Several types of temperature controllers are used with gas chromatographic ovens. Principally, these differ in terms of cost, accuracy, ad flexibility in programming. Column oven temperature controllers may be functionally divided into the classes below:

1. *Isothermal Controller.* Electronic feedback-type device for precision isothermal control; usually with nonlinear programming capability.

2. *Linear Temperature Programmer* (LTP). Electronic feedback-type device for precision isothermal control and selectable rates of linear temperature programming.
3. *Multilinear Programmer*. Sophisticated electronic system for precision temperature control, providing a multitude of temperature programming profiles. This version frequently includes automatic cooling and program recycle to allow more reproducible chromatograms.

3.6 COLUMNS

3.6.1 Purpose

It is the column which performs the separation, and separation is the primary objective of GC. Choosing the proper columns remains an important decision in all gas chromatographic techniques. We will discuss in order liquid phases, packed columns, solid supports, and capillary columns.

3.6.2 Liquid Phase

It is the liquid phase which must exhibit the differential solubility to effect a separation between components. This selectivity is a thermodynamic property and could be calculated from partition coefficients. Unfortunately, only limited thermodynamic data is available, so it is easier to experimentally determine liquid phase selectivity by chromatographic techniques.

1. *Relative Retention*. Choose the sample pair to be separated; chromatograph each as a standard, measure the adjusted retention times (from t_o) and calculate the ratio of adjusted retention times. This ratio is called "relative retention" or alpha (α). It measures the differential solubility in that liquid phase at that column temperature. Table 3.2 provides a valuable insight into the effect of α on the plates required to resolve two peaks 98% (Resolution, R = 1.5). Obviously, an α of about 1.1 is required to separate peaks by most packed column techniques, since plate counts higher than 6,000 are not easily obtained.

A good reference for helping to choose a liquid phase is an article by Supina and Rose (11) that tabulates the Rohrschneider Constants for 80 common liquid phases, enabling a decision to be made, almost by inspection alone, as to whether a particular liquid phase is worth trying.

Table 3.2 Solvent Efficiency and Plates Required

α	n[a]
1.015	165,000
1.075	7,400
1.157	2,000
1.245	930
1.375	484

[a] Assumes retention time is three times t_0 (i.e., $k' = 2$).

Equally important, the article identifies very similar liquid phases. For example, in terms of selectivity the following liquid phases have identical chromatographic properties: SE–30, OV–1 and OV–101; Celanese Ester No. 9 and diisodecyl phthalate; QF–1 and OV–210; OS–124 and OV–25; LAC–4–R–886 and ethylene glycol succinate; and LAC–2–R–466 and diethylene glycol succinate.

Solvent efficiency is not the sole criterion, however, for choosing a liquid phase. Also important are temperature limitations (both maximum and minimum), absolute solubility, cost, and availability.

2. *Minimum Temperature*. The minimum temperature limit is determined by the viscosity of the liquid phase. As the liquid phase viscosity increases, the mass transfer between the gas and liquid phases becomes so slow that excessive peak broadening occurs.

3.6.3 Packed Columns

The column tubing can be stainless steel, nickel, or glass in a straight, bent, or coiled form. Glass is recommended for amino acids, steroids, drugs of abuse, carbohydrates, and pesticides. The best solution is a large column oven which allows the use of U-shaped glass columns. These glass columns are easily packed, provide high efficiencies and are convenient to handle. Unfortunately, though, glass does break.

For this reason, stainless steel columns are commonly used, packed while straight to obtain a uniform packing density, and coiled to facilitate long lengths. Straight columns are more efficient, but can be cumbersome, particularly for work at high temperatures. If coiled, the spiral diameter should be at least ten times the column diameter to avoid diffusion and "racetrack effects."

Packed columns vary in length from a few inches to more than 20 feet. Common analytical columns are 3–10 feet in length. Longer lengths give more theoretical plates and better resolution. Long packed columns, however, require high inlet pressures. High pressures present problems in injection technique and in preventing gas leaks. An advantage of long columns, however, is that sample capacity is proportional to the amount of liquid phase present. This means that larger sample sizes may be injected onto longer columns.

Typical packed column diameters vary from 1.0–4.6 mm i.d. The smaller the column diameter, the higher the column efficiency. Standard analytical columns are $\frac{1}{8}$- and $\frac{1}{4}$-inch outside diameter. An obvious way to increase the column capacity is to increase the column diameter. Preparative scale separations are run on $\frac{3}{8}$-inch, $\frac{1}{2}$-inch o.d., and larger-diameter columns. Unfortunately, column efficiency decreases with increased column diameter, and less resolution is obtained.

3.6.4 Solid Support

Proper selection of the solid support is no longer a problem, thanks primarily to the efforts of the Johns-Manville Corporation, which has made available a variety of supports (both diatomaceous earth and porous polymer types) of different mesh sizes and different grades of activity.

The purpose of the solid support is to expose the sample to a thin, uniform film of liquid phase. An optimum support should have certain characteristics:

1. A large specific surface area of 1–20 m^2/g,
2. A uniform pore diameter in the range of 10 μm or less,
3. Inertness—a minimum of chemical and adsorptive interaction with the sample; this frequently requires chemical deactivation,
4. Regularly shaped particles, uniform in size for efficient packing,
5. Mechanical strength—should not crush on handling.

No material has yet been described that fills all these requirements; however, several suitable supports are commercially available. Usually, one must choose between inertness or efficiency (high surface area).

The raw material for most gas chromatographic supports is diatomite; also known as diatomaceous earth, or "Kieselguhr." Diatomite is composed of the skeletons of diatoms (microscopic unicellular algae), which are primarily microamorphous hydrous silica.

There are five forms or types of Chromosorb* in common use: A, G, P, W, and 750. Each is available either untreated or treated and in a variety of mesh ranges.

* Chromosorb is the registered trademark of the Johns-Manville Corporation.

Chromosorb A is for use in preparative-scale gas chromatography. It has a high capacity for liquid phase (25% maximum), a structure that does not readily break with handling, and a surface that is not highly adsorptive. It is available in mesh ranges of 10/20, 20/30, and 30/40. These allow the use of long preparative columns with low-pressure drop.

Chromosorb G is for the separation of polar compounds. Its low surface area, hardness, and good handling characteristics make it a possible replacement for Chromosorb W. Because of its lower surface area and higher density. Chromosorb G is employed with a lower liquid phase coating. A 5% liquid phase loading on Chromosorb G corresponds to 12% liquid phase on Chromosorb W.

Chromosorb P is prepared from Johns-Manville's Sil-O-Cel C-22 Firebrick. It is a calcined diatomite, pink in color and relatively hard. Its surface is more adsorptive than the other Chromosorb grades and is used primarily for hydrocarbon work. It has the best column efficiencies but cannot be used with polar samples.

Chromosorb W is a flux-calcined diatomite support prepared from the production of Johns-Manville Celite. Chromosorb W is white in color and friable. Its surface is relatively nonadsorptive and is used for the separation of polar compounds. When acid-washed and silane treated, high-performance Chromosorb W is the most inert solid support available.

3.6.5 Capillary Columns

The original capillary columns (wall-coated) had severe limitations. The liquid film adhering to a smooth metal or glass surface was so thin that only a small amount of liquid phase was present. To provide a reasonable β value (volume of gas/volume of liquid) meant restricting the column-inside diameter to a small dimension, commonly 0.01 inch. This small diameter in turn meant a slow flow (about 1 ml/min) and the final result was strict instrument requirements for sample splitting, slow flow rate, low dead volumes, limited column capacity, and the inability to do trace analyses. Still, the results obtained in resolution or speed of analysis were better than with packed columns.

In 1964, the second-generation capillary column appeared: support coated open tubular (SCOT), in which a layer of celite was adsorbed onto the tubing wall and liquid phase was adsorbed onto the celite. The prime advantage of an open tube (its low pressure drop makes long lengths feasible) is preserved and we now have a β approaching packed columns. Column capacity is greatly increased. The tubing is wider (0.02 inch), the flow rates faster (4–10 ml/min), and dead volume connections are less critical. Sample splitting is useful, but not required (use 0.5 μl or less), and useful trace analysis is possible.

Capillary columns are expensive. They require good technique and good instrumentation, but they provide separations and speeds unattainable by packed columns. If a separation demands more than 6,000 theoretical plates, a capillary column is probably the best solution. Capillary columns are not well suited for the separation of light, fixed gases due to the small amount of liquid phase present; however, some capillary columns containing Al_2O_3 do provide interesting separations of light gases.

3.7 DETECTORS

3.7.1 Detector Characteristics

The chromatographic detector measures the concentration of the sample component and generates an electrical signal proportional to the sample concentration. There are many different detectors in use, however space allows us to discuss only the major types used in inorganic analyses. The primary detector characteristics are as follows:

1. *Sensitivity* means the amount of signal generated for a given sample concentration or mass flow. A sensitive detector will generate a large electrical signal for a given sample size. Sensitivity can also be measured as the slope of the plot of detector response vs. sample concentration or mass flow.

2. *Noise* refers to random, short-term detector response determined by electrical properties, temperature, or flow sensitivity when no sample is present. Long-term noise (minutes to hours) is commonly called "drift." The level of noise determines how small a sample can be detected. That sample size which generates a signal two times the noise level is defined as the *minimum detectable quantity* (m.d.q.). Note that sensitivity, peak width, and noise determine the m.d.q.

3. *Universal response* means that the detector generates a response for all sample components. This is a desirable characteristic; it is very useful to have a sensitive detector with universal response. Only the thermal conductivity detector shows universal response.

4. *Selective response* means that the detector sees only certain types of compounds; for example, the flame photometric detector can only see compounds containing S or P atoms. This is also useful in certain restricted applications, that is, sulfur-containing pollutants in air.

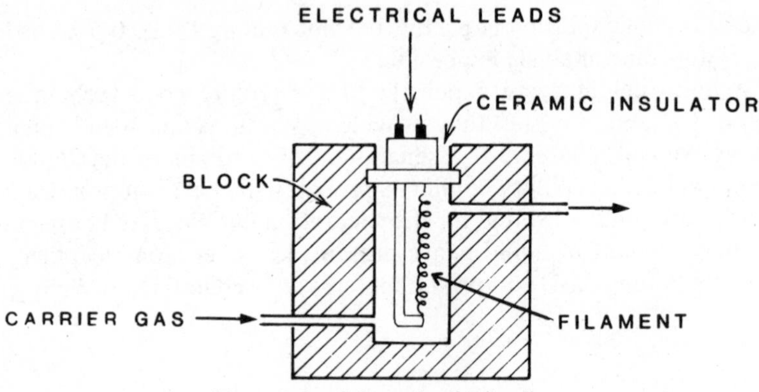

Figure 3.5. Schematic of TC cell.

5. *Linear range* means the region over which the detector signal is directly proportional to sample concentration. Stated another way, on a log–log plot of detector response and concentration, it means the range over which the curve is linear with a slope of 1.0 ± 5%. A wide linear range is very useful for quantitative analysis of both major and trace components in the same sample.

3.7.2 Thermal Conductivity

Theory of Operation

The thermal conductivity detector works on the principle that a hot body will lose heat at a rate that depends on the composition of the surrounding gas. Thus, the rate of heat loss can be used as a measure of the gas composition.

Figure 3.5 shows a typical TC cell consisting of a spiral tungsten filament supported inside a cavity. The cavity is inside a large stainless steel block to provide a constant reference temperature.

The heated filament can lose heat to the cooler block by the following processes:

1. Thermal conduction to the gas stream.
2. Convection (free and forced).
3. Radiation.
4. Conduction through the metal contacts.

The major heat-loss processes, however, are gaseous thermal conduction and forced convection. These two processes account for 75% or more

Table 3.3 Thermal Conductivities of Selected Compounds

Compound	Thermal Conductivity	Molecular Weight
Hydrogen	41.6	2
Helium	34.8	4
Methane	7.2	16
Nitrogen	5.8	28
Pentane	3.1	72
Hexane	3.0	86

of the total filament heat loss. Use of a light carrier gas, such as helium or hydrogen, will cause heat loss by thermal conductivity to predominate. It is assumed in the following discussion that thermal conduction by the carrier gas is the only mode of heat transfer.

Heat is transferred instantaneously by conduction when gas molecules strike the heated filament. Differences in thermal conductivity of gases are based on the mobility or speed at which the gas molecules diffuse; the smaller the molecule, the higher its mobility and the higher its thermal conductivity. Thus hydrogen and helium, which are the smallest molecules, have the highest thermal conductivity. Table 3.3 gives the thermal conductivities (in c.g.s. units and at 0°C) of several compounds, and shows how thermal conductivity decreases with increasing molecular weight.

Detector Cell

A TC cell contains a tungsten filament (see Fig. 3.5) whose electrical resistance varies greatly with temperature, that is, it has a high temperature coefficient of resistance. A constant current is passed through the filament causing its temperature to rise. With pure carrier gas flowing, the heat loss is constant and the filament temperature is constant.

When sample elutes from the columns, the sample molecules are larger than the carrier gas; they move more slowly and conduct less heat. The filament temperature increases, causing a corresponding increase in electrical resistance. It is this filament resistance change which is measured by a Wheatstone bridge circuit. Filaments are chosen on the basis of high temperature coefficient of resistance, and resistance to chemical corrosion. Common filament metals are platinum, tungsten, and tungsten alloys. The popular WX filaments from GOW-MAC are tungsten with 4% rhenium.

Table 3.4 Summary of Thermal Conductivity Detector[a]

M.D.Q.	10^{-8} g; about 50 ppm
Response	Universal, all components except carrier gas
Linearity	10^4
Stability	Good
Carrier gas	Helium or hydrogen
Temperature limit	400°C

[a] Nondestructive, moderate stability, moderate sensitivity, simple to operate. Requires good temperature, flow control. Used frequently for fixed gas analyses and preparative scale work.

Most detector blocks contain a pair of matched filaments in the sample flow channel, and a similar pair of matched filaments in the reference flow channel. The reference and sample flow channels are drilled into a single metal block having a high heat capacity to provide temperature stability. A 30-volt power supply is used to heat the filaments.

To increase the sensitivity of a TC detector, one should increase the filament current, decrease the block temperature, and choose a carrier gas having high thermal conductivity. In some cases, special filaments of higher resistance or specially designed cell blocks may increase sensitivity. The characteristics of the TC detector are in Table 3.4.

3.7.3 Flame Ionization Detector

Theory of Operation

Ionization detectors operate on the principle that the electrical conductivity of a gas is directly proportional to the concentration of charged particles within the gas. For the FID, a hydrogen flame is the ionizing source.

Carrier gas from the column flows into the flame, which ionizes some of the organic molecules in the gas stream. The presence of charged particles (positive ions, negative ions, and electrons) causes a current to flow across the gap and through a measuring resistor. The resulting voltage drop is amplified by an electrometer and fed to a recorder.

It is helpful to think of the electrode gap as a variable resistor whose resistance value is determined by the number of charged particles within the gap. With pure carrier gas flowing, a constant, very low concentration of charged particles will be present in the gap. When an organic component passes into the flame, it is combusted and charged particles are formed. This increases the number of charged particles, which in turn

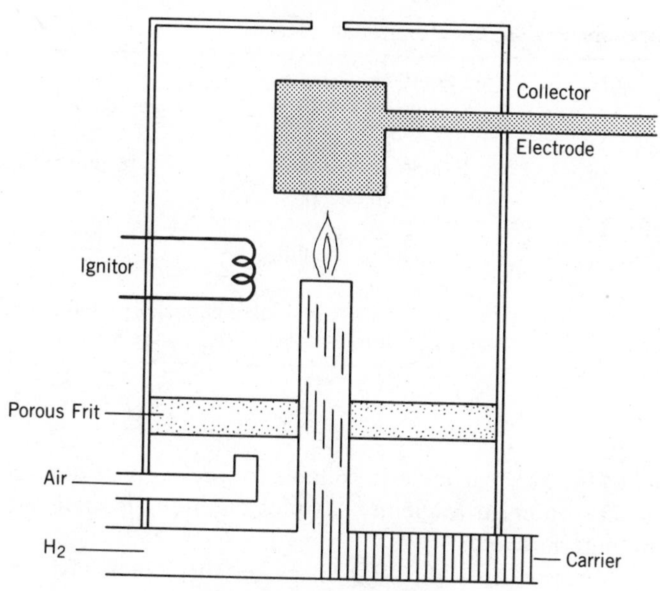

Figure 3.6. Schematic FID.

causes current to flow, producing a signal that is amplified and registered as a peak on the recorder.

Detector Cell

Figure 3.6 shows a schematic FID. Hydrogen and carrier gas are mixed together and pass up into the flame. Air passes through the detector base to support the flame. Oxygen may also be used. An ignitor coil is placed close to the flame jet to allow easy external ignition. In one common arrangement, the flame jet is grounded and serves as one electrode. The collection electrode is at 300 volts positive. Charged particles generated in the flame migrate to the polarized electrodes and cause a current to flow. This current is very small in a FID (10^{-8} to 10^{-12} amps), and it requires amplification by an electrometer before it goes to the recorder.

FID Response

The FID responds only to organic compounds, and is useful in inorganic analysis only when the atoms of interest are combined with an organic ligand. Some compounds which give no response are air, water, inert gases, CO, CO_2, CS_2, NO, SO_2, and H_2S. The lack of response to air and

Table 3.5 Summary of Flame Ionization Detector

M.D.Q.	5×10^{-12} g/sec
Response	Selective, sensitive only to organic compounds
Linearity	10^6
Stability	Excellent (relatively insensitive to temperature and flow changes)
Temperature Limit	400°C
Carrier Gas	Nitrogen or helium

Rugged; nonresponsive to water and air, making it especially useful for analyzing dilute aqueous solutions and air samples; destructive; widely used.

water makes the FID particularly suitable for the analysis of trace organic matter in air, water, or aqueous samples such as alcoholic beverages, biological materials, etc.

Table 3.5 summarizes the FID characteristics. It is a very sensitive detector with the widest linear range of any detector in common use. The combination of high sensitivity and wide linear range makes the FID an excellent detector for quantitative trace analysis.

3.7.4 Electron Capture Detector

Operating principles

The electron capture detector measures the loss of signal rather than an increase in electrical current. As the nitrogen carrier gas flows through the detector, a radioactive Ni^{63} foil ionizes the gas, and slow electrons are formed. These electrons migrate to the anode, which has a potential of 90 volts positive. When collected, these slow electrons produce a steady current of about 10^{-8} amps, which is amplified by an electrometer. If a sample containing electron-capturing molecules is then introduced to the detector, this current will be reduced.

Recent EC detectors have pulsed-voltage power supplies that maintain a constant current. With no sample, the pulse frequency is very low; as the sample enters the detector, the frequency increases to offset the current loss due to the electron-capturing species. The pulse frequency is proportional to the sample concentration and can be used for quantitative analysis.

Detector Selectivity and Sensitivity

The electron capture detector is extremely sensitive to certain molecules, such as alkyl halides, conjugated carbonyls, nitriles, nitrates, and organometals. It is virtually insensitive to hydrocarbons, alcohols, and ketones. Stated another way, the detector selectively responds to molecules containing electronegative atoms, since these atoms easily attach or attract an electron and thus produce an electrical signal. Selective sensitivity to halides makes this detector especially valuable for the analysis of pesticides; certain pesticides can be detected at picogram levels.

Radioactive Source

A radioactive source is an essential part of the EC detector. It supplies the primary ionization of the carrier gas. Both H^3 (tritium) and Ni^{63} have been employed. The tritium design was used first; however, the tritium foil is limited to 220°C and thus becomes easily contaminated with high boiling samples or with column bleed. When the radioactive surface becomes coated, the primary emission decreases and the detector loses sensitivity. It is then necessary to remove the radioactive foil and clean it in an ultrasonic bath. If facilities are not available for handling radioactivity, it is recommended that the detector be returned to the manufacturer for cleaning.

The Ni^{63} source is more expensive, but it can be heated to 350°C. It can be maintained at a higher temperature, and therefore can operate for months without the foil needing to be cleaned, making it the popular design today.

Linearity

All electron capture detectors suffer from a limited linear range. The detectors are easily saturated, and very dilute samples should be injected. Samples must be dry, since traces of water destroy the normal detector response. The characteristics of the detector are in listed Table 3.6.

3.7.5 Gas Density Balance

The gas density balance (G.D.B.) invented by A. J. P. Martin (12) was one of the first chromatographic detectors. The difficulties encountered in construction of this complicated detector prevented its widespread use. In 1960, Nerhein (13) described a gas density balance that differed little

Table 3.6 Summary of Electron Capture Detector

M.D.Q.	Nano- and picograms
Response	Very selective; sees only electronegative atoms
Linearity	500–10,000
Stability	Fair
Temperature limit	220°C (H^3), 350°C (Ni^{63})
Carrier gas	Nitrogen or argon + 10% CH_4

Detector is easily contaminated; carrier gas must be dry; is very selective; can be extremely sensitive; nondestructive; requires DOE License for radioactive source.

in principle but was of greatly simplified construction. This model is now commercially available from Gow Mac.

Gas density detectors have several advantages:

1. Calibration is not required for quantitative analysis.
2. Molecular weight of components can be determined by using two different carrier gases.
3. They can be used with corrosive gases, since the sample does not pass over the filaments.

Referring to Fig. 3.7 reference gas enters at *A* while gas from the column enters at *C*; both gas streams exit at *D*. Measuring elements are mounted in the reference stream at R_1 and R_2 and connected in opposite arms of a Wheatstone bridge. The sample never contacts the measuring elements, making the G.D.B. useful for the analysis of corrosive materials. If the gas eluted at *C* has the same density as the reference gas, the gas flows are at equilibrium and no unbalance is detected by the bridge.

If the gas entering at *C* carries a sample component of higher density, gravity pulls more of the gas stream down, and this increased flow from *C* retards the lower gas flow AR_2 while the upper flow increases. This flow imbalance causes a variation in resistance of the measuring elements R_1 and R_2, causing an umbalance in the bridge. Solutes of lower density than the carrier gas tend to rise, producing the opposite effect on the flow paths.

Sensitivity depends, in part, on the difference in density between the carrier gas and the sample component. Nitrogen is the preferred carrier gas except when CO, C_2H_4, and C_2H_2 are being determined; then, carbon

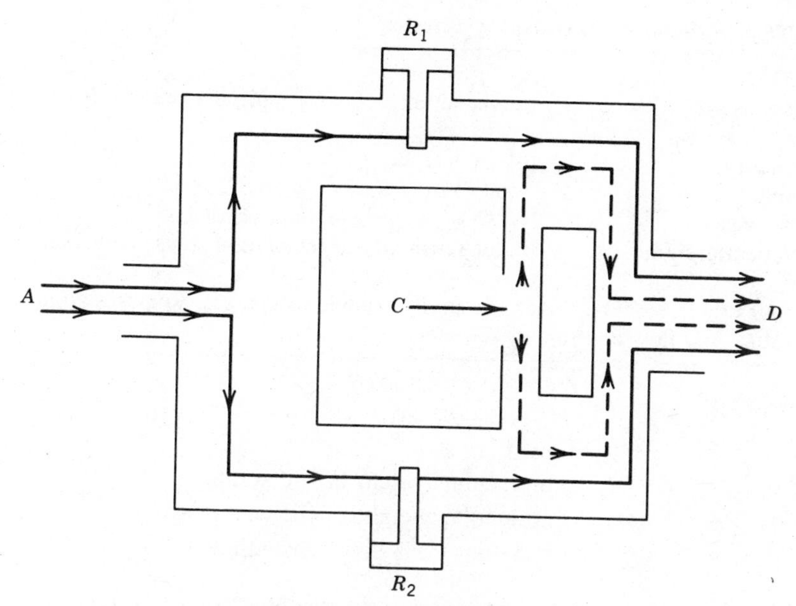

Figure 3.7. Schematic of GDB.

dioxide or argon should be used as carrier. Sulfur hexafluoride has been preferred by Guillemin and Auricourt as carrier gas (14). Hydrogen and helium should not be used with a gas density balance, since diffusion can occur, allowing the sample component to enter the passages containing the measuring elements.

A gas density detector can be used over a wide range of flow rates. However, it is important that the reference flow be 15–20 ml/min faster than the column flow. This prevents sample components from contacting the detector elements. There is an optimum reference flow rate for maximum response. Column flow rate should be selected for the sample and the column in use.

The outstanding feature of a gas density detector is that quantitative analyses may be made without calibration. The gas density balance gives responses directly in weight percent if peak areas are multiplied by a factor derived from the molecular weights of the solute and the carrier gas. Weight of a component = $X \cdot A \cdot K$ where:

$$K = M_s/(M_s - M_{cg})$$

where:

Table 3.7 Summary Gas Density Detector

M.D.Q.	10^{-6} g
Response	Universal, except same molecular weight as carrier gas
Linearity	10^3
Stability	Good
Carrier gas	N_2, CO_2, Ar, SF_6 and not H_e or H_2
Temperature limit	Better sensitivity at temperatures less than 150°C

Nondestructive; stable; poor sensitivity; simple to operate. Requires good temperature and flow control.

$$A = \text{peak area}$$
$$M_s = \text{molecular weight of the solute}$$
$$M_{cg} = \text{molecular weight of the carrier gas}$$
$$x = \text{constant for a given instrument}$$

However, peaks must be completely resolved for the above to hold true, and, of course, the identity of the peak must be known. Gas density detectors can also be used for the determination of molecular weight. Phillips and Timms (15) were able to determine molecular weights of a variety of organic compounds with errors of ± 2%. Table 3.7 summarizes this detector.

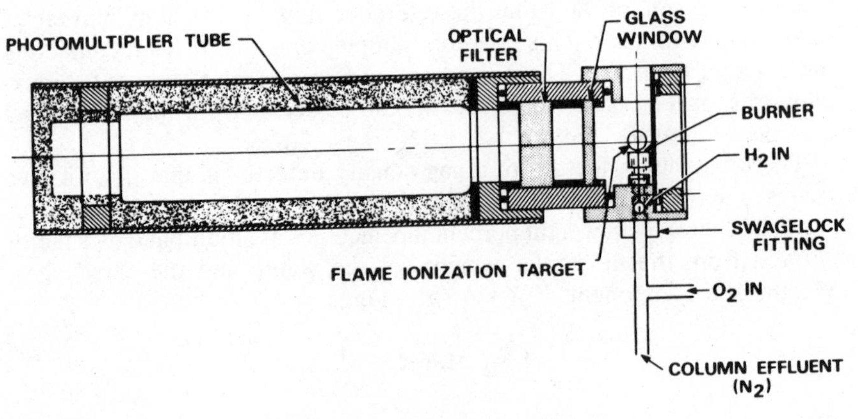

Figure 3.8. Schematic of FPD.

Table 3.8 Summary Flame Photometric Detector

M.D.Q.	10^{-10} g P; 10^{-9} g S
Response	Very selective; only S and P
Linearity	10^4
Stability	Good
Temperature limit	400°C
Carrier gas	Nitrogen

A very selective, sensitive detector for compounds containing S and P. Flame requires very clean H_2 and air. Calibration at very low levels requires thermostated permeation tube. Expensive, but useful for pesticide and air pollution work.

3.7.6 Flame Photometric Detector

The flame photometric detector (FPD), developed by Brody and Chaney (16), is shown schematically in Fig. 3.8. In this detector, the column effluent is mixed with oxygen and combusted in a flame surrounded by an envelope of hydrogen. The flame is seated within a high-walled cup to avoid background emission. Organic species in the sample are fragmented and excited to higher energy states in the reducing atmosphere of the flame. A high ratio of hydrogen to oxygen assures a reducing atmosphere. The plasma above the cup is viewed by a conventional photomultiplier tube through a narrow bandpass filter of the appropriate wavelength.

The interference filters insure maximum sensitivity toward sulfur and phosphorus emissions without significant interference from hydrocarbon species. The interference filters are chosen to have maximum transmittance at either 536 mm, corresponding to the wavelength of phosphorus HOP species in the hydrogen-rich flame; or 394 mm, which is the prime wavelength of sulfur $S_{\overline{2}}$ emissions. The detection limit of the FPD is 10 pg of phosphorus and 40 pg of sulfur. The characteristics of the FPD are in Table 3.8.

REFERENCES

1. L.S. Ettre and A. Zlatkis (eds.), *The Practice of Gas* Chromatography, Wiley-Interscience, New York, 1967.
2. H.M. McNair and E.J. Bonelli, *Basic Gas Chromatography*, Varian Aerograph, Walnut Creek, California, 1969.

3. O.E. Schupp *Gas Chromatography*, Wiley, New York, 1968.
4. J. Krugers, (Ed.), *Instrumentation in Gas Chromatography*, Centres, Eindhoven, Holland, 1968.
5. G. Schomburg, In *Proceedings of the Fourth Hindeland Symposium*: R.E. Kaiser (Ed.), Institute of Chromatography, Bad Durkheim, BRD, 1981.
6. K. Grob and K. Grob, Jr., *J. HRC and CC*, **1,** 263 (1978).
7. K. Grob and K. Grob, Jr., *J. Chromatogr.,* **151,** 311 (1978).
8. M. Galli, S. Trestianu, and K. Grob, Jr., *J. HRC and CC,* **2,** 366 (1979).
9. M. Galli and S. Trestianu, *J. Chromatogr.,* **203,** 193 (1981).
10. J.C. Giddings, *J. Chem. Educ.,* **69,** 569 (1962).
11. W.R. Supina, and L.P. Rose, *J. Chromatogr.,* **8,** 214 (1970).
12. A.J.P. Martin and A.T. James, *Biochem. J.,* **63,** 138, 1956.
13. A.G. Nerheim, *Anal. Chem.,* **25,** 1640 (1963).
14. C.L. Guillermin and F. Auricourt, *J. Chromatogr.* **2,** 156 (1964).
15. C.S.G. Phillips and P.L. Timms, *J. Chromatogr.,* **5,** 131 (1961).
16. S.S. Brody and J.E. Chaney, *J. Gas Chromatogr.,* **4**(2), 42 (1966).

CHAPTER

4

INSTRUMENTATION FOR HIGH-PERFORMANCE LIQUID CHROMATOGRAPHY

HAROLD M. McNAIR

Department of Chemistry
Virginia Polytechnic Institute and State University
Blacksburg, Virginia

4.1 INTRODUCTION

The requirements for the basic parts of liquid chromatographs are discussed briefly. These parts include: a solvent reservoir, pumping systems, injection systems, column oven and columns, detectors, and recorders. The most popular versions of these components are discussed in the order presented.

4.2 INSTRUMENTATION

Figure 4.1 is a schematic of the instrumentation required for HPLC. The instrumentation is comprised of seven components:

1. Solvent reservoir that contains the mobile phase
2. High-pressure pump used to push the mobile phase through the tightly packed column
3. Pressure gauge for monitoring the pump pressure
4. Injection device, usually containing a sample loop, used to introduce the sample into the moving mobile phase
5. Column, typically a tube of stainless steel 10–30 cm in length, 3 or 4 mm i.d., which has been tightly packed with small particles (~10 μm) of the material used to effect the separation
6. Detector, frequently a UV photometer (254 nm), used to measure the concentration of the sample components as they elute from the column
7. Potentiometric recorder that produces the chromatogram.

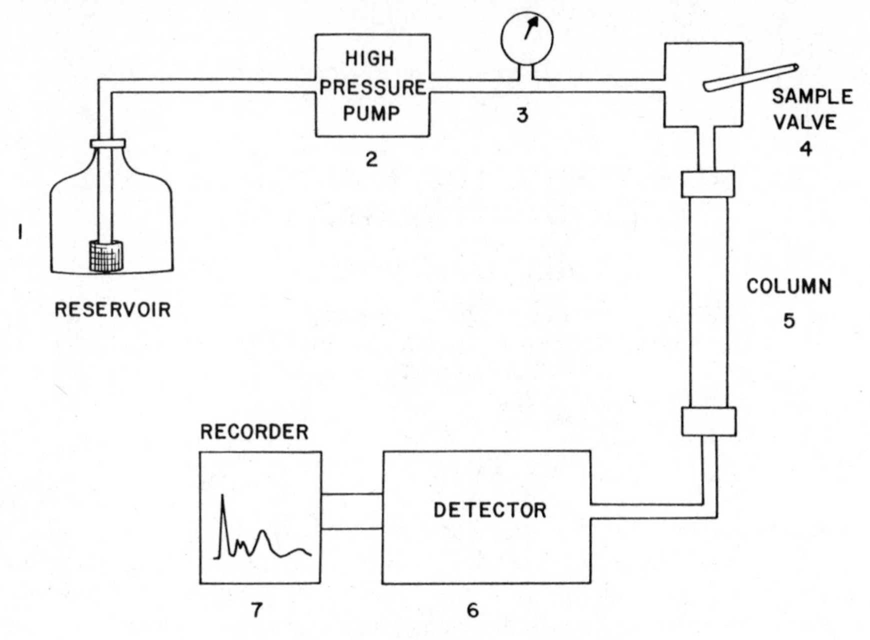

Figure 4.1. Schematic HPLC instrumentation.

4.3 SOLVENT RESERVOIR/SOLVENT

The solvent reservoir for HPLC is determined by the type of pumping system used, for example, syringe pumps contain a limited reservoir equal to the volume of the fully displaced piston. Here the solvent is degassed externally and added to the pump chamber. However, other types of pumps such as reciprocating piston types contain a separate solvent reservoir. Ideally, the reservoir must meet several requirements:

1. It must contain a volume adequate for repetitive analysis.
2. It must provide solvent degassing either by heating or applying vacuum or allowing sparging with He.
3. It must be inert with respect to the solvent.

Frequently, glass or stainless steel containers of 0.5 to 2 liters make suitable solvent reservoirs. Where space allows, the use of the glass bottle in which the solvents are purchased makes an excellent solvent reservoir. These should be carefully capped to avoid contamination from the laboratory atmosphere.

Table 4.1 Commonly Used HPLC Solvents

Normal phase (silica gel, cyano, amino)—Hexane or iso-octane, methylene chloride, chloroform, ethyl acetate.
Reverse phase—water, methanol, acetonitrile.
Ion exchange—aqueous buffers (usually 0.01–0.1 *M*).
Steric exclusion chromatography
Gel filtration—water (occasionally small amounts of alcohols)
Gel permeation—toluene, chloroform, trichlorobenzene, THF, pyridine, decalene.

The solvent is the mobile phase; its purpose is to carry the sample through the column and produce a reasonable distribution (capacity factor) between the mobile and stationary phases. The solvent must dissolve the sample. In addition, the solvent must be of high purity, often HPLC quality. Other desirable factors involved in solvent selection include low cost, low viscosity, low toxicity, and low boiling point. Depending on the type of column employed, the solvents will vary. Table 4.1 illustrates the most commonly used solvents.

4.4 PUMPING SYSTEMS

One of the most important components in HPLC is the pumping system. By producing reproducible high pressures, the pump is a major factor in obtaining high resolution, high-speed analyses, and reproducible quantitative analyses. The requirements of a good pump include:

1. A stable flow without pulsations to minimize detector noise
2. The solvents delivered must have a range of flow rates suitable for the various HPLC modes (usually 0.5–10 ml/min)
3. A constant volume delivery to facilitate qualitative and quantitative analysis
4. Amenability to high pressure (6000 psi)
5. Easy adaptability to gradient operation.

Several types of pumping systems are available including:

1. Liquid displacement by compressed gases (holding coil)
2. Liquid displacement by a piston driven by a compressed gas (pneumatic amplifier)

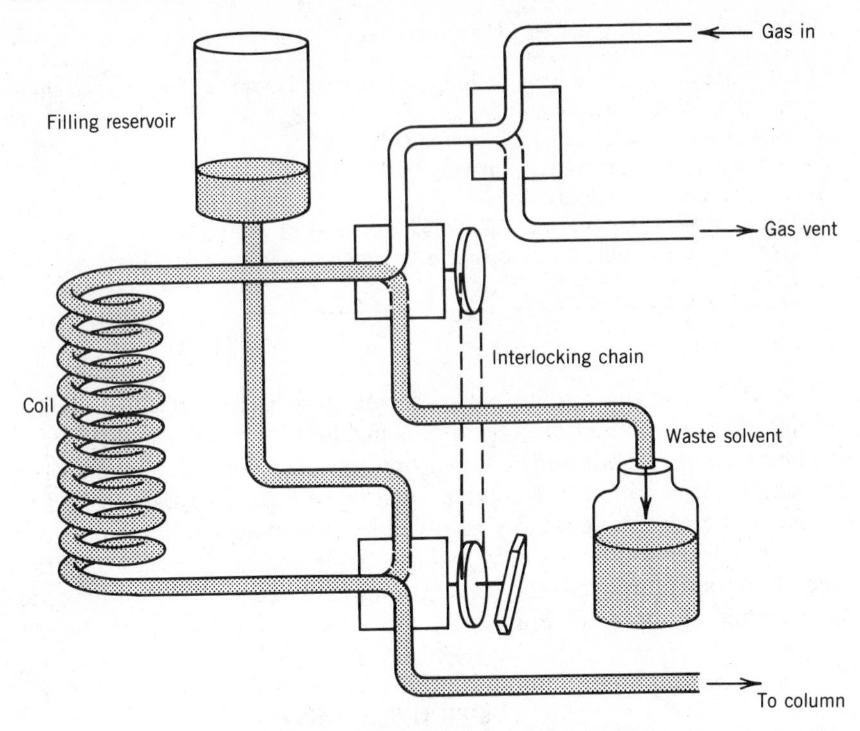

Figure 4.2. Holding coil solvent delivery system.

3. Piston or diaphragm driven by a moving fluid
4. Reciprocating piston
5. Syringe pumps.

4.4.1 Holding Coil

Liquid displacement by compressed gas is sometimes available in less expensive HPLC units. It consists of filling a large holding coil, usually of stainless steel tubing, with the solvent (see Figure 4.2). Compressed gas from a cylinder forces the liquid at constant pressure from the holding coil into the chromatographic column. Flow rates are dependent upon column permeability and the gas pressure used. These low-cost units usually operate isocratically, that is, gradient accessories are not available or are very complex. Pressures are usually limited to 1500 psi and care must be used to avoid dissolving the driving gas in the solvent that enters

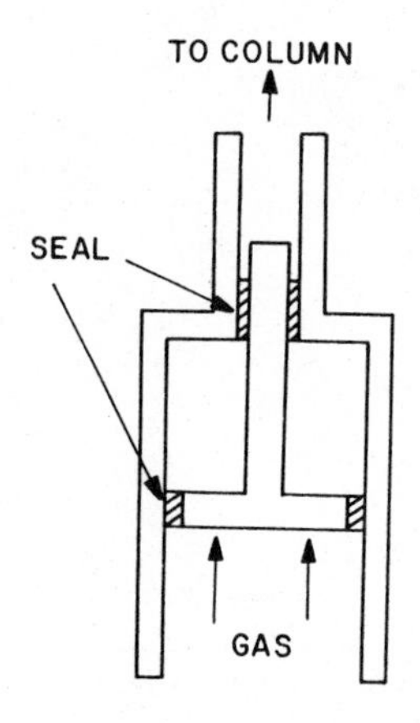

Figure 4.3. Schematic pneumatic amplifier pump.

the column. These types of pumps are not very popular and today are used mainly for teaching purposes or simple quality-control applications.

4.4.2 Pneumatic Amplifier

A pump utilizing a piston driven by a compressed gas is commonly referred to as a pneumatic amplifier. Low pressure gas (usually less than 200 psi) contacts a large surface area piston in contact with a smaller surface area of solvent (see Figure 4.3). The surface area ratios of the gas-piston/solvent-piston form the amplification factor of the gas pressure. Common ratios are 10- and 20-fold increased outlet pressures. Some models allow rapid filling of the solvent cylinder without affecting chromatographic performance. Flow is essentially pulseless, making this a good pump for quantitative analysis. Flow rate is dependent on column permeability and pressure available. Some models have optional flow controllers, which allow a choice of constant pressure or constant flow operation. High-pressure (20,000 psi) versions of this type pump are frequently used to pack HPLC columns.

4.4.3 Moving Fluid Type

Some HPLC units utilize a piston or diaphragm driven by a moving liquid. Generally, both models are driven from a single hydraulic pump. Pulseless operation is obtained by utilizing two or more pistons or diaphragms (see Fig. 4.4). This approach easily converts to gradient operation. The diaphragm type is essentially analogous to the piston type.

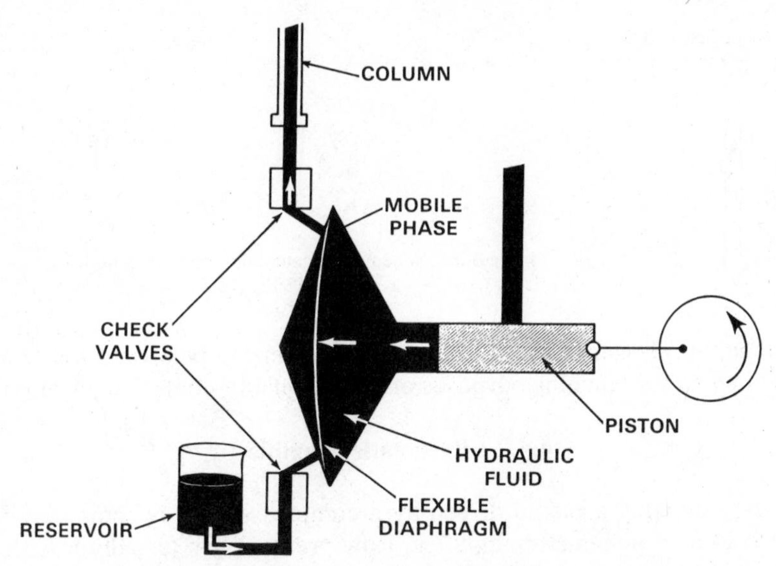

Figure 4.4. Schematic of hydraulic driven diaphragm pump.

4.4.4 Reciprocating Piston

These pumps utilize a piston in direct contact with the solvent. The piston is driven either mechanically with motors and gears or by solid-state pulsing circuits.

A simple version of the reciprocating piston pump is shown in Fig. 4.5. A motor-driven piston moves rapidly back and forth in a hydraulic chamber. By means of check valves, on the backward stroke the piston sucks in solvent from a reservoir. At this time the outlet to the column is closed, to preserve the operating pressure inside the column. On the forward stroke, the pump pushes solvent to the column and the inlet from the reservoir is closed. An eccentic cam drives the piston in three steps: (1) rapid reset; (2) rapid displacement until operating pressure is reached; (3) and, finally, a smooth constant volume displacement to provide a uniform flow rate. This is an inexpensive pump that allows a wide range of flow rates. Unfortunately, it does produce flow pulses and a pulse-dampening system must be employed.

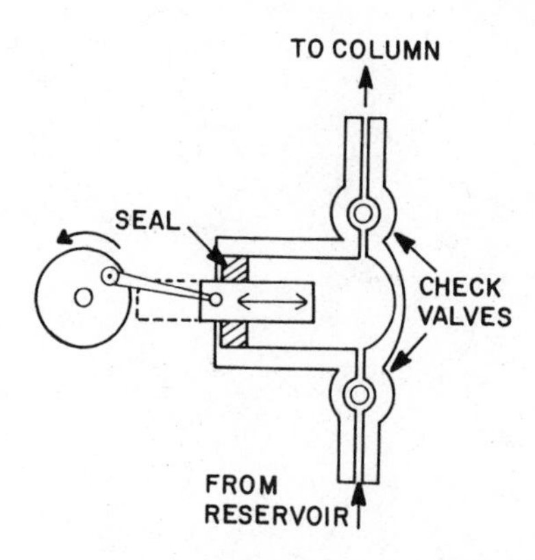

Figure 4.5. Schematic reciprocating piston pump.

One approach in eliminating pulsations is with multiheaded pumping systems. Two or three pumping assemblies offset in time can produce essentially pulseless flow. Other models electronically sense the pressure between pump pulses and automatically change the speed of the stepping motors to minimize flow pulses. Gradient operation is achieved either by premixing the solvents at low pressure prior to pumping or by the addition of a second pump and mixing at operating pressures.

4.4.5 Syringe Pumps

These pumps operate by a screw gear displacing a plunger through the solvent reservoir (see Fig. 4.6). They are expensive, but usually produce stable flow rates and high pressures. These advantages are sometimes overshadowed by the inconvenience of the several washings required when changing solvent systems. Syringe pumps are particularly well suited to gradient operation, since each solvent is contained in separate reservoirs and the speed of each syringe plunger can be electronically controlled to produce the desired gradient. Of course, two syringe pumps must be used for gradient operation.

4.5 INJECTION SYSTEMS

Samples are introduced by one of three methods: (1) syringe injection; (2) sample valve; and, (3) automated sample valves. Sample valves are most widely used, since they provide a more reproducible volume injected (and hence better quantitative analysis). They can be easily automated for unattended operation, and they are not expensive.

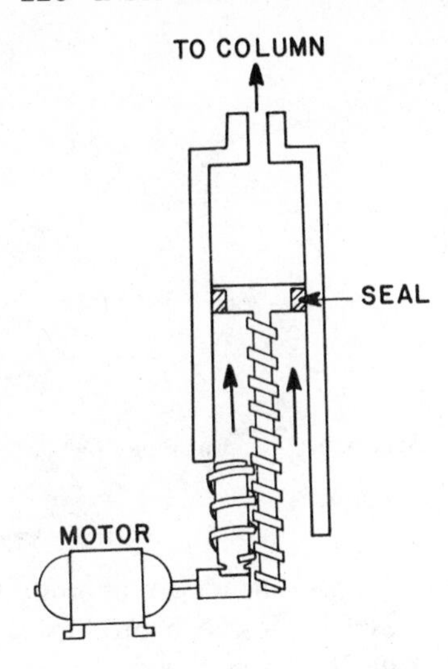

Figure 4.6. Schematic syringe pump.

4.5.1 Syringe Injection

Low-pressure syringes for gas chromatography (GC) can be successfully used up to 1500 psi, however, precautions must be taken. At higher pressures, specially designed syringes are available. Requirements for septum materials are more rigorous in HPLC operation than in GC. The elastomer must have sufficient strength to enable successive needle penetrations without total rupture or polymer extrusion through the needle hole. This requirement is difficult with elastomers, which swell in solvents. Generally, silicone polymers such as those used in GC are the most popular for LC systems operating with aqueous mobile phases, aqueous-alcohol, or other polar phases. However, with nonpolar organic solvents, swelling of the silicone elastomer precludes their usage. Data are available from LC manufacturers concerning compatibility of elastomers with nonpolar solvents. Perfluoroelastomers are ideal with hydrocarbons and other nonpolar mobile phases. Septum material covered with a thin sheet of Teflon or other insert polymer is also available.

To obtain reasonable septum life, it is advantageous for the needle to penetrate the same hole on successive injections. This can be obtained with needle guides.

Septumless syringe injection at full operating pressure is available on several units. One approach utilizes a syringe whose needle is inserted and sealed into a Teflon cylinder to form a pressure-tight seal. The syringe can then be pushed forward and injection made when a transverse seal

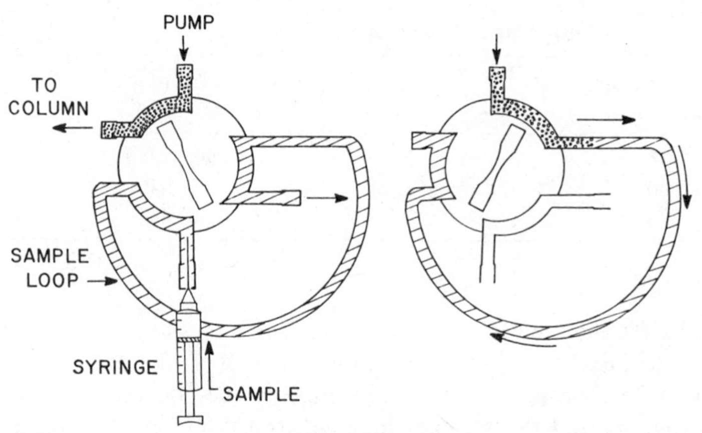

Figure 4.7. HPLC sample valve: (*a*) sample loading position; (b) sample introduction position.

is opened. A second approach uses a syringe with a side injection aperture which is transversely inserted through the mobile phase stream.

4.5.2 Sample Valve

A fixed-volume loop is filled with the sample, and by turning a valve this loop is placed into the solvent stream before the column. In Fig. 4.7, on the left, is a sample loop of stainless steel that has been filled with sample from a syringe. For analytical work, the sample loop volume is usually 10 or 20 μl. Note that in this filling position the mobile phase at high pressure is passing directly through this sample valve to the column. On the right in Fig. 4.7, the valve has been rotated so that the high-pressure mobile phase now sweeps through the sample loop carrying the sample with it onto the column. This is a simply reproduced technique and sample valves can be easily automated. Sample sizes can be changed by changing the sample loop volume. Unlike syringes, they do not suffer the lack of repeatability due to operator variability, or downtime due to septum deterioration. Sample valves must be constructed with minimum dead volumes and of suitable materials to eliminate reaction between sample or solvent. Some valves contain a large volume sample loop and allow "variable volume" injection by syringes. This is a convenience in method development, but does not provide as good quantitative results as with fixed volume loops that are completely filled with sample.

The sample-valve approach has been automated for unattended HPLC operation. In several versions, the sample is placed in small glass vials.

Table 4.2 Effect of Column Temperatures

Temperature (°C)	R	PLATES/M N	α
25	2.4	11,000	1.20
50	1.60	12,400	1.14
75	0.85	12,500	1.11

The vials are indexed to a position where an excess volume of sample is pumped through a high-pressure injection valve. The excess volume is necessary to thoroughly clean the sample loop. The sample is automatically injected and the sampler indexed to a new vial. Microprocessors can be used to repeat the injection of the same sample, to rinse the sample loop and connecting tubing between each sample, or simply to inject in sequence the samples as loaded in the sampler tray.

4.6 COLUMN OVEN

Many HPLC separations can be performed at ambient temperatures without the aid of a column oven. However, elevated temperatures are useful either to reduce retention volumes or to decrease the mobile-phase viscosity and thereby increase the column efficiency. Lower mobile-phase viscosity also enables operation at lower pumping pressures or higher flow rates at the same pumping pressure. Three approaches are currently used in temperature control:

1. Using an external heat exchanger enclosing the column through which water from a thermostatically controlled water bath is circulated
2. Submerging the column in a water bath
3. Using a hot-air circulation oven as in GC.

Useful temperature ranges are from 30 to 100°C.

Recent work has emphasized the change in column selectivity caused by changes in temperature (see Table 4.2). This data is for the separation of PNA's on a RP–18 reverse phase column. This is true even for adsorbents such as silica gel. Thermostats were always required for ion exchange, gel-permeation, and liquid partition chromatography; they now appear highly advantageous even for adsorption and bonded-phase chromatography.

4.6.1 Columns

Columns for HPLC vary, depending primarily on the type of separation desired and the physical characteristics of the packing materials. The most popular column length ranges from 10 to 30 cm. Analytical columns have internal diameters of 3 or 4 mm and are precision-bore stainless steel. Glass (maximum 500 psi) and glass-lined stainless steel tubing can also be used. Column packings of small diameters (3 and 5 μm) use shorter columns (5–15 cm) due to the higher efficiency and larger pressure drops. The column shape is usually straight; however, with slurry-packing techniques, bent or coiled columns have been shown to be equally efficient.

Packing is usually retained by inserting nickel or stainless steel frits into the fittings at the end of column. An alternative procedure is to have the frit accommodated in the nut attached to each end of the column. The frit is particularly effective in reducing dead volume as well as producing uniform flow profiles.

Columns for preparative separations are usually wider. Preparative scale columns have diameters ranging from 6 to 10 cm. They are useful for milligram up to gram quantities of sample. Larger flow rates are required and most often automatic sample collectors are used in preparative scale work. There are also available dedicated preparative scale systems using columns with diameters of 4 inches and larger. These columns can accommodate sample loads of up to 50 grams, but require special high-volume pumps.

4.7 DETECTORS

The purpose of the detector is to measure the sample concentration in the solvent and produce an electrical signal proportional to the sample concentration. The two most common detectors are the UV and the differential refractive index (RI).

4.7.1 UV/VIS Photometers

Shown schematically in Figure 4.8 is the most widely used detector, a simple fixed wavelength double-beam UV photometer. On the left is the source, a low-pressure mercury lamp that emits a sharp (essentially monochromatic) line spectrum with a strong line at 253.7 nm (254). A quartz lens focuses the UV radiation on the sample and reference cells. The sample cell usually contains about 10–20 μl of the continuously flowing column effluent. The reference cell is usually filled with air. A UV filter

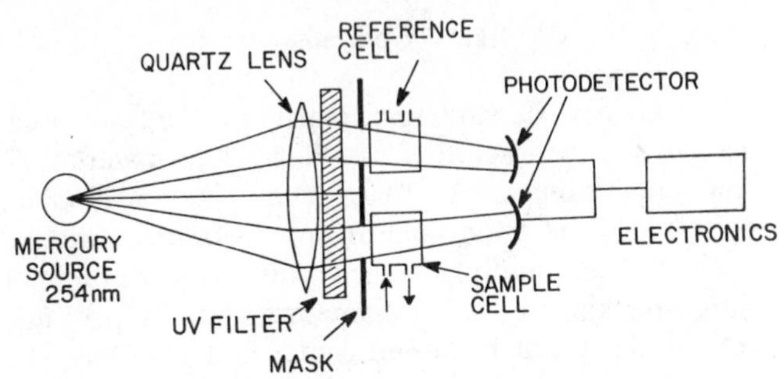

Figure 4.8. Schematic UV photometer, 254 nm.

removes unwanted radiation. The radiation passing through the reference and sample cells falls on two photodetectors. The outputs of these two detectors are passed through a preamplifier to a log comparator that produces the electric signal for the recorder. The log comparator is necessary to convert the photons measured at the photodetector (transmittance) into absorbance units that are directly proportional to concentration.

This UV detector is inexpensive, sensitive, insensitive to normal flow and temperature fluctuations, and well suited to gradient elution. It is, however, a selective detector. Only sample molecules which absorb at 254 nm can be detected. It cannot be used for lipids, hydrocarbons, most polymers, carbohydrates, and fatty acids.

Since many compounds do not absorb at 254 nm, a variety of UV/VIS detectors offering other wavelengths have become available. With the medium-pressure mercury lamp and appropriate filters, wavelengths available include 254, 280, 312, 365, 436, and 546 nm. Also available are detectors offering 254 and 280 nm simultaneously, as well as 254 minus 280, and 280 minus 254 (i.e., differential detectors).

4.7.2 UV/VIS Spectrophotometers

Spectrophotometers adapted for HPLC detection are also very popular. Some are true recording spectrophotometers, which allow a UV/VIS spectrum to be generated on an eluting peak that is trapped in the flow cell. Others have only manual selection of wavelength (usually 200–800 nm). This allows selection of a wavelength to maximize sensitivity or minimize interferences, but does not allow a spectrum to be recorded.

Figure 4.9 shows a schematic spectrophotometer used as an HPLC detector. This instrument uses a grating, thus allowing selection of any wavelength of 200–800 nm. Light from a continuum source is focused on the entrance slit of a grating monochromator. By the appropriate optics,

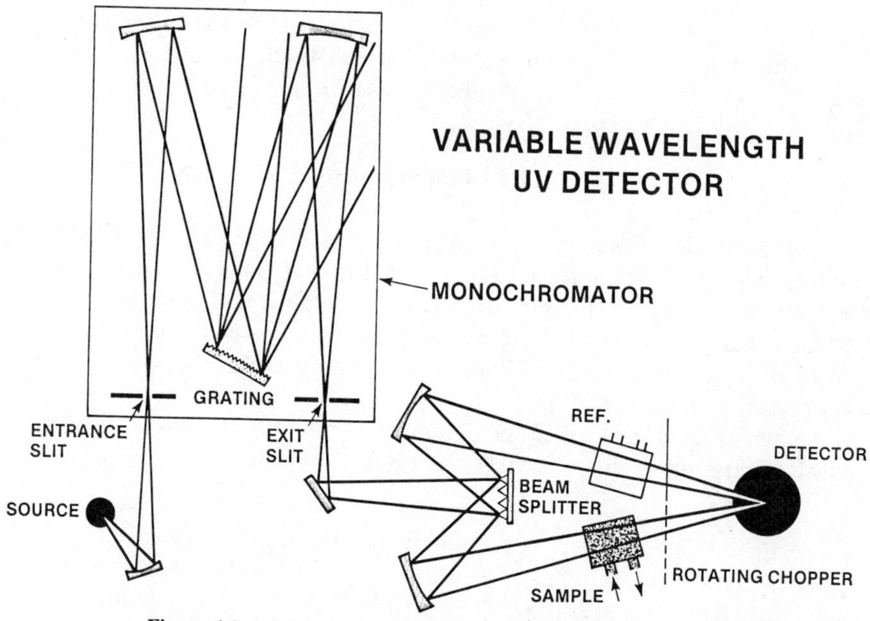

Figure 4.9. Schematic spectrophotometer for HPLC detector.

this "white light" is focused on the grating where it is dispersed into the various wavelengths. By varying the position of the grating, the desired wavelength is focused on the exit slit and then passed through sample and reference cells by means of a beam splitter. The detector measures the difference in light intensity passing between sample and reference cells. The detector signal is converted into absorbance by a logarithmic comparater.

Spectrophotometers allow the selection of any useful wavelength in the UV and visible regions. The primary benefit is the increased sensitivity for the compounds of interest. Many compounds would not absorb well at 254 nm, and therefore greater sensitivity is possible with a spectrophotometer. In some cases, noise or interferences from the sample matrix are also reduced by using a spectrophotometer.

4.7.3 Refractive Index

This is a universal detector and will respond to all sample types. It is more expensive than the simple UV detector, but less expensive than a spectrophotometer; it requires very good temperature and good flow control; and it is not amenable to gradient elution. Sensitivity is limited to about 1 μg of sample. The RI is a useful detector in GPC work, preparative separations, and routine quality control where trace analysis or gradient elution is not required. It is not usually sensitive nor stable enough for routine analytical work.

4.7.4 Fluorescence

Fluorescence detectors are becoming more popular due to their selectivity and sensitivity. Selectivity frequently means that the compounds of interest can be readily distinguished from a complicated matrix of compounds that do not exhibit fluorescence. Sensitivity has been shown to be 10–11 g or low parts per billion. Both fixed wavelength and scanning fluorescence units are offered.

There are a variety of other detectors available, but space does not allow discussion in this brief introduction.

4.8 RECORDERS

Standard strip chart recorders (1 and 10 mv) such as are used in GC are also employed in HPLC. Chart speeds should be reproducible so that retention times (chart distance) can be used to identify peaks. Quantitative analysis is performed as in GC; that is, peak areas or peak heights are measured and compared with standards. Disk integrators, electronic integrators, and computers are also used to measure peak areas.

4.9 FACTORS IN CHOOSING HPLC INSTRUMENTATION

Several major factors must be considered when purchasing an HPLC system. In addition, there are usually several factors unique to each laboratory. The comments concerning each factor are the subjective opinion of the author.

It is essential to know if HPLC will solve the analytical problem. Fortunately, most manufacturers have experienced applications laboratories that can run samples. Prospective buyers are urged to take the time to run the sample before making a purchase. For additional information, consult one of the recent textbooks (1,2).

Most samples can be run isocratically; however, for an initial purchase, gradient elution is a strongly recommended capability. It is essential in research and method development, less useful in routine quality control.

A fixed wavelength (254 nm) UV photometer is recommended as a standard component for everything except GPC work. Most sample types show some absorbance at 254, and this is a robust, inexpensive, and sensitive detector. Sample types that may require a variable wavelength UV or refractive index detector include sugars, fatty acids, triglycerides, most polymers, and nonaromatic hydrocarbons. The author's second choice is a variable wavelength UV that can be used for both sensitivity (choose a wavelength of maximum sample absorbance) or selectivity (choose a wavelength to suppress the background of undesired components or mobile phase).

Modular systems are more flexible and usually less expensive than integrated systems. They allow start-up with a modest investment and can be upgraded later. Modular systems usually require some experience on the part of the chemist and may require more "playing in the lab" initially to set up and optimize the modules. Integrated systems use matched components in a fully assembled unit; start-up time is minimal and the manufacturer often provides installation, start-up assistance, and some training. Integrated systems are usually more expensive and easier to automate, but frequently cannot be taken apart for swapping parts or troubleshooting.

The buyer must finally choose a specific instrument manufacturer. In many cases, the previous considerations will have narrowed the field to two or three candidates. The most important factors at this point will be the manufacturer's reputation in HPLC and the after-sales service. A good picture of the manufacturer's reputation can be obtained by talking to several local LC customers who have purchased equipment in the last few years. It is well to try to avoid "good" or "bad" opinions created years ago, often in a different product line and often associated with a particular person.

A minor factor (in most cases) should be the relative cost, provided there is no more than a 20% variation. The real criterion is the price/value; frequently, higher-priced units offer more value.

REFERENCES

1. L.R. Snyder and J.J. Kirkland, *Introduction to Modern Liquid Chromatography*, 2nd ed., Wiley, New York, 1979.
2. E.L. Johnson and R. Stevenson *Basic Liquid Chromatography*, Varian, Palo Alto, 1978.
3. H.M. McNair, "High performance liquid chromatography equipment IV," *J. Chromatogr. Sci.*, 588 (1978).
4. J.F.K. Huber, (Ed.), *Instrumentation for High Performance Liquid Chromatography*, Elsevier, Amsterdam, 1978.

CHAPTER

5

GAS CHROMATOGRAPHY OF INORGANIC COMPOUNDS, ORGANOMETALLICS, AND METAL COMPLEXES

PETER C. UDEN

Department of Chemistry
University of Massachusetts
Amherst, Massachusetts

5.1 INTRODUCTION

It cannot be denied that in the 30 years since its inception as a practical analytical separatory technique, gas–liquid chromatography in all of its forms has had the greatest impact in the quantitative resolution of volatile organic compounds. This impact has been reinforced and extended in recent years by the explosive growth in the application of high-resolution capillary columns to the separation of complex multicomponent samples. Although the sister discipline of gas–solid adsorption chromatography had its primary field of application in permanent and light inorganic gas analysis along with that of metalloid halides and hydrides, the potential of both GLC and GSC for analytical separation of the more "definitive" inorganic chemical compounds, metal complexes, and organometallics has been less realized than might have been expected.

The reasons for the relative obscurity of inorganic gas chromatography are varied. Perhaps the most pervasive has been the expectation that inorganic compounds are inherently nonvolatilizable and/or thermally unstable under conventional gas chromatographic column conditions. Concurrent with this is the notion that there is an inherent incompatibility of inorganic or organometallic compounds with column substrates, and a certainty of undesirable reactivity that will make gas chromatography difficult if not impossible. Another aspect of the unwillingness of some analytical chemists to explore inorganic gas chromatography concerns the assumption of sample incompatibility with conventional detectors. While there are certainly cases where some of these many strictures apply, there are nontheless numerous literature examples of very valuable and rigorous gas chromatography of metallic and metalloid compounds; it is to the review and commentary of such studies that this chapter is devoted.

In this chapter, the conventional limitations of definition of organometallics and metal complexes will be adhered to. Thus, the former will include compounds in which sigma or pi carbon to metal bonds are present; appropriate examples will also be given of closely analogous organometalloid compounds. Under the definition of metal complexes will be covered compounds in which a coordinated bond is formed between a ligand atom, such as oxygen, sulfur, nitrogen, phosphorus, or halogen, and the central metal atom. The latter class of compounds will comprise primarily metal chelate systems. Many metallic compounds are necessarily excluded from consideration since they cannot be volatilized unchanged at conventional gas chromatographic column temperatures; however, there are good examples of successful elution of such compounds as metal oxides and halides at high temperatures. The ionic nature of many inorganic compounds in solid or solution phases precludes direct gas chromatography, but in some instances suitable derivatization methods may enable successful elution to be achieved. There are cases where the great 'reactivity' of metallic compounds would be thought to forbid viable gas chromatography, but where in fact the nonoxidative and nonhydrolytic conditions prevailing under typical gas chromatographic situations has enabled qualitative and quantitative separations to be achieved.

There is often considered to be a minimal group of chemical properties to be possessed by an inorganic compound before gas chromatography can be successful; although not all-embracing, these certainly give a good guideline for experimental development. They are: adequate volatility; thermal stability; monomeric form; neutrality; low molecular weight (certainly a matter of perspective); coordinative saturation; and, perhaps most importantly, adequate structural shielding of the metal atom(s) by bulky and inert organic functional or coordinating groups. It is perhaps appropriate to consider that the inorganic complex or organometallic compound should appear to the GC column as much like a simple organic compound as possible and that the free metal atom should never be exposed to reactive chromatographic substrates or materials. As will be noted later, the majority of successful quantitative GC separations of inorganic compounds honor most if not all of these criteria.

The nature of the materials in the GC system are more than normally important in defining whether inorganic gas chromatography is viable. The criteria of inertness of column materials, injection and detection pathways and ancillary flow devices is particularly important for this type of gas chromatography. As is noted later the advent of inert capillary column materials promises to reopen the study of many inorganic chromatographic applications previously impeded through the lack of such facility.

Similarly, the advent of novel detector devices that are more compatible, selective, or specific for inorganic compounds has opened the way for quantitative and sensitive analysis that was previously unattainable. In many ways, the time is ripe for the reappraisal of many of the early gas chromatographic separations and analyses described in this chapter in the light of the advancing technology of the past few years, and it is hoped that the following discussion and appraisal of information will serve to stimulate such studies.

The general methods of inorganic analysis involve the removal of interfering substances from those to be determined and increasing selectivity and sensitivity by sample treatment and preparation methods such as extraction and derivatization. The classes of inorganic substances for which gas chromatographic analysis is viable, either directly or by indirect derivatized modes, are noted as follows:

1. Elemental gases and vapors.
2. Binary inorganic compounds such as selected halides, hydrides, and oxides.
3. Sigma-bonded organometallic and organometalloid compounds such as metal and metalloid alkyls and aryls.
4. Pi-bonded organometallics such as metal carbonyls and metallocenes.
5. Chelated metal complexes having nitrogen, oxygen, sulfur, phosphorus, etc. as ligand atoms.
6. Derivatized species for metal, metalloid, nonmetal, or functional group determinations.

The areas to be reviewed in this chapter are primarily the last four classes, although observations on simpler inorganic applications are also considered briefly.

5.2 GAS CHROMATOGRAPHIC COLUMNS, SUBSTRATES, AND MATERIALS

While the first quarter century of gas chromatography can realistically be considered as the era of packed column separations, the subsequent period has seen rapid movement towards open tubular capillary column emphasis. The impetus for this trend, which has culminated in the wide adoption of fused silica open tubular capillary columns, has been in the necessity for the highest available resolution of multicomponent volatile mixtures in complex matrices. Inorganic gas chromatographic separation problems have not typically had high resolution as their primary objective, but rather the lowest possible detection limits achievable for frequently intractable samples. Even so, there is a marked recent trend towards open tubular column applications, a procedure which will undoubtedly revolutionize inorganic GC as it has other analytical areas.

Historically, inorganic GC separations have relied on both adsorption and partition processes. The chemical reactivity of volatile inorganic compounds has frequently necessitated the most rigorous control of columns and substrates in order to prevent undesirable on-column decomposition or unfavorable interactions.

Adsorption GC has been widely employed for the separation of gases; Kiselev and Yashi reviewed applications to inorganic separations in their 1969 monograph (1). A wide variety of adsorptive column packings have been investigated, and detailed applications are given later. Alumina, silica, other metal oxides, and inorganic and carbon molecular sieves have predominated for the separation and quantitation of stable permanent atomic and molecular gases. Organic polymers, notably polystyrene copolymer based materials, have been much favored for more reactive and polar analytes such as hydrides, low molecular weight organometallics, and polar compounds such as water and ammonia (2). Inorganic materials ranging from quartz and diamond to metal salts and eutectics have been used for GC of reactive species at very high temperatures. The greater part of viable quantitative analysis has been by means of molecular sieves and porous polymers, which show minimal residual adsorption.

The choice of suitable partitioning stationary phases for separation of organic compounds has remained a challenge. With the exception of the chromatography of a few particularly stable compounds, such as tetraalkyl silanes, polar substrates such as esters, polyesters, and polyalkylene glycols such as Carbowaxes™ have had little utility. The main aim has been analyte stability and absence of on-column degradation rather than aiming for subtleties of resolution.

Methyl silicone oils and waxes have provided the widest application, although other substituted silicones such as phenyl, cyano, and particularly fluoroalkyl have also been favored. Nevertheless, elution problems and nonideal chromatography have always plagued inorganic GC, and many approaches have been necessary to reduce them. All chromatographic parameters have been varied, including column material, solid support, treatment and deactivation, injection port design, and system pretreatment. Most studies have been conducted with glass or stainless steel columns, and the former are preferred for most inorganic separations. Since metals present on internal column surfaces or supports are often responsible for on-column catalytic decomposition, there has always been danger of increasing sample degradation with continued column use as more metals and oxides are deposited.

The increasingly rapid adoption of capillary column GC techniques promises an advance for inorganic separations as much by virtue of the absence of decomposition sites on supports or columns as by high resolution capabilities. This is particularly true for fused-silica capillary columns, which are characterized by extremely low residual trace metal concentrations in the silica matrix. It is probable that the majority of packed column partition or adsorption separations and the traditional stainless steel or glass capillary GC described in this chapter may benefit from reassessment on fused silica columns where thermal stability of the analytes will prove to be the most important criterion for successful quantitative elution, particularly at trace levels.

5.3 GAS CHROMATOGRAPHY OF INORGANIC GASES

Although it was first published in the early 1970s, the text by Guiochon and Pommier entitled *Gas Chromatography in Inorganics and Organometallics* still provides an excellent summary of this topic (3). This may be usefully augmented by Schwedt's coverage in his 1981 text, *Chromatographic Methods in Inorganic Analysis* (4). Gas–liquid partition chromatography (GLC) has very limited application for the analysis of permanent gases, since their solubilities in liquid stationary phases are minimal at ambient and subambient temperatures. This, in combination with their low boiling points, minimizes retention. Although there have been examples of GLC of some polar gases, such as ammonia with ester, and silicone oil phases, gas–solid adsorption (GSC) is the preferred method for such analysis as well as those of nonpolar atmospheric gases. Inorganic zeolite-based molecular sieves possessing concentrated surface charge generate surface adsorption based on polarizability of adsorbed molecules (5) and have found most favor for atmospheric gases. Aubeau et al. (6) showed how the separation and elution order of hydrogen, oxygen, nitrogen, carbon monoxide, methane, krypton, and xenon can be changed by altering the degree of hydration of molecular sieve 5A (Fig. 5.1). Reproducible activation of the adsorbing substrate is clearly of the greatest importance in this type of analysis. There may be scope for capillary columns for GSC wherein the adsorbent is coated as small particles on the inner column walls (the support-coated open-tubular column—

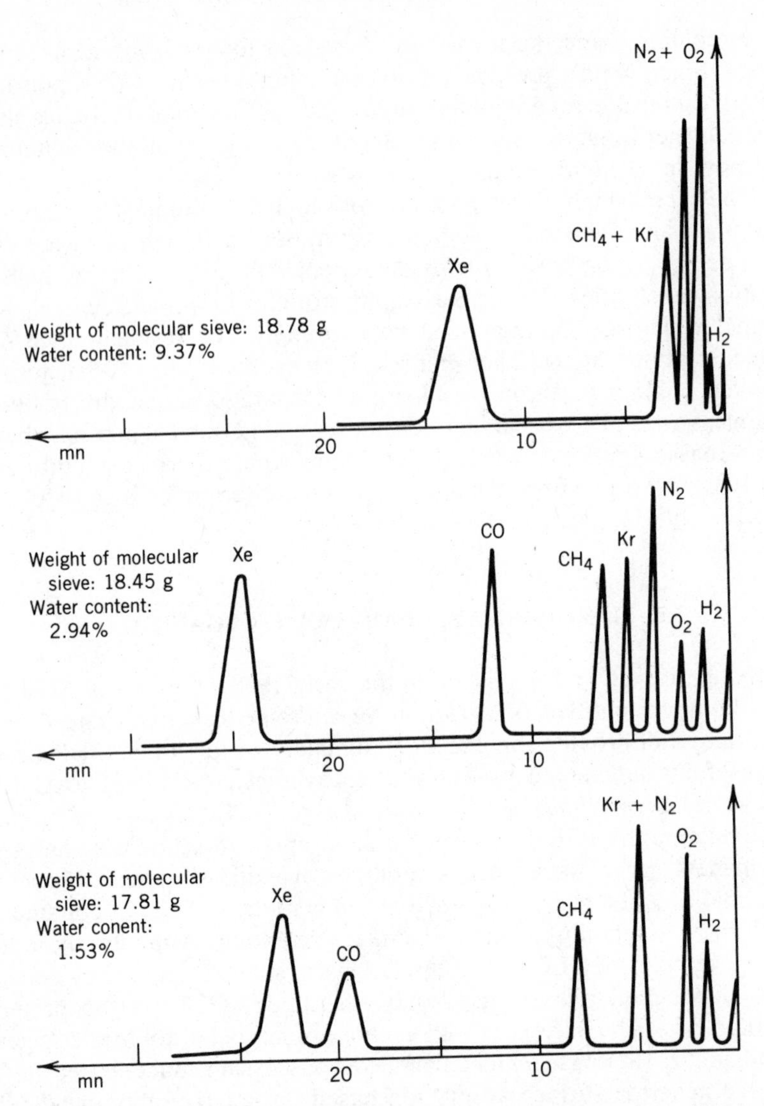

Figure 5.1. Influence of the degree of hydration of molecular sieve 5A on its retention properties of permanent gases. Column 2 m, 60°C helium carrier gas at 30 ml/min. Reprinted with permission from R. Aubeau, L. Leroy, and L. Champeix, *Journal of Chromatography*, **19**, 245 (1965). Copyright 1965 Elsevier Science Publishers.

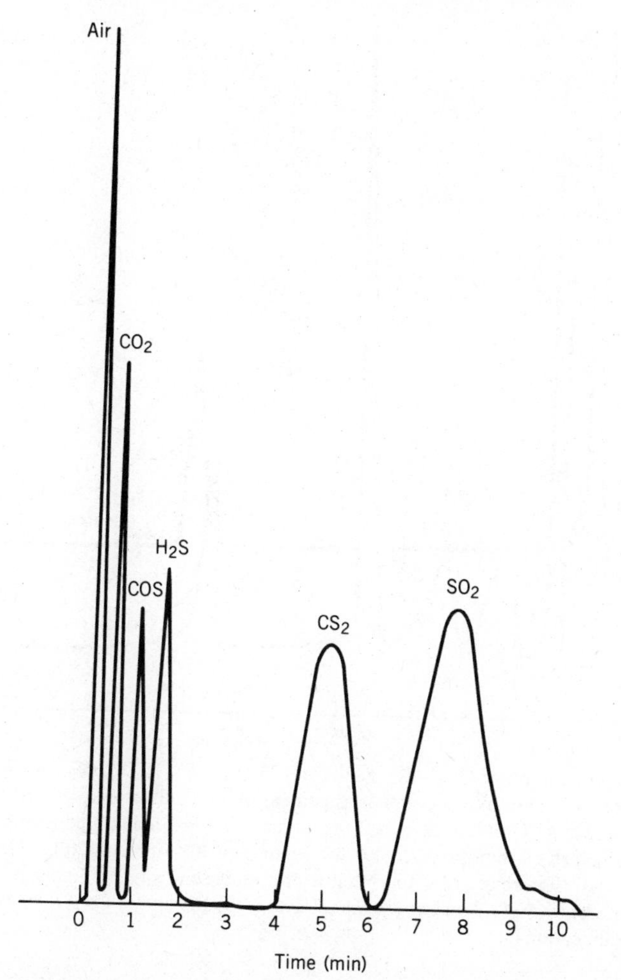

Figure 5.2. Analysis of sulfur compounds on silica gel. Column temperature 100°C, helium carrier gas at 40 ml/min. Reprinted with permission from C. T. Hodges and R. F. Matson, *Analytical Chemistry,* **37,** 1065 (1965). Copyright 1965 American Chemical Society.

SCOT) (7). The situation in which carbon dioxide is also present in such gas mixtures is more complicated, since it is irreversibly retained under normal conditions. Silica will permit resolution of all these components except oxygen and nitrogen. The full resolution of the mixture may be achieved by setting silica and molecular sieve columns in series separated by a low-volume, nondestructive detector. Backflushing of the molecular sieve column periodically will maintain reproducibility.

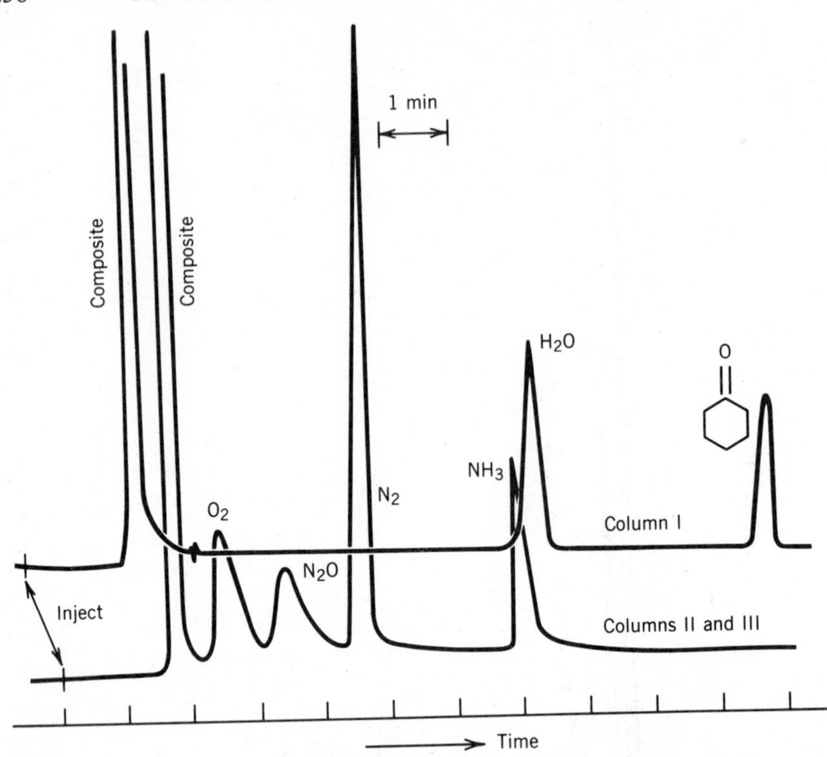

Figure 5.3. Separation of O_2 (14%), N_2 (28%), N_2O (14%), H_2O (10%), and cyclohexane (2%), balance as helium by a parallel arrangement of molecular sieve 5A and Chromosorb 104 (isothermal at 45°C) followed by a 2.4-m, 4 mm i.d. glass column containing Carbowax 20M on 40/60 mesh Chromosorb T (40°C for 4 min then 30°C/min to 220°C). Helium carrier gas at 60 ml/min (Reference 12). Reprinted with permission from W. Gates, P. Zambri and J. N. Armor. *Journal of Chromatographic Science,* **19,** 183 (1981) Copyright 1981. Preston Technical Publications.

Another noteworthy application of silica has been for separation of sulfur gases. Figure 5.2 shows such a separation (8). Water presented a major problem in early gas chromatography, since large amounts of adsorptive tailing are observed with most substrates, and liquid phases exhibit poor partition characteristics. Porous polymers of the Porapak™ and Chromosorb Century™ types of polystyrene or similar polymer (9,10) have extremely low affinity for water, which is eluted rapidly with efficient peak shape. Other polar compounds such as ammonia, hydrogen cyanide, and sulfur dioxide are also eluted rapidly before low molecular weight organic compounds. These polymers may also be used advantageously in series with molecular sieves or as column pairs at different tempera-

Table 5.1 Separation Materials for Adsorption Gas Chromatography of Inorganic Substances

Separation Material	Typical Separations	Ref.
Alumina	O_2, N_2, CO_2	13
Beryllium oxide	H_2S, H_2O, NH_3	14
Silica gel	O_2/N_2, CO_2, O_3, H_2S, SO_2	15–17
Chromium(III) oxide	O_2, N_2, Ar, He	18
Clay minerals (Attapulgite, Sepiolite)	O_2, N_2, CO, CO_2	19
Kaolin	He, O_2, N_2, CO, CO_2	20
Sodium-, lithium fluoride, alumina	MoF_6, SbF_5, UF_6, F	21
Quartz granules	Ta, Re, Ru, Os, Ir: Oxides, hydroxides	22
Chromosorb 102	Element hydrides	23
Graphite	NH_3, N_2, H_2	24
Synthetic diamond	CF_2O, CO_2	25
Molecular sieve	Hydrogen isotopes	26
Carbon molecular sieve	O_2, N_2, CO, CO_2, N_2O, SO_2, H_2S	27
XAD resins	NH_3, SO_2, H_2S, CO, CO_2, H_2O	28
Porapak Q	GeH_4, SnH_4, AsH_3, SbH_3, $Sn(CH_3)_4$	29,31
Porapak QS polymers	H_2S, CH_3SH, $(CH_3)_2S$, $(CH_3)_2S_x$, SO_2	30
Porapak P	Chlorides of Si, Sn, Ge, P, As, Ti, V, Sb	32
Teflon	F, MoF_6, SbF_6, SbF_3	33

Condensed from Ref. 4, *Chromatographic Methods in Inorganic Analysis,* G. Schwedt, p. 43.

tures. Two Porapak Q™ columns in series at 75°C and −65°C gave a complete resolution of CO_2, H_2S, H_2O, COS, SO_2, N_2, O_2, Ar, and CO. Helium carrier gas containing 100 ppm of SO_2 was employed for this separation (11). An added dimension to this type of analysis can be gained by the use of parallel configured columns for the same analysis. Thus Gates et al. (12) utilized three columns with automatic valve switching, parallel molecular sieve 5A and Chromosorb 104™ followed by Carbowax 20M on Chromosorb T to separate O_2, N_2, and N_2O as well as various volatile organics (Fig. 5.3). A summary of adsorption materials and applications is given in Table 5.1.

Table 5.2 Boiling Points (1 atm) of Metalloid Hydrides

Compound	Boiling Point (°C)
SiH_4	−111.8
GeH_4	− 88.5
SnH_4	− 52
PH_3	− 87.7
AsH_3	− 55
SbH_3	− 17.1
SeH_2	− 41.5
TeH_2	− 2.2

Detection for inorganic gas analysis has been almost entirely by means of thermal conductivity since ionization detector response is minimal. The prospect for enhanced detection by specific element methods involving atomic plasma emission spectroscopy is an attractive one, since elemental response is independent of molecular form.

5.4 BINARY METAL AND METALLOID COMPOUNDS

The principal simple binary inorganic compounds that have adequate vapor pressures and thermal stabilities at normal gas chromatographic temperatures are main group hydrides and halides. If high-temperature applications with temperatures to 1000°C or above are also considered, a number of metal oxides also show adequate physical characteristics for undecomposed elution.

Considering first metal and metalloid hydrides, the primary candidates are those of Groups IIIA, IVA, and VA elements. Boron hydrides were among the earliest of inorganic compounds to be successfully gas chromatographed; silicon, germanium, and tin hydrides have been widely studied; and in Group V arsenic, antimony, bismuth, and also phosphorus may be considered. Representative boiling points of a range of these hydrides are listed in Table 5.2.

A number of analytical results for boron hydrides appeared in the 1950s and 1960s, one of the earliest being a separation of the borane homologs from diborane to pentaborane (34). Low column temperatures and an inert system were needed to minimise on-column degradation, but some interconversion of species was seen due to catalysis by the substrate. Chloroboranes such as B_2H_5Cl and $BHCl_2$ were chromatographed on highly

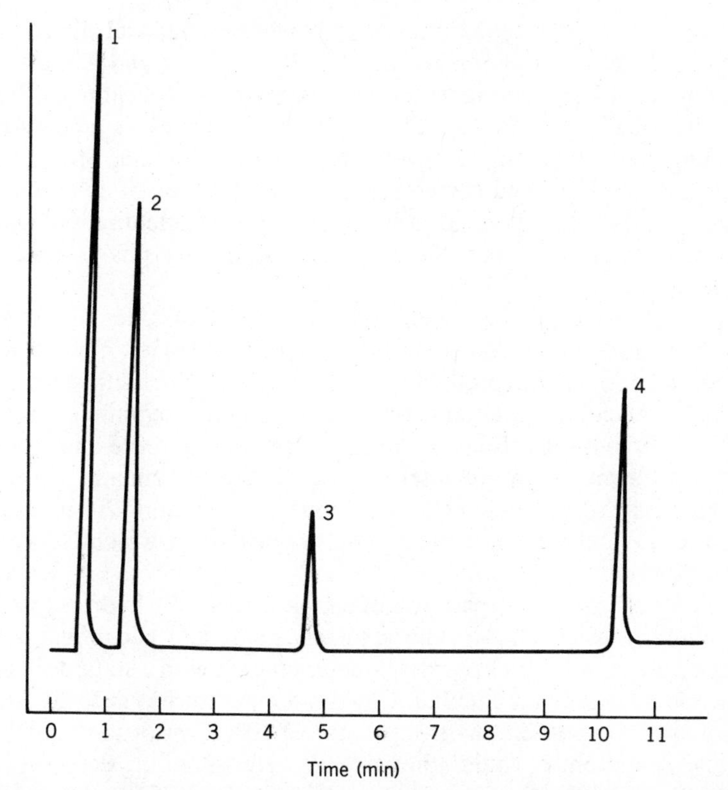

Figure 5.4. Separation of a boron hydride mixture. Column 6% of OV 17 on 80/100 Chromosorb W. Temperature program, 40°C isothermal for 2 min, 20°C/min to 180°C. Helium carrier gas at 80 ml/min. Peak identities: 1—B_2H_6; 2—B_5H_9; 3—B_6H_{10}; 4—$B_{10}H_{14}$. Reprinted with permission of E. J. Sowinski and I. H. Suffet, *Analytical Chemistry,* **44,** 2237 (1972). Copyright 1972 American Chemical Society.

inert substrates such as Teflon™, fluorocarbon oils and paraffins by Myers and Putnam (35). These and similar separation methods are attracting renewed attention in view of the need for trace-level determinations in electronic grade organometallics. Sowinski and Suffet achieved successful boron hydride separations with programmed temperature GC (36,37) (Fig. 5.4) and it seems possible that new inert capillary column materials may further aid such separations. The reactivity of aluminum, gallium, and indium hydrides suggests that subambient capillary GC may be the only feasible approach to their separation.

Silicon and germanium hydrides were also studied early in GC development. Phillips reported separation of silanes with up to eight silicon atoms on ester and silicone phases (38), while Feher and Strack chromatographed compounds up to $Si_{18}H_{18}$ (39). Phillips and Timms characterized silanes and germanes by plotting (40) the logarithm of the relative retention times and those of mixed siligermanes as a function of total combined silicon and germanium atoms in the chain. Straight lines of various slopes allowed correlation and identification of compounds of various classes. Pollard et al. utilized gas density detection for alkyltin mono- and dihydrides; only the most inert stationary phases precluded on-column reaction (41).

In an extensive study, Soderquist and Crosby (42) developed a GC method for the simultaneous determination of triphenyltin hydroxide and its potential degradation products, tetraphenyltin, diphenyltin oxide, benzenestannoicacid, and inorganic tin. The organotin compounds were converted to phenyltin hydrides by lithium aluminum hydride and were detected by electron capture after GC separation. Minimum detectable quantities for Ph_3SnH, Ph_2SnH_2, and $PhSnH_3$ were around 200 pg. Among the Group IVA elements, lead alone has hydrides for which GC has not proved possible.

An early example of separation of simple Group VA hydrides was the work of Zorin et al. (43,44) who resolved arsine and phosphine as well as silane and germane on a packed 8-meter column with a stationary phase of silicone oil coated on alumina. Column temperature was 30°C. Arsine, AsH_3 was decomposed to hydrogen, etc., at 1000°C prior to thermal conductivity detection; it would otherwise degrade within the detector channels as do some other organometallics; $Ni(CO)_4$ exemplifies this behavior also. More recently, hydrides of arsenic and antimony along with those of germanium and tin have been successfully resolved on porous polymer column phases at or above ambient temperatures. Such a separation obtained by Kadeg and Christian (45) is shown in Fig. 5.5. This study involved the utilization of established hydride generation procedures with a balloon trapping of analytes prior to introduction onto a Porapak Q™ column. Temperature programming proved very important in gaining complete resolution; atomic absorption spectroscopy was used successfully for detection, and analyte integrity was confirmed mass spectrally. Hydride determinations have been the focus of a number of recent investigations. Vinsjansen and Thrane reported procedures for the determination of trace phosphine in ambient air (46). A packed squalane column at 80°C proved effective and an alkali metal thermionic detector (NPD) used in the phosphorus specific mode permitted a detection limit of 0.03 parts in 10^9 and a useful working range of 0.5–5 parts in 10^9. Dumas and Bond carried out a similar study on Chromosorb 102™ to measure 0.01 ppm of phosphine after absorption on Tenax™ or Chromosorb 102 (47).

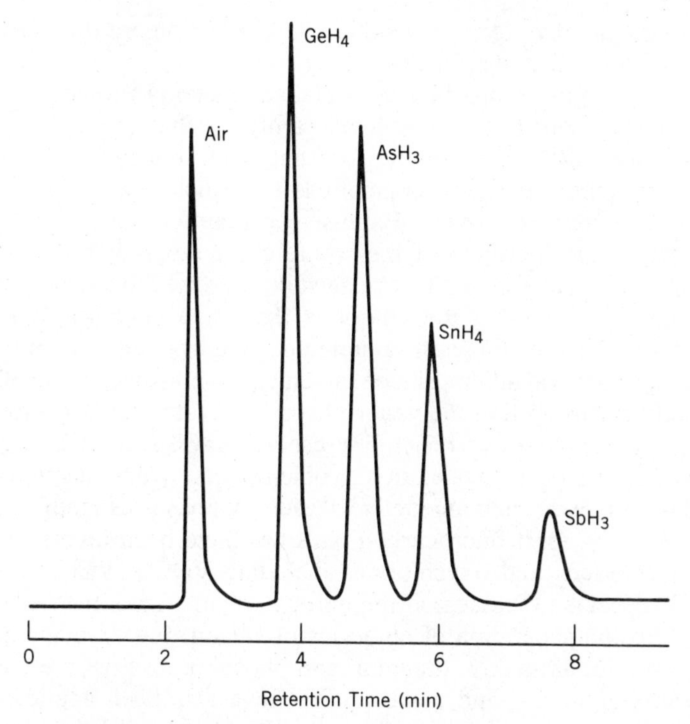

Figure 5.5. Separation of metalloid hydrides. Column 3 ft. Porapak Q. Temperature program from 75 to 120°C at 8°C/min (Reference 45).

The GC of metalloid hydrides has been aided by specific methods of detection and other novel procedures. Caruso et al. have used interfaced atomic absorption spectrophotometry (48) and microwave induced plasma atomic emission (49) detection for specific element monitoring of arsenic, antimony, selenium, germanium, and tin hydride separations on Chromosorb 102 columns. Further reference is made to these techniques later in this chapter. In a very innovative detection study, Gifford and Bruckenstein used a gas porous gold electrode detector for electrochemical detection of arsenic, antimony, tin, and mercury hydrides (50). Analysis of environmentally significant arsenic and antimony compounds with microwave plasma emission detection was also reported by Talmi et al. (51,52). For example, alkylarsonic acids were reduced with sodium borohydride to alkylarsines prior to quantitative GC separation. Reports of GC of hydrogen selenide and hydrogen telluride have been limited to that noted earlier (44,49) for the former. It seems reasonable that methods suited to quantitative hydrogen sulfide determination are the most likely to succeed for these compounds.

The other major group of binary metal and metalloid compounds to be gas chromatographed at normal temperatures is that of halides, certain of which are sufficiently volatile for analytical application. The major practical difficulty for these compounds often lies in high reactivity in vapor and condensed phases. Extensive precautions must be taken to ensure maximum inertness of the whole chromatographic and sample-handling system. In particular, care must be exercised to avoid hydrolysis during injection. Among the chlorides that have been gas chromatographed are those of titanium, aluminum, mercury, tin, antimony, germanium, gallium, vanadium, silicon, arsenic, and phosphorus, all of which are readily hydrolyzed in the vapor phase at elevated temperatures. The first reported analysis by Freiser (53) resolved $SnCl_4$ and $TiCl_4$. A general observation has been that elution problems arise from reaction of the chlorides with even such unreactive stationary phases as methyl silicone oils; frequently, inert fluorocarbon packings have been favored (54) for reactive chlorides and oxychlorides including $VOCl_3$, VCl_4, PCl_3, and $AsCl_3$. Parissakis (32) assessed the porous polymer stationary phase Porapak P for the separation of chlorides of silicon, tin, germanium, vanadium, arsenic, antimony, titanium, and phosphorus. Low melting inorganic salts, eutectics, and metal phases have also been applied in this field with considerable success; for example, Juvet and Wachi separated $TiCl_4$ and $SbCl_3$ on a column of $BiCl_3$-$PbCl_2$ eutectic to 240°C (55). Pommier et al. subsequently reported the separation of $NbCl_5$ and $TaCl_5$ at 444°C on a LiCl-KCl eutectic (56). Later Tohyama and Otazai (57) made an extensive study of eutectic and single chloride phases at column temperatures from 450–1000°C using electrical conductivity detection. The chlorides eluted ranged from $BiCl_3$ (B.P. 441°C) to $CdCl_2$ (B.P. 960°C) and included the zinc, thallium (I), and lead (II) compounds. A recent application of this type of analysis involved determination of traces of plutonium in soil (58). The matrix was decomposed by hydrofluoric acid and the residue chlorinated to either $PuCl_3$ or $PuCl_4$; alpha spectroscopic detection was used. The mechanism of separation operating in these cases is believed to involve the formation of chloro complexes with free chloride ions present in the liquid phase.

The more volatile silicon tetrahalides of chlorine and bromine may be effectively separated on silicone oils coated on a polytetrafluorethylene support (59). Hydrogen carrier gas and a thermal conductivity detector allowed detection at low microgram levels. Crompton (60) provides a comprehensive survey of chlorosilane separations; a typical example is that reported by Green (61) shown in Fig. 5.6 of a range of polychlorod-

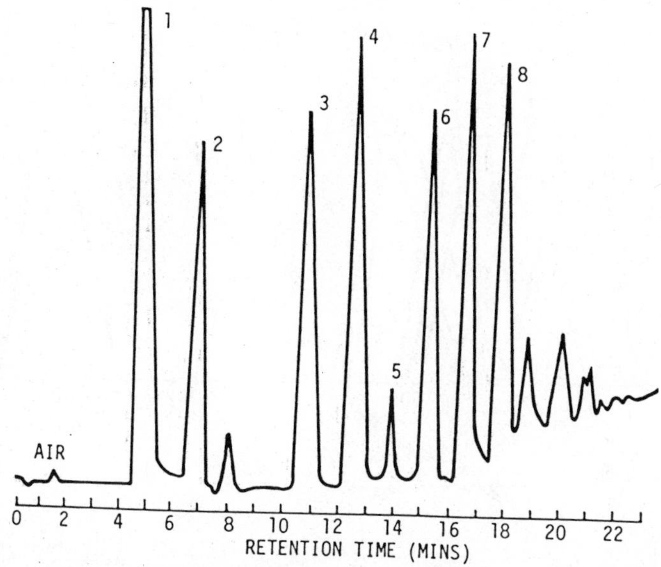

Figure 5.6. Separation of chlorosilanes and chlorosiloxanes. Column 10% SF-96. Peak identities; 1—trichlorosilane; 2—tetrachlorosilane; 3—1,1,3,3-tetrachlorodisiloxane; 4—1,1,1,3,3-pentachlorodisiloxane; 5—hexachlorodisiloxane; 6—1,1,2,3,3-pentachlorotrisiloxane; 7—1,1,1,2,3,3-hexachlorotrisiloxane; 8—1,1,1,2,3,3,3-heptachlorotrisiloxane (Reference 61). Reprinted with permission from L. F. Green, L. J. Schmaush and J. C. Worman *Analytical Chemistry,* **36,** 1512 (1964).

isiloxanes and trisiloxanes. In contrast, the less volatile metal bromides provide a very challenging task; high-temperature stationary phases are generally necessary. The most comprehensive study is that of Tsalas and Bächmann (62), who used alkali bromide salts coated on silica as stationary phases. The use of bromine/nitrogen and boron tribromide/bromine/nitrogen mobile phases enabled quantitative elution and separation of the bromides of zinc, niobium, molybdenum, technetium, indium, antimony, tin, bismuth, tellurium, and iodine. Detection was by means of gamma-ray spectroscopy of radiochemically labeled eluates. Although not all of the bromides could be completely resolved in a single experiment, adequate separations could be obtained by variation of elution parameters. Typical chromatographic profiles are shown in Fig. 5.7.

Some metal fluorides have low boiling points; examples are those of tungsten (17.5°C), molybdenum (35°C), tellurium (35.5°C), rhenium (47.6°C), and uranium (56.2°C), for which low column temperatures are feasible. Thus Juvet and Fisher (63) obtained good peak shapes and separation of the above fluorides (excepting tellurium) at 75°C on a column

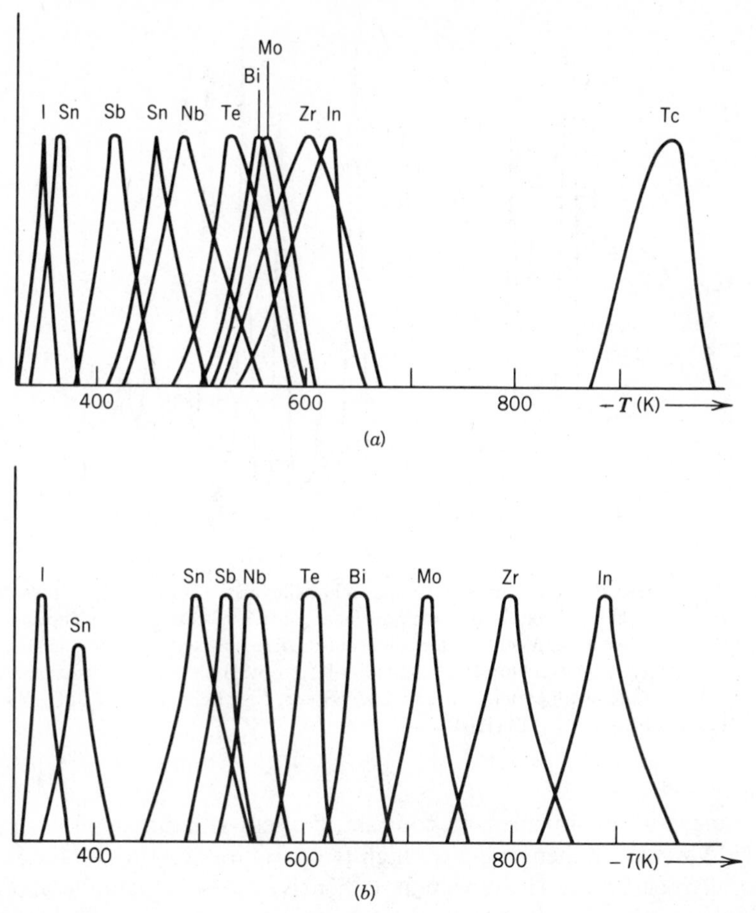

Figure 5.7. Separation of inorganic bromides. Carrier gas; 60 mm Hg of Br_2 and 4 mm Hg of BBr_3 in N_2. Stationary phases (*a*) NaBr; (*b*) KBr on quartz granules. Temperature program 400–900°C (Reference 62). Reprinted with permission from S. Tsalas and K. Bačhmann, *Talanta,* **27**, 201 (1980). Copyright 1980 Pergamon Press, Ltd.

of Kel-F oil™ coated on Chromosorb T™ (Teflon™). Gas solid chromatography has also been examined on alkali metal fluoride and alumina substrates, although Kel-F modification was found helpful in improving column efficiency (33,64). Total absence of water was vital in all such studies in order to preserve columns, detectors, and other equipment. The determination of alloys and metal oxides, carbides, etc. after conversion to their analogous fluorides by fluorination appears feasible (63). Kel-F was also used as a stationary phase for the GC of a range of fluorinated compounds including ClF, ClF_3, OF_2, ClO_4F, ClO_3F, and UF_6 (65).

One of the most extreme modifications of conventional gas chromatographic concepts has been in the area of "thermochromatography," very high (column) temperature GC. The applications for chlorides—and particularly bromides, mentioned earlier—lead to consideration of this type of separation (62), but perhaps the most striking examples are those for metal oxides, hydroxides, and oxychlorides. Bächmann and colleagues have been the principal investigators of these compound classes; they have employed temperatures as high as 1500°C for the separation of oxides and hydroxides of technetium, rhenium, osmium, and iridium (66). Quartz granules were used as substrate and oxygen and oxygen/water mixtures as carrier gases with the necessary equipment modifications. Silica gel and graphite were employed as column packings in another study at temperatures to 900°C (67). The mechanisms for these separations are by no means certain, since a differential volatilization/distillation process may predominate over adsorption or partition mechanisms. Since very carefully constructed instrumentation is required for these applications, it is likely that their application will be mainly in physicochemical rather than analytical studies.

5.5 ORGANOMETALLIC COMPOUNDS

The fact that organometallic compounds have attracted a considerable amount of gas chromatographic study is stressed by Crompton (60), who in the introduction to his 1981 text, *Gas Chromatography of Organometallic Compounds*, notes that during the preceding decade more than one thousand papers were published on this topic, relating to compounds of more than 50 elements. The reader is referred to Crompton's study for an exhaustive survey of this topic, and only a selection of the more significant examples is discussed here.

By definition, organometallic compounds are those which contain at least one metal to carbon bond, other than metal carbides. Conventionally, metal carbonyls and their derivative compounds are included. Metal complexes chelated with ligands through oxygen, nitrogen, sulfur, or phosphorus donor atoms are not included, but are considered subsequently in a separate section. There are various ways by which organometallics may be classified: according to the ligand type, period metal group, or bonding type(s). Perhaps the most useful of these ways, from the gas chromatographic standpoint, is to define as main group those organometallics with primarily sigma carbon-metal bonds, and as transition group those compounds with mainly pi carbon to metal bonding.

5.5.1 Sigma-Bonded Compounds

The elements forming sigma-bonded organometallics that have been gas chromatographed or for which GC may be feasible are B, Al, Ga, and In from Group III; Si, Ge, Sn, and Pb from Group IV; (P), As, Sb, and Bi from Group V; and Se and Te from Group VI. Also included are Hg and possibly Zn and Cd. Groups I and II form no sigma-bonded organometallics of adequate stability for GC. The elements in this group that have attracted most analytical interest have been silicon and lead, with tin, arsenic, and mercury also being prominent.

The organic functionalities bonded to these metals and metalloids are most commonly alkyl, aryl, and substituted aryl groups; perfluoroalkyl and aryl groups, which typically impart enhanced volatility; and mixed alkyl-chloro and aryl-chloro systems. Alkoxy groups also are common for silicon and in this group may be placed siloxanes, stannoxanes, carboranes, and compounds containing such groups as $R_3Si\text{-}S\text{-}SiR_3$, the chalcogenides.

5.5.2 Group III Sigma-Bonded Compounds

Organoboron Compounds

Although they are considerably more reactive than some other metalloid alkyls, alkylboranes were the subject of early GC separation studies. Schomberg et al. (68) investigated alkyl group redistribution reactions between triethyl, tri-*n*-propyl, and tri-isobutyl boranes, while Koster et al. (69) extended the study to alkyl group interchange among trialkylboranes and the analogous organoaluminum compounds although only the former compounds could be chromatographed. Gas chromatography was also used to investigate alkylborane reactions catalyzed by diborane (B_2H_6) (70). The reactive hydrogen-bridged alkyldiborane derivatives and isomers could be resolved at ambient temperatures. Higher alkylborane homologs up to decaborane derivatives have also been examined (71). As with many other areas of GC analysis, it appears likely that modern inert capillary columns may play a role in renewed study of such compounds.

A group of compounds that has had a notable impact on gas chromatography is the Dexsil™ range of carborane-silicone polymers, whose high temperature stability has enabled their use as stationary phases to 400°C and above. The electron-rich carbon boron cage structure of the carborane moiety contributes to this high stability. The GC characteristics and heats of solution in a variety of stationary phases have been measured for six carborane monomers of different structure (72).

Organoaluminum and Organogallium Compounds

The thermal and chemical stability of the alkyl derivatives of Group III elements decreases with increasing atomic number; very little effective GC has been reported for metal alkyls in this group and none for other carbon-bonded compounds. Nevertheless, interest in their chemistry is increasing, in view of applications in such areas as polymerization catalysis (aluminum alkyls) and electronic material precursors (gallium and indium alkyls), and thus there is a clear present need for reexamination of their GC applicability.

The two reported successful separations of aluminum alkyls involved somewhat different approaches. Bortnikov et al. (73) used a conventional packed methyl silicone column at 110°C with thermal conductivity detection. It was not clear what detection limits could be achieved or whether any decomposition occurred on-column. Longi et al. (74) used a reduced pressure system with an outlet pressure of 350 torr, the objective being to decrease compound decomposition during elution. The stationary phase was a paraffin wax/triphenylamine mixture (17:3). Triethylaluminum was claimed to be eluted, but there was extensive peak tailing indicative of undesirable column interactions. Specific element monitoring of eluate would be helpful in further investigations of such challenging separations. Trimethylgallium was also reported to be eluted by Bortnikov (73) and by Fukin et al. (75), who used similar GC conditions. The latter investigators identified $(CH_3)_2GaCl$, CH_3GaCl_2, and $(CH_3)_2AlCl$ by means of post-column analysis.

No organometallic indium or thallium compounds have been reported to be gas chromatographed.

5.5.3 Group IV Sigma-bonded Compounds

The largest amount of gas chromatography of sigma-bonded organometallics has been reported for compounds of Group IVA elements. The very extensive application of GC in organosilicon chemistry is made evident by Crompton, who devotes more than 160 pages in his text (60) to alkyl and aryl silane and siloxane separation and reaction analysis. Tetraalkylsilanes have been chromatographed frequently, their stabilities and elution behavior paralleling alkane hydrocarbons. Thus logarithmic retention plots are typically linear with respect to compound parameters of boiling point and molecular weight with modification depending on molecular symmetry. Good retention linearity was noted for symmetrical tetraalkylsilanes from tetramethyl (C_4) to tetra(*n*-octyl) (C_{32}) (76). Figure 5.8 depicts a capillary column separation of an alkyl group redistributed

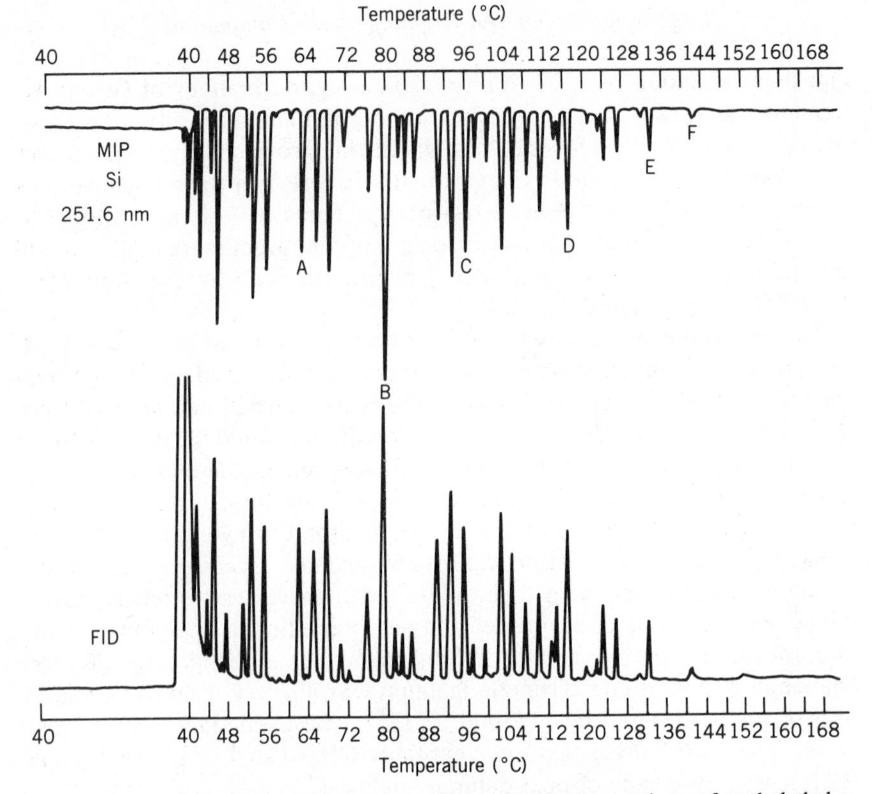

Figure 5.8. Dual-detection chromatogram of the redistribution products of methylethyl-*n*-propyl-*n*-butylsilane. Column 100 m × 0.25 mm i.d. OV 225 glass capillary support-coated open tubular column. Temperature program 40–170°C at 4°C/min. Lower chromatogram, FID; upper chromatogram MIP for silicon at 251.6 nm. Peak identities; A—methylethyldi-*n*-propylsilane; B—methylethyl-*n*-propyl-*n*-butylsilane; C—methylethyl-di-*n*-butylsilane; D—ethyl-*n*-propyl-di-*n*-butylsilane; E—*n*-propyltri-*n*-butylsilane; F—tetra-*n*-butylsilane (Reference 77). Reprinted with permission from S. A. Estes, C. A. Poirier, P. C. Uden, and R. M. Barnes, *Journal of Chromatography,* **196,** 265 (1980). Copyright 1980 Elsevier Science Publishers.

tetraalkyl silane mixture of 35 compounds formed by the aluminum chloride catalyzed reaction of methylethyl-*n*-propyl-*n*-butylsilane (77).

Dual detection is by a flame ionization (FID) and a microwave induced and sustained helium plasma (MIP) with monitoring at 251.6 nm for silicon. The upper trace also shows a second higher sensitivity trace for the earliest eluting silanes, $(CH_3)_4Si$ and $(CH_3)_3SiC_2H_5$. Statistical alkyl group interchange predicts peak-area ratios of 12, 24, 12, 12, 4, and 1 for

$CH_3C_2H_5Si(C_3H_7{}^n)_2$, $CH_3C_2H_5SiC_3H_7{}^nC_4H_9{}^n$, $CH_3C_2H_5Si(C_4H_9{}^n)_2$, $C_2H_5C_3H_7{}^nSi(C_4H_9{}^n)_2$, $C_3H_7{}^nSi(C_4H_9{}^n)_3$ and $Si(C_4H_9{}^n)_4$; these ratios are clearly observed in the chromatogram.

The group substitution parameters for tetra-substituted organosilicon compounds were very carefully considered by Semylen and Phillips (78) for the alkylsilanes and more recently by Peetre and Smith (79,80) for alkyl and alkoxysilanes. Very precise instrument control afforded highly reproducible measurements; equations were developed to predict retention indexes for different systems.

The literature of alkyl, aryl, and vinyl chlorosilane GC is very extensive. The great importance of methylchlorosilanes in silicone production ensured that methods for their qualitative and quantitative separation were thoroughly investigated. The facile hydrolysis of chlorosilanes to siloxanes necessitates a careful choice of stationary phases and column conditions. Silicone oils have been generally preferred, sometimes in combination with phosphate or phthalate esters (81). Thermal conductivity detectors give similar sensitivities for the chlorosilanes with different ratios of methyl to chlorine groups unlike flame ionization detectors. Burson and Kenner (82) demonstrated the quantitative determination of methylchlorosilanes at the 100 ppm level and below in silicon tetrachloride and trichlorosilane. In a similar manner the GC of phenylchlorosilanes has been successfully accomplished, temperatures above 200°C being needed (83). Wurst and Churacek (84) presented retention data for alkyl and arylchlorosilanes, vinyl chlorosilanes, and many substituted siloxanes.

Alkylgermanes and Alkylstannanes

The GC characteristics of alkyl compounds of germanium and tin are similar to those of silicon, although there is some tendency towards decreasing thermal stability, which becomes more pronounced for the alkyllead compounds. Phillips et al. (40,78) and Pollard, Nickless, and Uden (76) carried out extensive studies of alkylgermanes deriving from alkyl redistribution of groups as large as *n*-butyl; retention behavior could be predicted effectively. These studies were extended to mixed trialkylchlorine, bromine and hydrogen compounds but greater deviations from predicted retention was observed (85). Bortnikov et al. (86) separated a range of germanium compounds which had bonds between germanium, tin, silicon, and sulfur heteroatoms; they used Apiezon L and polyethylene glycol stationary phases and also employed gas solid chromatography on graphitized carbon. The chromatographic properties of alkylstannanes are similar to the germanes except that greater care must be excercised to prevent on-column oxidation, hydrolysis, or thermal degradation. Pollard et al. (41,87) and Putnam et al. (88) carried out extensive studies of alkyl, vinyl, and aryl stannanes, including the reaction products of trimethylstannane. An "inert" gas density balance detector proved suitable for detection of alkyl tin mono and dihydrides $(CH_3)_3SnH$, $(CH_3)_4Sn$, and $(C_3H_7^n)_2SnH_2$ being resolved although there was the possibility of on-column hydrogen abstraction from hydrides to form tetraalkylstannanes (41). Redistribution reactions of substituent groups on tin and among adjacent Group IVA elements have been investigated (76,77), plasma emission specific element detection proving helpful in examining reaction products.

The range of GC study of organostannanes has greatly exceeded that of the organogermanes. The butylbromostannanes were analyzed quantitatively as methylbutylstannanes after derivatization with a methyl Grignard reagent (89), and perfluorovinylchlorostannanes were chromatographed at 60°C (90). Preparative GC of tetra-alkylstannanes gave fractions with as little as $10^{-3}\%$ of impurities (91). As noted earlier (42) the degradation chemistry of triphenylhydroxystannane was studied extensively by GC.

Selective detection has been used with advantage for organotin compounds. Flinn and Aue (92) utilized flame photometric detection (FPD) and gas phase luminescence for organogermanium and tin analysis. The response factor found for these elements was as much as six orders of magnitude greater than for hydrocarbons. Kapila and Vogt (93) carried out a study of tetraalkylstannanes with FPD detection, monitoring in both blue and red spectral regions. The former (360–450 nm) was broad and was influenced by siliceous surfaces in the flame area; the latter (610 nm) was less affected. For optimal performance a 610 nm optical filter without quartz enclosure was recommended. Hill has made an extensive study of the hydrogen atmosphere flame ionization detector (HAFID) for selective detection of metal compounds. The HAFID, which is readily set up as a modification of a standard FID, possesses selectivity of three orders of magnitude or more over carbon while retaining typical FID sensitivity (94). Applications of this system included GC of bis(tributyltin)oxide (TBTO) in marine paints, triphenylhydroxystannane pesticide after derivatization to triphenylmethylstannane, and tricyclohexylhydroxystannane (Plictran™) in apple leaves after derivatization to tricyclohexylbromostannane. An example of the latter determination is shown in Fig. 5.9, which depicts FID and HAFID detection of a dilute Plictran sample and a tricyclohexylbromostannane standard. Another interesting application of tetraalkyltin chemistry in GC was noted by Rodriguez-Vazquez

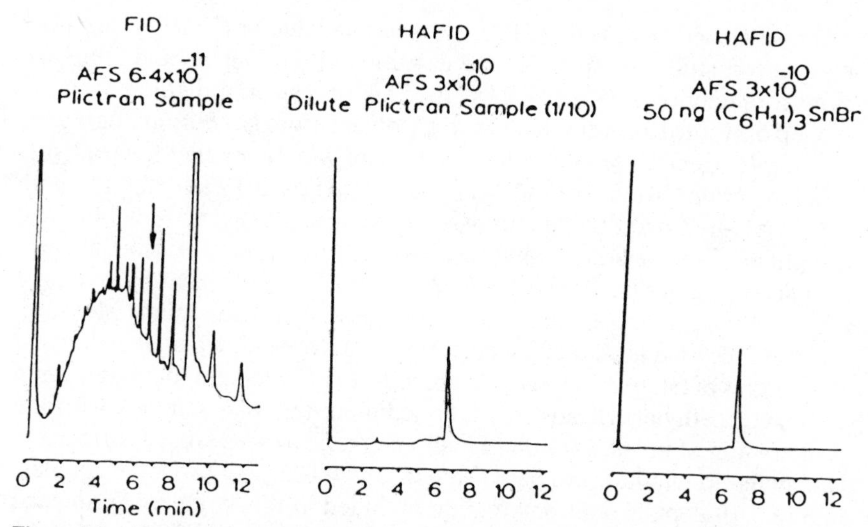

Figure 5.9. Hydrogen atmosphere flame ionization (HAFID) detection of "Plictran"™ as tricyclohexyltinbromide. Column 6 ft, 3% SE 30 (Reference 94). Reprinted with permission from D. R. Hansen, T. J. Gilfoil, and H. H. Hill, *Analytical Chemistry,* **53,** 857 (1981). Copyright 1981 American Chemical Society.

(95) who used tetramethylstannane to derivatize mercury as methylmercuric chloride prior to GC analysis.

Alkyl Lead Compounds

The literature on the GC of tetraalkyl lead compounds is extensive and covers some 20 years of the developments in analysis of these environmentally significant compounds. Crompton (60) devotes 90 pages to discussion of organolead GC. These compounds are generally more toxic than inorganic lead compounds, the degree of toxicity varying with the degree of alkylation. The question of bioalkylation of inorganic to organic lead in the aquatic environment by microorganisms remains one of considerable debate (96,97). GC separation of tetramethyllead (plumbane) from analogous Group IV compounds was reported in 1960 by Abel et al. (98) at 80°C on Apiezon L. An early selective method for GC determination of alkylleads in hydrocarbon matrixes such as gasoline was that of Parker and Hudson (99), who absorbed eluate in iodine solution and used quantitative spectrophotometric measurement with dithizone. Lovelock and Zlatkis (100) showed the electron capture detector to be highly selective over hydrocarbons for tetraalkylleads, but detector contamination proved to be a major problem.

The hydrogen atmosphere flame ionization detector (HAFID) has been used successfully for specific lead detection (101). Dupuis and Hill (102) showed detection limits of 7.2 $\times$ 10^{-12} g/s of lead in a method that employed only a 1:10 dilution of leaded gasoline before gas chromatography.

Atomic spectral detection has been quite widely applied for lead specific detection. Mutsaars and Van Steen (103) used FPD detection with an oxygen-hydrogen flame and measurement at 405.8 nm; they obtained a rectilinear response for alkylleads from 10–1000 ppm. Atomic absorption detection has been much favored, flame AA being used by Ballinger and Whittemore (104), who also used pressure programming. Measurement at 283.3 nm gave a 20 ng detection limit for lead. Chau et al. carried out a series of GC-AAS studies on procedure development for quantitation of mixed methylethylleads in water sediment and fish samples with the primary aim of investigation of bioalkylation processes (105, 106); typical lead detection limits were between 0.01 and 0.025 μg/g for solid samples. Figure 5.10 depicts a chromatogram obtained in this study wherein each peak corresponds to 5 ng of the tetraalkyllead compound.

Radzuik et al. (107) also used GC–AAS for the determination of alkylleads in air, with improved detection results. They used graphite furnace electrothermal atomization at 1500°C, sampling being for 70-liter air volumes condensed at −72°C prior to chromatography. The effluent was transferred to the furnace through Teflon-lined aluminum tubing. An alternative tantalum connector allowed furnace operation to 2500°C, but the lower temperature proved best for quantitative analysis. Typical total alkyllead concentrations were found to be around 70 ng per cubic meter of sampled air. Robinson et al. (108) have also used GC-GFAAS with direct interfacing of the column into the furnace atomizer. In view of the extensive application of AAS detection for alkyllead GC this method appears well established.

With the increasing interest in plasma atomic emission detection for specific GC, applications for alkylleads have been noteworthy. Such detectors are discussed in more detail later. The primary mode of plasma employed has been the microwave induced and sustained helium plasma (MPD, MIP, or MED), the energetics of which facilitate excitation and emission from both metals and nonmetals. Early applications focused on reduced pressure plasma, and Reamer et al. (109) reported use of a $\frac{3}{4}$ Wave Evensen cavity as used by McCormack (110) for atomospheric alkyllead determinations. Air sampling was carried out in diverse locations and the determined levels of tetramethyllead ranged from 7–650 ng/m^3, the highest being, predictably, in automobile exhaust. The atmospheric pressure microwave cavity devised by Bennakker (111) permits sample direct GC interfacing and was explored by Quimby et al. (112), who obtained a

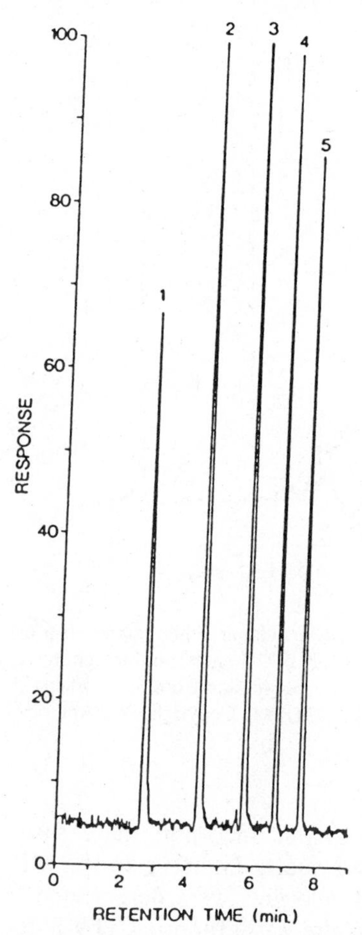

Figure 5.10. Atomic absorption detection of alkyl lead compounds. 5 ng of lead. Peak identities: 1—Me_4Pb; 2—Me_3EtPb; 3—Me_2Et_2Pb; 4—$MeEt_3Pb$; 5—Et_4Pb (Reference 106). Reprinted with permission from Y. K. Yau, P. T. S. Wong, G. A. Bengert, and O. Kramar, *Analytical Chemistry,* **51,** 186 (1979). Copyright 1979 American Chemical Society.

detection limit of 0.49 pg/s of lead in tetraethyllead, measured at 283.3 nm. The selectivity of this plasma was emphasized in the determination of mixed alkylleads in different gasolines (113), a detection limit of 0.17 pg/s and a selectivity over carbon of 246,000 being reported.

The high toxicity of tetraalkylleads is attributable to their ability to undergo the following decomposition in the environment: $R_4Pb \rightarrow R_3Pb^+ \rightarrow R_2Pb^{2+} \rightarrow Pb^{2+}$. The formation of alkyllead salts such as chlorides, carbonates, or nitrates, probably associated with proteins, arising in tissues from rapid metabolic dealkylation of tetraalkylleads is of toxicological importance. The gas chromatographic determination of trialkyllead as the chloride is difficult because of its reactivity; however, Estes et al.

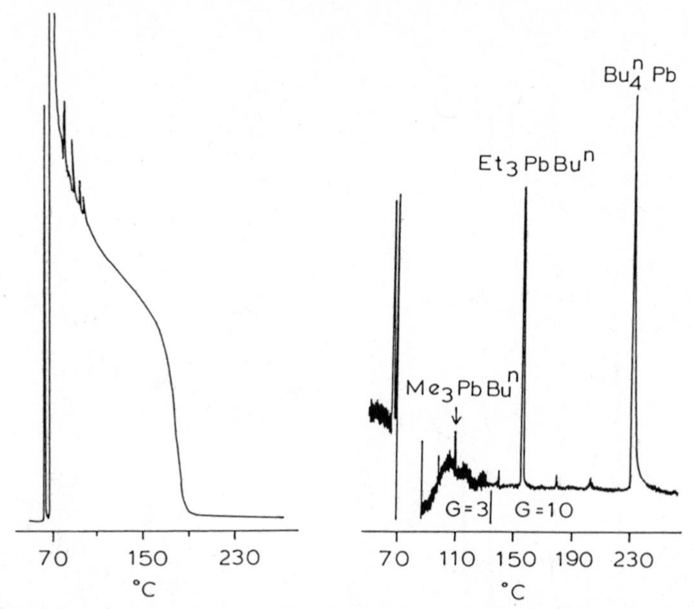

Figure 5.11. Microwave induced plasma (MIP) detection. 300-microliter sample of *n*-butylated benzene extract from industrial plant effluent. (*a*) 247.9 nm carbon detection; (*b*) 283.3 nm lead detection (Reference 115). Reprinted with permission from S. A. Estes, P. C. Uden, and R. M. Barnes, *Analytical Chemistry,* **54,** 2402 (1982). Copyright 1982 American Chemical Society.

devised a sufficiently inert capillary GC system incorporating a fused silica silicone oil column and a deactivated quartz interface to the MIP detector (114). Trimethyl and triethyllead chlorides were determined in water in the range of 10 ppb to 10 ppm. An alternative approach by Estes et al. (115) involved derivatization by butyl Grignard reagent to form the trialkylbutylleads. A precolumn Tenax™ trap enrichment procedure enabled low ppb determinations of the trialkylead moiety to be carried out. Figure 5.11 illustrates the high selectivity for lead of the MIP detector, the trimethyllead ion being estimated as present in the industrial plant effluent at the 5 ppb level.

5.5.4 Group V Sigma-Bonded Compounds

In general, the alkyl and aryl derivatives of arsenic and antimony are more labile than those of group IV elements, and require carefully controlled GC conditions for successful elution. Bismuth compounds have not been chromatographed successfully. Gudzinowicz et al. chromatographed methyl, ethyl and butylbromoarsines, trimethyl, triphenyl and trivinyl arsine, although high column temperatures were needed for phenyl derivatives, up to 290°C (116). On column reactions were investigated and retention data relationships were set up. Parris et al. used GC-GFAAS to determine trimethylarsine in gases generated microbiologically (117). A packed silicone column operated at 40°C was suitable for this analysis. Schwedt and Ruessel (118) determined arsenic in biological materials by derivatization to triphenylarsine; 2 ppm of arsenic could be determined. Perfluorinated organoarsenic compounds may also be successfully chromatographed. Gudzinowicz and Driscoll reported quantitative separation of such compounds as ethylbis(trifluoromethyl)arsine and phenylbis(heptafluoropropyl)arsine; fluorine substitution increases volatility and lowers the column temperature needed (119). In general, sigma-bonded organoarsenicals should be quite readily eluted on insert fused silica capillary columns. Although much less GC has been carried out for organoantimony compounds, the good peak characteristics found for triethylantimony (74) suggested that alkylantimony compounds should be amenable to quantitative separation. Talmi and Norvell (52) determined As^{3+} and Sb^{3+} in environmental samples by cocrystallization with alpha-mercapto-N,2-naphthylacetamide followed by reaction of the precipitate with phenylmagnesium bromide to form triphenylarsine and triphenylstibine. GC separation was followed by low pressure microwave plasma detection at 228.8 nm and 259.8 nm, respectively, giving detection limits of 20 pg and 50 pg. Alkylarsenic acids in commercial pesticides and environmental samples were also determined as the corresponding alkylarsines generated by sodium borohydride reduction (51).

5.5.5 Group VI Sigma-Bonded Compounds

The chromatography of dialkylselenides and dialkyldiselenides was studied by Evans and Johnson (120), the latter compounds showing high response to electron capture detection. Thus dipropyldiselenide had a ECD/FID response ratio 1.5×10^4 times higher than dimethylselenide. Parris et al. (117) used GC-GFAAS for the determination of dimethylselenide, as did Talmi (121), who used GC-MIP, and Estes et al. (122), who noted a detection limit of 62 pg/s for selenium. One of the rare records of GC of organotellurium compounds is that of Bortnikov et al., who chromatographed bis(triethylsilyl)telluride (123).

Alkyl and Aryl Mercury Compounds

The GC of these environmentally significant organometallics has attracted similar attention to that of organolead compounds. In addition to their agricultural uses, biomethylation of inorganic mercury is a significant natural process. Such compounds as alkylmercuric bromides have also found synthetic utility.

Packed column GC was used by Baughman et al (124), diethyleneglycolsuccinnate stationary phase being preferred for methylmercurials and methylsilicones for phenylmercurials. $(CH_3)_2Hg$ and $(C_6H_5)_2Hg$ were stable, but some decomposition of organomercury halides was shown by GC-MS. Bache and Lisk measured $(CH_3)_2Hg$ and CH_3HgCl on short packed glass columns using MIP detection (125), such columns being generally preferred for organomercurials. Risby and Talmi note in their review of MIP detectors for GC (126) that a GC-MIP system has been in regular use for many years in routine determination of methylmercurials (52) with subpicogram detection limits.

Nonflame "cold vapor," atomic absorption spectrophotometry at 254 nm has proved to be a useful procedure; Longbottom (127) converted organomercurials to elemental mercury catalytically over Cu_2O at 800°C or in a flame ionization detector. Detection limits down to 20 pg were obtained for dimethyl, diethyl, dipropyl, and dibutylmercury. Another novel spectroscopic detector found to be effective for organomercurials is the Atmospheric Pressure Active Nitrogen Afterglow System (APAN) described by Rice et al. (128). A detection limit for dimethylmercury of 2 pg is reported and a typical chromatogram is shown in Fig. 5.12 for dialkylmercurials at the 2–5 ng level.

Shaviat (129) converted dialkylmercurials to alkylmercuryiodides in order to utilize the sensitivity of the electron capture detector (ECD) for iodine-containing compounds.

Experimental procedures for the determination of environmental organomercurials are extensively reviewed by Crompton (60), with emphasis on procedural development in this ara by Westoo (130). Diphenylmercury and phenylmercury halides have been quantitatively determined on a diethyleneglycol adipate column, quantitative conversion of phenylmercuric nitrate to the chloride being utilized for trace chloride measurement (131). Care must be taken to avoid disproportionation of the chloride to diphenyl mercury in the hot GC injection port.

Alkylation and Arylation Reagents

A summary of reagents and substrates that have been used to generate volatile alkyl and aryl metal or metalloid derivatives suitable for GC de-

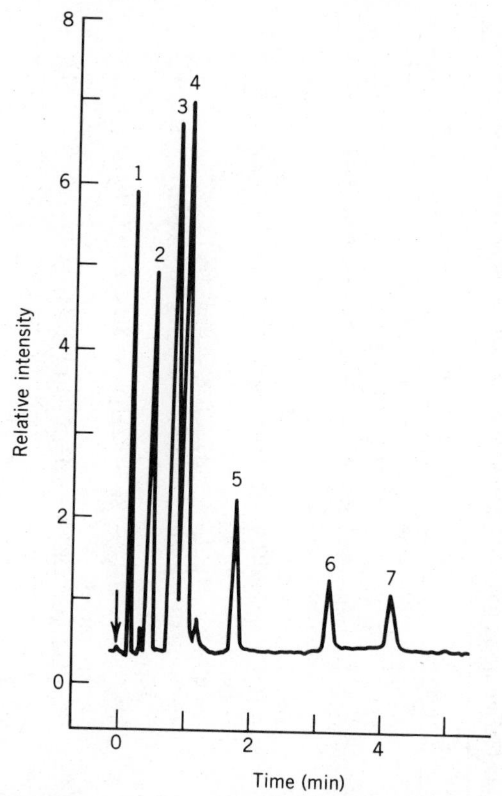

Figure 5.12. Atmospheric Pressure Active Nitrogen Afterglow (APAN) detection for organomercury compounds. Column 4 ft. 4% SE30/6%Sp2401 on 100/120 Supelcoport. Temperature program, 50–200°C at 32°C/min. 2–5 ng peak levels. Peak identities: 1—$(CH_3)_2Hg$; 2—$(C_2H_5)_2Hg$; 3—$(C_3H_7)_2Hg$; 4—$(C_4H_9)_2Hg$; 5—$(C_6H_{13})_2Hg$; 6—$(C_6H_5)_2Hg$ (Reference 128). Reprinted with permission from G. W. Rice, J. J. Richard, A. P. D'Silva, and V. A. Fassel, *Analytical Chemistry*, **53**, 1919 (1981). Copyright 1981 American Chemistry Society.

termination is given by Schwedt (4). A version of this summary is seen in Table 5.3.

Metal and Metalloid Alkoxides and Oxy-salts

Although they are not formally classified as organometallics, many alkoxides share similar GC characteristics with alkyl compounds. The siloxanes and silicones, the GC of which is very widely established, will not be discussed here; the area is well reviewed by Crompton (60) and in the text by Smith (139) on the analysis of silicones. Mention may be

Table 5.3 Alkylation and Arylation Reagents for Gas Chromatography of Organometal and Organometalloid Compounds

Reagent	Derivatization	Ref.
Magnesiumdiphenyl	As diethyldithiocarbamate to triphenylarsane	118
Tetramethyltin	$Hg^{2+} + (CH_3)_4Sn \rightarrow CH_3HgCl + (CH_3)_3SnCl$	132
Phenylmagnesium bromide	Se-, Te-, Hg-, As-, Sb-, Bi diethyldithiocarbamates to element phenyls	133
Phenylmagnesium bromide	As-, Sb 'thionalid' to triphenyl cpds.	52
Sodium tetraphenylborate	$4\ HgCl_2 + NaBPh_4 + 3\ H_2O \rightarrow 3\ HCl + NaCl + B(OH)_3 + PhHgCl$	134, 135
Phenylsulfinic acid	$PhSO_2H + HgCl_2 \rightarrow PhHgCl + SO_2 + HCl$	136
Lithium Pentafluorobenzenesulfinate	$Ph - SO_2Li + Hg^{2+} \rightarrow (Ph - SO_2Hg^+) \rightarrow SO_2 + Ph - Hg^+$	137
4,4-Dimethyl-4-silapentane 1-sulfonate	$Hg^{2+} + (CH_3)_3 - Si - (CH_2)_3 - SO_3Na$ to CH_3Hg derivatives	138
Pentacyanomethylcobaltate (III)	$Hg^{2+} + (Co(CN)_5 - CH_3)^{3-}$ to $HgCH_3^+$	135

From *Chromatographic Methods in Inorganic Analysis,* G. Schwedt, (4), pg. 53).

made, however, of some other separations which have been carried out. Group IV elements other than silicon: that is, germanium, tin, titanium, zirconium, and hafnium, form stable volatile alkoxides, as do some Group III elements, notably aluminum. Brown and Mazdiyasni (140) investigated the isopropyl and tert-pentyl alkoxides ($M(OR)_n$) on Teflon™ columns lightly loaded with Apiezon or silicone oil phases. Great care was necessary to exclude moisture. Some separations were achieved although column efficiencies were low; mass and infrared spectroscopy confirmed elution of the eluted compounds undecomposed.

There have been GC separations reported of oxycarboxylate salts of beryllium and zinc. The former have been extensively studied; Barrett et al. (141) showed beryllium oxyacetate, $Be_4O(COOCH_3)_6$ to elute satisfactorily from an Apiezon L column at 150°C but chromatography of the zinc analog and beryllium oxypropionate was less effective. Cardwell and Carter (142) extended this study to monitor solution reactions between beryllium oxyacetate and oxypropionate, noting resolution of mixed oxycarboxylates.

5.5.6 Pi-Bonded Compounds

The classes of transition metal organometallics that have received most gas chromatographic attention have been those having carbonyl, arene, and cyclopentadienyl ligand moieties. A notable feature of these compounds in general has been that their elution characteristics have been very favorable with only rare on-column degradation or adsorption in most instances.

Metal Carbonyls and Related Compounds

An effective separation of $Fe(CO)_5$, $Cr(CO)_6$, $Mo(CO)_6$, and $W(CO)_6$ was reported by Pommier and Guiochon (143). They employed the hydrocarbon phases, Apiezon L and squalane at 90°C. It seems likely that satisfactory methods may be employed for the highly toxic $Ni(CO)_4$ species, since its high volatility will permit elution at low column temperatures. Ultratrace determination by electron capture detection appears feasible.

Many more results have been reported for arene, cyclopentadienyl, and related derivatives of metal carbonyls. Among early studies were those of Veening et al. (144), who resolved alkyl substituted derivatives of benzenechromiumtricarbonyl. Segard et al. developed these separations further (145), obtaining effective separation of isomeric compounds. In Fig. 5.13 there is shown a typical separation obtained by VandenHeuvel et al. (146), who used GC–Mass Spectroscopy for eluate confirmation.

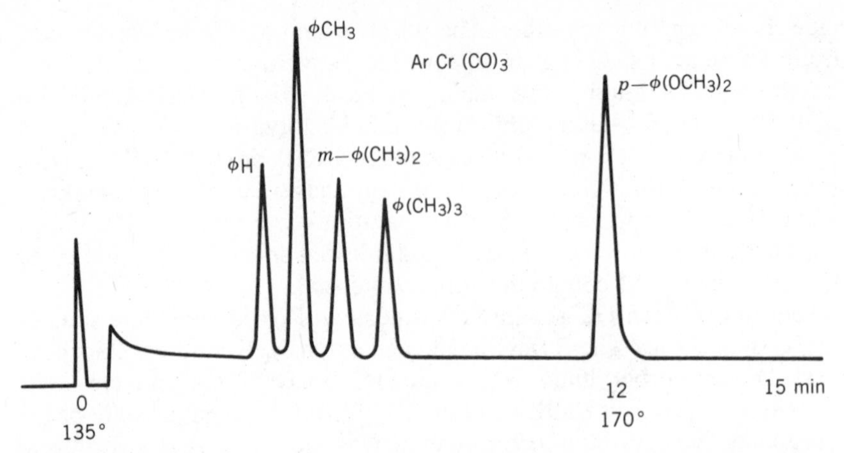

Figure 5.13. Separation of arene tricarbonyl iron compounds. Peak designation indicate ring substitution. Column 12 ft. 2% DC 560 and 1.5% SE 30 on 80/100 mesh Gas-Chrom P (Reference 146) Temperature program 135–170°C. Reprinted with permission from W. J. A. VandenHeuvel, J. S. Keller, H. Veening and B. R. Willeford. *Analytical Letters,* **3,** 279 (1970) Copyright 1970. Marcel Dekker Inc.

Later developments included GC of thiophenechromiumcarbonyls, which, however, proved rather less chromatographically stable than the arenes (147). Veening also showed that the analogous arenemolybdenumtricarbonyls could be effectively gas chromatographed (144).

Manganese group elements also form volatile thermally stable carbonyl derivatives, the properties of such compounds as cyclopentadienylmanganesetricarbonyl ($C_5H_5Mn(CO)_3$)–cymantrene being particularly well suited to quantitative GC. Uden et al. (148) reported a GC method for the determination of methylcyclopentadienylmanganesetricarbonyl, ($CH_3C_5H_4Mn(CO)_3$–MMT) in gasoline by packed column separation with DC argon plasma emission specific element detection. Cymantrene was used as an internal reference and low nanogram determination was possible. A dual detector chromatogram of an unleaded gasoline containing MMT additive is shown in Fig. 5.14. Effluent is split 1:1 to a flame ionization detector and to the DC argon plasma, which monitors manganese atomic emission at the 273.83 manganese line. Other detection methods have permitted the determination of lower levels of MMT. Depuis and Hill (149) used a hydrogen atmosphere flame ionization detector (HAFID) to monitor to levels of 1.7×10^{-14}g/s of manganese. Quimby et al. (112) obtained a detection limit of 2.5×10^{-13} g/s of manganese and a selectivity of 1.9×10^{6} over carbon using an atmospheric pressure microwave induced helium plasma detector (MIP). These very low detection limits confirm the absence of any on-column losses for this compound.

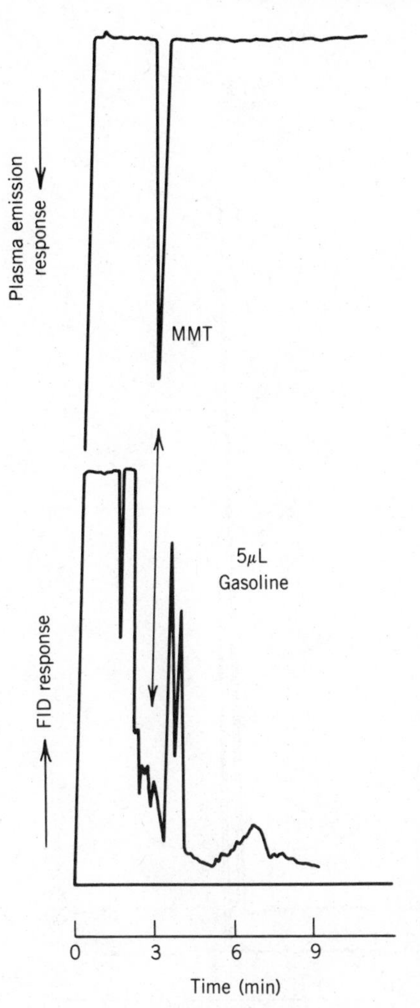

Figure 5.14. Dual-detection chromatogram of unleaded gasoline containing MMT. Column 2 m × $\frac{1}{8}$ in. o.d. 2% Dexsil 300GC on 100–120 mesh Chromosorb 750. 130°C. 1.5 μL injected; effluent split 1:1, FID and DC argon plasma detectors. Mn 279.83 nm line monitored (Reference 148). Reprinted with permission from P. C. Uden, R. M. Barnes, and F. P. DiSanzo, *Analytical Chemistry*, **50,** 852 (1978). Copyright 1978 American Chemical Society.

Capillary GC has also proved very effective for the resolution of compounds of this class; specific element detection simplifies such separations and provides qualitative and quantitative proof of complete elution. Such a separation of a range of organometallics with differing metals and functionalities, as obtained by Estes et al. (150), is shown in Fig. 5.15. The column was a 12.5-meter methylsilicone wall-coated fused-silica capillary. Chromatogram 5.15i shows six peaks, monitoring the carbon channel of a MIP detector to afford "universal" carbon detection. The identities of the eluted peaks are: (a) $C_5H_5Mn(CO)_3$, (b) $CH_3C_5H_4Mn(CO)_3$, (c)

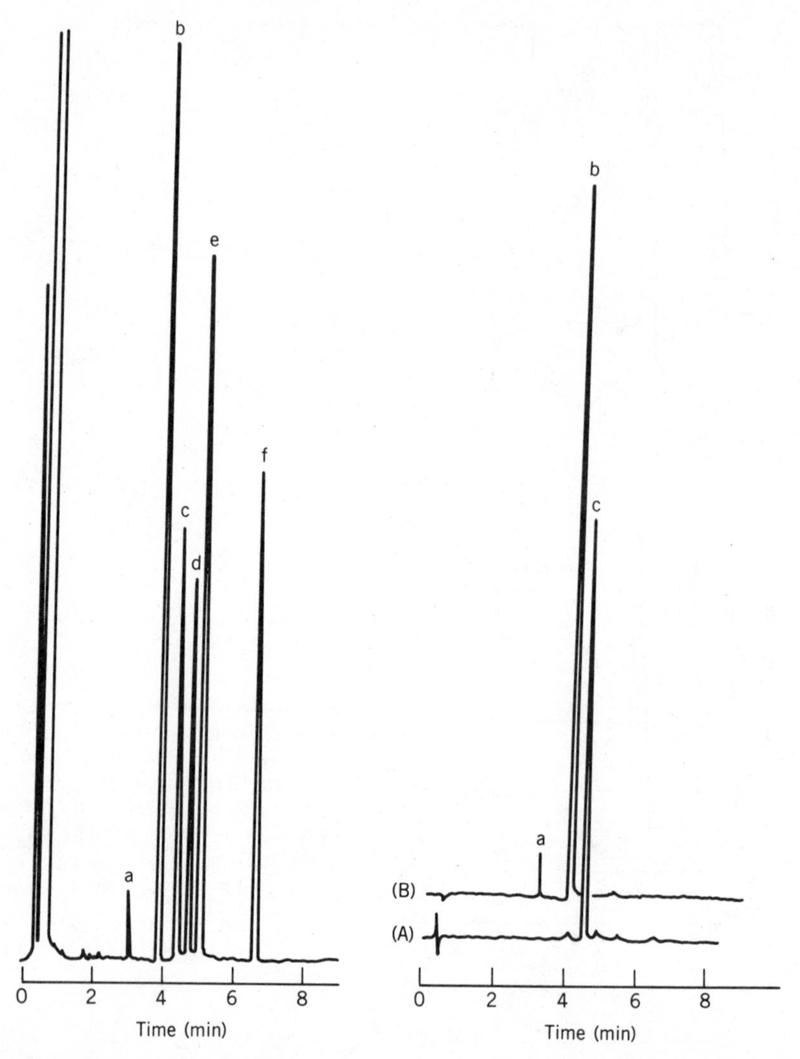

Figure 5.15. Specific element atmospheric pressure microwave induced plasma detection of organometallics. Left chromatogram: Carbon monitored at 247.9 nm. Right chromatogram(s): A, Chromium monitored at 267.7 nm; B, Manganese monitored at 257.6 nm. Peak identities: (*a*) $CpMn(CO)_3$, (*b*) $MeCpMn(CO)_3$, (*c*) Cp_2Ni and $CpCr(NO)(CO)_2$, (*d*) $CpV(CO)_4$, (*e*) Cp_2Fe, (*f*) $(Me)_5CpCo(CO)_2$ (Reference 150). Reprinted with permission from S. A. Estes, P. C. Uden, M. D. Rausch, and R. M. Barnes, *Journal of High Resolution Chromatography and Chromatographic Communications*, **3,** 471 (1980). Copyright 1980 Dr. Alfred Hüthig Publishers.

$C_5H_5Cr(NO)(CO)_2$ and $(C_5H_5)_2Ni$ (unresolved), (d) $C_5H_5V(CO)_4$, (e) $(C_5H_5)_2Fe$, and (f) $C_5H_5(CH_3)_5Co(CO)_2$. Chromatogram 15 iiA shows the same sample with spectral monitoring at 267.7 nm for chromium and 15 iiB shows the two manganese compounds detected specifically by monitoring emission at 257.6 nm. It may be noted that the column behavior of even the previously unchromatographed vanadium and cobalt compounds is excellent, and that the relatively unstable nickel compound $(C_5H_5)_2Ni$–nickelocene is also eluted undecomposed. This type of separation opens up the possibility for GC of a wider range of compound types than had been recognized earlier.

Another study of iron carbonyl derivatives emphasized the ability of GC to separate relatively unstable compounds. Forbes et al. (151) separated dienetricarbonyliron compounds including those incorporating cyclopentadiene, cyclohexa-1,3-diene, cyclohepta-1,3-diene, cycloocta-1,3-diene, cyclooctal-1,5-diene, cycloheptatriene, tropone, and other substituents. Column temperatures from 50–80°C proved best for these thermally labile compounds. Other recent experiments have resolved the cyclopentadienyldicarbonylnitrosyl compounds of tungsten, molybdenum, and chromium (152). Since many of these organometallics are attracting synthetic interest with respect to their potential as polymerization catalysts, etc., it appears probable that their GC determination in various matrices may become more widely utilized.

Metallocenes and Derivatives

Parallel to the developments in the GC of metal carbonyl compounds has been that of separation of the metallocenes, among which ferrocene (biscyclopentadienyl)iron-$(C_5H_5)_2Fe$ is the most familiar example. It has frequently been noted that these compounds have very favorable GC properties, ferrocene proving an ideal organometallic probe for determining column and system efficiency.

Tanikawa and Arakawa (153) and Ayers et al. (154) were among the early investigators to resolve ferrocene derivatives; among the compounds in the latter study were alkyl, vinyl, dialkyl, acetyl, diacetyl, and hydroxymethyl derivatives. In general, these and other investigators found that GC behavior was determined largely by the substituents and that the ferrocene moiety acted similarly to an inert alkane group. Little capillary GC has been reported for these compounds, but there is clearly great potential for its application. One example of a relatively low resolution capillary separation of metallocenes utilizes a 33-ft methylsilicone porous layer open tubular (PLOT) column (152). The separation of fer-

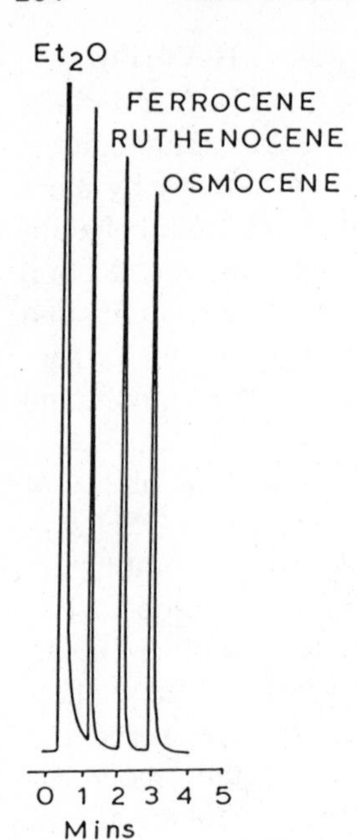

Figure 5.16. Metallocene separation. Column 33 ft × 0.03 in. i.d. stainless steel PLOT OV 101 at 150°C (Reference 152). Reprinted with permission from P. C. Uden, D. E. Henderson, F. P. DiSanzo, R. J. Lloyd and T. P. Tetu. *Journal of Chromatography,* **196,** 403 (1980) Copyright 1980. Elsevier Science Publishers.

rocene and its analogs, ruthenocene and osmocene, at 150°C is shown in Fig. 5.16.

5.6. METAL CHELATES

Neutral metal complexes in which the charge on the central metal atom is countered by equal and opposite charge deriving from a number of anionic ligands with oxygen, nitrogen, sulfur, or phosphorus donor atoms, comprise a widely studied chemical class. From the gas chromatographic standpoint, the critical features that determine whether the separation of such compounds is possible are an appropriate combination of volatility and thermal stability together with an adequate shielding of the metal atom from unwanted column interactions. The range of organic ligands that have been shown to be suitable for GC analysis has been relatively limited, but within these classes a considerable amount of development and application has been done.

5.6.1 Beta-diketonates

In 1955, Lederer (155) suggested that neutral chelate complexes in particular possessed adequate thermal stability in the vapor phase to make them amenable to GC. The acetylacetonates attracted the first experimental attention by reason of their facile formation and enhanced stability gained from the presence of multiple chelate rings. Those metallic ions having coordination numbers equal to twice their oxidation state such as Al(III), Be(II), and Cr(III) form coordinatively saturated neutral complexes which chromatograph well (156). Acetylacetonates of metals such as Cu(II) and Ni(II), which are not coordinatively saturated, however, tend to solvate, polymerize, or undergo undesirable adsorption on column or in the GC system. In fact, as in much metal complex GC, the nonfluorinated beta-diketonates are generally of marginal thermal and chromatographic stability; they usually require column temperatures too high for thermal degradation to be completely absent. These problems undoubtedly gave rise to unsubstantiated early claims that a wide range of acetylacetonates could be eluted. Little quantitative work was accomplished with these complexes.

The major breakthrough that transformed metal chelate GC to give useful analytical results was the introduction of fluorinated beta-diketone ligands, which formed complexes of greater volatility and thermal stability. Moshier and Sievers gave the major impetus to this development, and a major portion of their monograph published in 1965 (157) summarizes the analytical progress made to that time. Trifluoroacetylacetone (1,1,1-trifluoro-2,4-pentanedione-HTFA) and hexafluoroacetylacetone (1,1,1,5,5,5-hexafluoro-2,4-pentanedione-HHFA) have been the most widely studied and analytically developed of the fluorinated beta-diketonates. HTFA, in particular, extended the range of metals that may be gas chromatographed with little or no evidence of decomposition to include Ga(III), In(III), Sc(III), Rh(III), and V(IV) (158). Some HTFA complexes, however, have suspect elution below the microgram level, thus $Cu(TFA)_2$ has been shown to undergo partial on-column degradation in packed columns. Sokolov used tagging with radioactive ^{64}Cu to follow anomalous partly irreversible sorption (159). The problems for HTFA chelates have been further demonstrated for $Th(TFA)_4$, $U(TFA)_4$, and $Fe(III)(TFA)_3$, which degrade under normal column conditions but show improved peak shapes and lower elution limits if a continuous level of HTFA ligand vapor is maintained in the carrier gas (160).

HTFA chelates of trivalent hexacoordinate metals such as Cr(III), Co(III), Al(III), and Fe(III) exhibit geometrical isomerism with facial (cis) and meridonal (trans) forms present and interconverting at various rates dependent upon temperature. At GC temperatures below 100°C, interconversion of chromium chelates occurs slowly enough for each chelate to be eluted independently, but exchange of aluminum complexes is rapid and only the more stable meridonal form is observed. These effects have clear relevance to quantitative analysis.

The above phenomena are all clearly depicted in a high resolution capillary separation shown in Fig. 5.17. This study by Jennings et al. (160a) uses a 5-meter methyl silicone fused silica capillary column and hydrogen carrier gas. The presence of the cis and trans $Cr(TFA)_3$ chelates is evident, as is the single peak for the aluminum complex. The perturbation in baseline prior to the elution of $Cu(TFA)_2$ confirms on-column degradation even in the relatively inert fused silica wall coated column.

Numerous analytical applications of HTFA chelates are reviewed in detail by Mosher and Sievers (157), Uden and Henderson (161), Rodriguez-Vazquez (162), and in the texts of Guiochon and Pommier (3) and Schwedt (4). An example of a recent application is the analysis for beryllium in ambient air particulates (163). After filter sampling and extraction/chelation, packed column GC with electron capture detection allowed ppm level beryllium quantitation in collected particulates, which corresponded to levels of $2\text{–}20 \times 10^{-5}$ $\mu g/m^3$ in the sampled air. In Fig. 5.18 is seen the chromatogram of extract from a glass fiber filter after recovery of ambient air particulates for 24 h at an urban site.

Among other fluorinated beta-diketonates that have been extensively investigated are the HHFA complexes. Their great volatility and very limited solubilities in most stationary phases dictate very short retention times at low column temperatures. Further, their great electron capturing ability has permitted extremely low detection limits for $Cr(HFA)_3$ and $Be(HFA)_3$ (164). Quantitative derivatization is made difficult, however, by hydrate formation of the ligand and complexes.

Since modifications of the basic beta-diketone structure may be made readily, various modified chelating ligands have been evaluated for GC applications. The two major adaptations have been the replacement of the methyl group by higher branched alkyl groups, notably tert-butyl, and the incorporation of longer chain perfluoroalkyl groups in the ligand. While some fully fluorinated longer chain ligands such as 1,1,1,2,2,3,3,7,7,7-decafluoroheptanedione (HDFHD) have been employed in synergic combination with dibutylsulfoxide for extraction and GC separation of $U(VI)O_2$, Th(IV), and the lanthanides (165), their difficulty of preparation and hydration problems have prevented their wide adoption.

The widest development in ligand modification has been for those diketones containing a tert-butyl group and a long chain fluoroalkyl moiety. Thus Sievers et al. used 1,1,1,2,2,3,3-heptafluoro-7,7-dimethyl-4,6-octane dione [heptafluoropropanoylpivalylmethane (HFOD or HHPM)] for lanthanide separations (166). This and other similar ligands with different

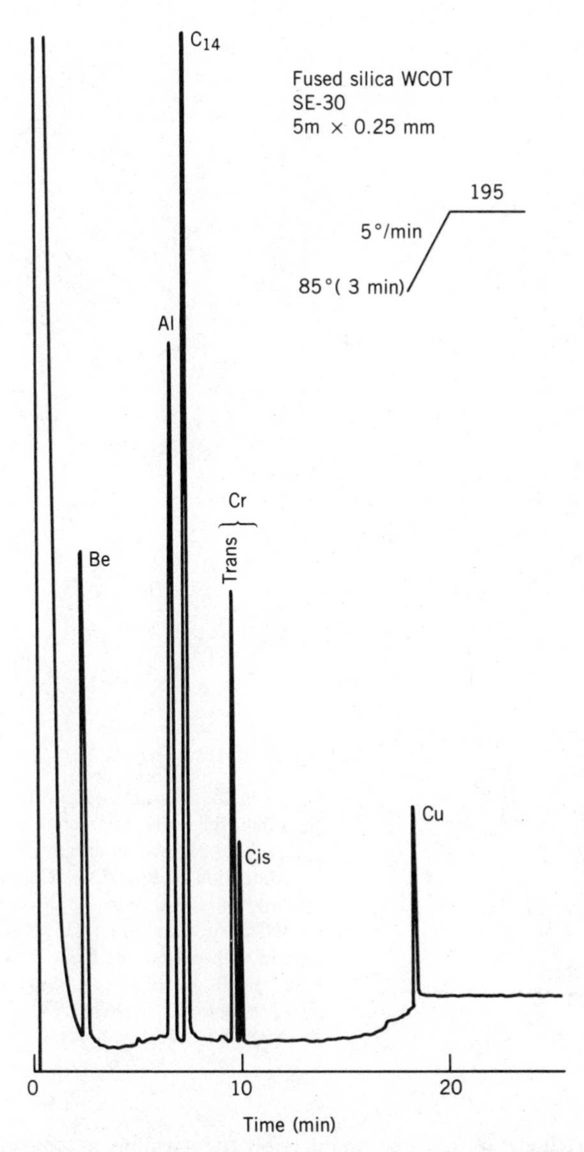

Figure 5.17. Capillary separation of Be, Al, Cr, and Cu trifluoroacetylacetonates and *n*-tetradecane. Inlet split 100:1. Fused silica WCOT SE-30, 5 m × 0.25 mm i.d. Temperature program 85–195°C at 5°C/min. Hydrogen carrier gas (Reference 160A). Reprinted with permission from L. Sucre and W. Jennings, *Journal of High Resolution Chromatography and Chromatographic Communications,* **3,** 452 (1980). Copyright 1980 Dr. Alfred Hüthig Publishers.

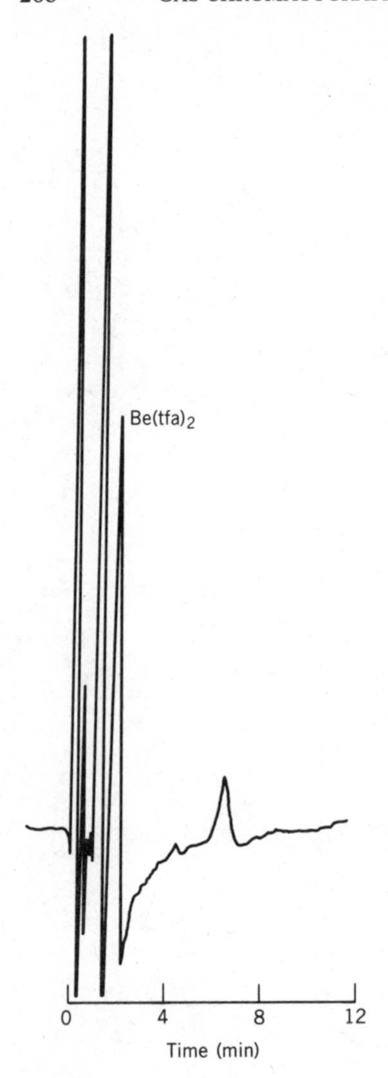

Figure 5.18. Chromatogram of extract from glass fiber filter after recovery of ambient air particulates for 24 hr at an urban site. Electron capture detection of $Be(TFA)_2$. Column 1.2 m × 4 mm i.d. glass; UC W98(3.8%) on Gas-Chrom Z 80–100 mesh. Column 110°C (Reference 163). Reprinted with permission from W. D. Ross, J. L. Pyle, and R. E. Sievers, *Environmental Science and Technology,* **11,** 469 (1977). Copyright 1977 American Chemical Society.

fluoroalkyl substituents have been used for a range of analytical procedures and preparative separations. Guiochon et al. (167) separated $Th(FOD)_4$, $U(FOD)_4$ and $U(FOD)_6$. Lead chelates were examined but exhibit some on-column degradation (168). Rather similar behavior was observed for alkali metal (169), alkaline earth (170), and mixed chelates of the form $MM'L_4$ where M is an alkali metal and M′ a rare earth (171).

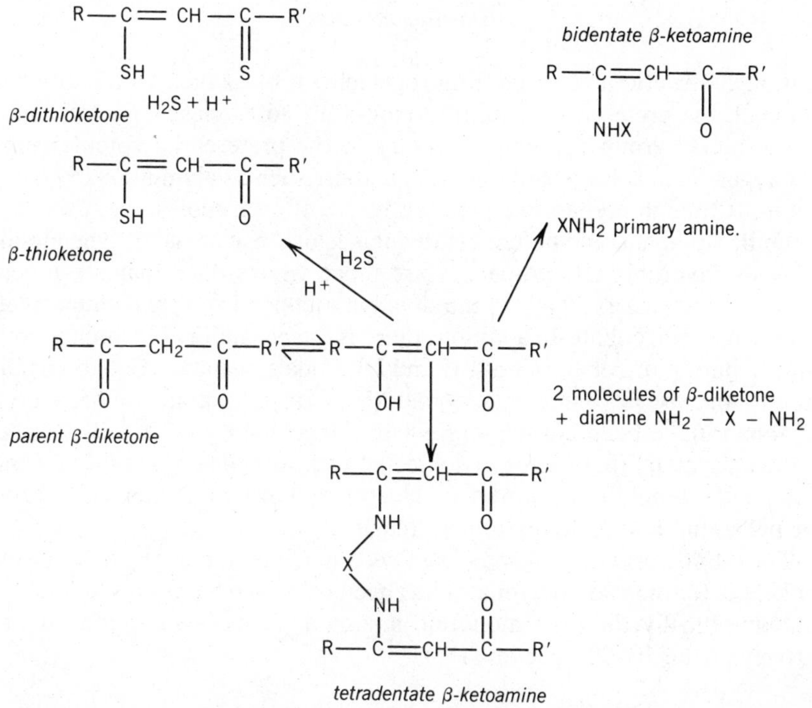

R—C═CH—C(═S)—R′ (SH on C)

β-dithioketone

$H_2S + H^+$

R—C═CH—C(═O)—R′ (SH on C)

β-thioketone

H_2S, H^+

bidentate β-ketoamine

R—C═CH—C(═O)—R′ (NHX on C)

XNH_2 primary amine.

R—C(═O)—CH_2—C(═O)—R′ ⇌ R—C(OH)═CH—C(═O)—R′

parent β-diketone

2 molecules of β-diketone + diamine $NH_2 - X - NH_2$

R—C(NH—)═CH—C(═O)—R′
X
R—C(NH—)═CH—C(═O)—R′

tetradentate β-ketoamine

Figure 5.19. Formation of major beta-difunctional ligands for metal complex GC (Reference 161). Reprinted with permission from P. C. Uden and D. E. Henderson, *The Analyst*, **102,** 889 (1977). Copyright 1977 Royal Society of Chemistry.

Although these studies showed only partial success, they do emphasize the versatility of these ligands for GC of metal compounds in many periodic groups.

5.6.2 Alternative Beta-difunctional Chelates

Although the beta-diketonates were very widely studied, it became clear that a logical approach to their chemical problems as regards stability, monomer integrity, and selective metal reactivity was to turn to alternative ligands related by the substitution of donor oxygen atoms by sulfur or nitrogen atoms. The major classes of ligand so produced are summarized (161) in the schematic diagram in Fig. 5.19. This indicates the mode of formation of the main alternate bidentate ligands with sulfur donors and bidentate and tetradentate ligands with nitrogen donors.

Beta-thioketonates

Although structural changes from replacement of oxygen with sulfur are minimal, the preference of sulfur donors for "soft" class B metals, such as the nickel group, is complementary to the preferential complexation of oxygen ligands for "hard" class A metals, such as aluminum and chromium. Although preparative procedures are often troublesome, there are a number of actual and potential advantages of these chelates. The metals showing favorable GC properties are those whose diketonates are generally unsatisfactory, such as the divalent metals nickel, palladium, platinum, zinc, and cobalt. In addition, there is the possibility of simultaneous determination of cobalt in both II and III oxidation states. Cobalt (II) bis monothioacetylacetonate and cobalt (III) tris monothiotrifluoroacetylacetonate have been chromatographed independently (172, 173). Nickel is the only metal that has been subjected to a complete quantitative analysis, as the monothiotrifluoroacetylacetonate but parallel determinations for palladium and platinum appear feasible.

The dithio derivative ligands have been much less investigated, but the nickel bis (dithioacetylacetonate) has been shown to be more stable chromatographically than the monothio analog and has been eluted quantitatively at the 10–20 ng level (174).

Beta-ketoaminates

Both bidentate and tetradentate chelates of a number of transition metals have been effectively chromatographed; in general, the presence of the nitrogen donor atom in the ligand determines for them an intermediate place between betadiketonates and monothioketonates in terms of the metals which are readily complexed. For GC purposes, there is closer correspondence with the sulfur ligands. The nickel group of metals is favorably complexed, but in addition stable chelates of copper(II) are formed, as are those of vanadyl (V(IV)O). The bidentate ketoamines are of only marginal GC stability, submicrogram elution levels often being the limit of chromatography (175). A review of their behavior is presently in order, however, since the new inert capillary columns may allow quantitative elution at low levels, enabling the rapid formation rates for these complexes to be exploited.

The tetradentate beta-ketoamine ligands form a group of particular utility for GC of divalent transition metals. The addition of the extra five-membered ring adds much stability to the complexes, which more than offsets the lowered volatility as compared with the analogous bidentate chelates. Table 5.4 lists the tetradentate betaketoamine ligands that have

Table 5.4 Tetradentate β-Ketoamines

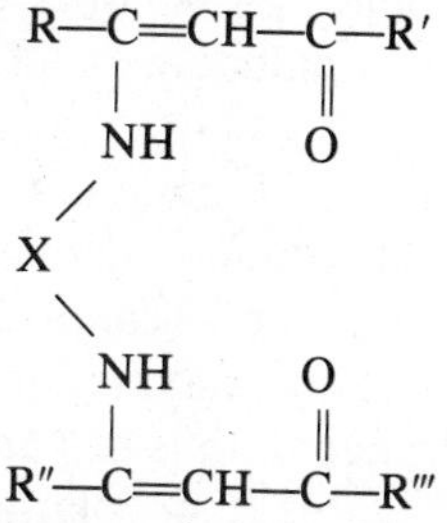

```
R—C=CH—C—R′
  |     ‖
  NH    O
 /
X
 \
  NH    O
  |     ‖
R″—C=CH—C—R‴
```

R	R′	R″	R‴	X	Symbol	Ref.
CH_3	CH_3	CH_3	CH_3	en[a]	$H_2(enAA_2)$	175–178
CH_3	CH_3	CH_3	CH_3	pn	$H_2(pnAA_2)$	175–178
CH_3	CH_3	CH_3	CH_3	bn	$H_2(bnAA_2)$	179
CF_3	CH_3	CH_3	CF_3	en	$H_2(enTFA_2)$	175–176, 178–180
CF_3	CH_3	CH_3	CF_3	pn	$H_2(pnTFA_2)$	175–176, 178–180, 184
CF_3	CH_3	CH_3	CF_3	bn	$H_2(bnTFA_2)$	179–181
$C(CH_3)_3$	CH_3	CH_3	$C(CH_3)_3$	en	$H_2(enAPM_2)$	178–179, 182
$C(CH_3)_3$	CH_3	CH_3	$C(CH_3)_3$	pn	$H_2(pnAPM_2)$	178–179, 182
CF_3	$C(CH_3)_3$	$C(CH_3)_3$	CF_3	en	$H_2(enTPM_2)$	183, 185
CF_3	CH_3	CH_3	CH_3	en	$H_2(enAATFA)$	176, 179
CF_3	CH_3	CH_3	CH_3	pn	$H_2(pnAATFA)$	179
CF_3	CH_3	CH_3	CH_3	bn	$H_2(bnAATFA)$	179

en = $CH_2 - CH_2$; pn = $CH(CH_3) - CH_2$; bn = $CH(CH_3) - CH(CH_3)$.

been evaluated for GC of copper, nickel, palladium, and vanadyl complexes. An important distinction occurs in volatility; in contrast to all bidentate beta-difunctional chelating ligands, for the tetradentates, the fluorinated chelates show uniformly lower volatilities than do their nonfluorinated analogs; this factor allows the latter to be more readily utilized analytically, since lower column temperatures may be used (176). Relative retention of $Ni(pnAA_2)$ to $Ni(pnTFA_2)$ changes from 0.35 on a QF 1 fluorosilicone stationary phase to 2.5 on Apiezon L.

The fluorinated chelates have very great electron capturing abilities (180) affording pg level detection, but the complexes form more slowly than those of the nonfluorinated complexes, which themselves possess adequate volatility and stability for analytical use. The latter are also detectable by ECD to the 100 pg level. The ligands $H_2(enAA_2)$ and

$H_2(pnAA_2)$ were thus employed successfully for trace determination of palladium in nickel- and copper-based ores (177). Belcher et al. (182) described a procedure for simultaneous determination of copper and nickel by solvent extraction and GC of complexes of $H_2(enAPM_2)$. Dilli and coworkers have examined the chemistry and GC of this series of complexes in great detail, their studies having included detailed analysis of preparative procedures (185) and reaction chemistry (184,186). The formation of tetradentate ligands having unsaturated bridging groups formed by oxidative dehydrogenation is of particular interest, as the chelates formed may be more stable and volatile than those of the parent ligands (184). A sensitive analytical study of the simultaneous determination of copper, nickel, and vanadium at the 5 pg level as $H_2(pnTFA_2)$ complexes formed a part of the same study. There have been a number of unpredicted features of the GC of these chelates. Of potential analytical interest is the very high relative volatility of the $H_2(enTPM_2)$ complexes; Dilli and Patsalides (185) showed that this change in property corresponded to structures which place trifluoromethyl substituent groups adjacent to the diamine bridge, and not remote from it as is the case for other ligands of this class.

GC has provided useful experimental methodology in chelate structural studies as exemplified by the straightforward resolution of the racemic and meso geometrical isomers of chelates of $H_2(bnTFA_2)$. The separation of a series of copper chelates including the $Cu(bnTFA_2)$ isomers is shown in Fig. 5.20 (152). Moderate resolution on a 33-ft PLOT column is adequate to emphasize the wide separation between isomers.

A related series of ligands which combine the properties of nitrogen, sulfur, and/or oxygen donor atoms is that examined by Patsalides et al., the beta-thionoenamines (187). They achieved separation of the copper, nickel, zinc, palladium, and platinum chelates of 4,4′-(ethane-1,2-diyldiimino)bis(pent-3-ene-2-thione) (DT-AAED). A further example of the continuing interest in novel volatile metal chelate systems is the report of an analytical determination of nickel with various alkyl and fluoroalkyl substituted amino-vinylketones (188).

5.6.3 Dialkyldithiocarbamates

Although these complexes have been successfully used for decades for the extraction of metal ions and for their spectrophotometric determination and other analytical procedures (189), it was not until 1975 that the GC of these chelates was first reported (190–192). Most attention was given to nickel and zinc complexes, although palladium, platinum, cadmium, and lead complexes were also reported to be eluted. One problem

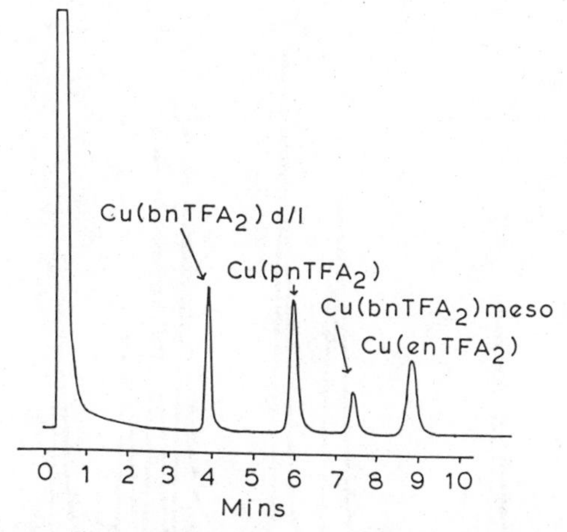

Figure 5.20. Separation of copper(II) *N,N'*-alkylenebis(trifluoroacetylacetoneimine) chelates. Column as Fig. 5.16. 225°C (Reference 152). Reprinted with permission from P. C. Uden, D. E. Henderson, F. P. DiSanzo, R. J. Lloyd and T. P. Tetu. *Journal of Chromatography* **196,** 403 (1980). Copyright 1980. Elsevier Science Publishers.

encountered was the need for high column temperatures, often above 250°C, which created concern for complete elution without some loss from on-column decomposition. In general, the metals complexed by these ligands are parallel to those chelated by the tetradentate beta-ketoamines for GC applications. The quantitative application of nonfluorinated ligands has been investigated by Radecki et al., who have determined zinc, copper, and nickel in marine bottom sediments (193) and in sea sands and muds (194). Metal levels in the 1–100 ppm range were readily determined. The stereochemical properties of nickel and zinc complexes with alkyl substituents of differing chain lengths has been investigated by capillary GC, but no analytical benefits of substituents longer than ethyl was evident (195). Riekkola et al. have also shown capillary column separations (196) of selenium, zinc, cadmium, lead, palladium, arsenic(III), cobalt(III), rhenium, and indium. Cobalt was also determined in human tissue by capillary GC (197).

In a parallel fashion to the development of fluorinated beta-diketone ligands, fluorinated dialkyldithiocarbamates have shown considerable analytical promise in view of their increased volatility over nonfluorinated analogs. Tavlaridis and Neeb first investigated di(trifluoroethyl)dithiocarbamates, eluting zinc, nickel, cadmium, lead, antimony, and bismuth complexes at 185°C from a packed SE 30 methylsilicone column (198). An example of the packed column separations possible with these chelates

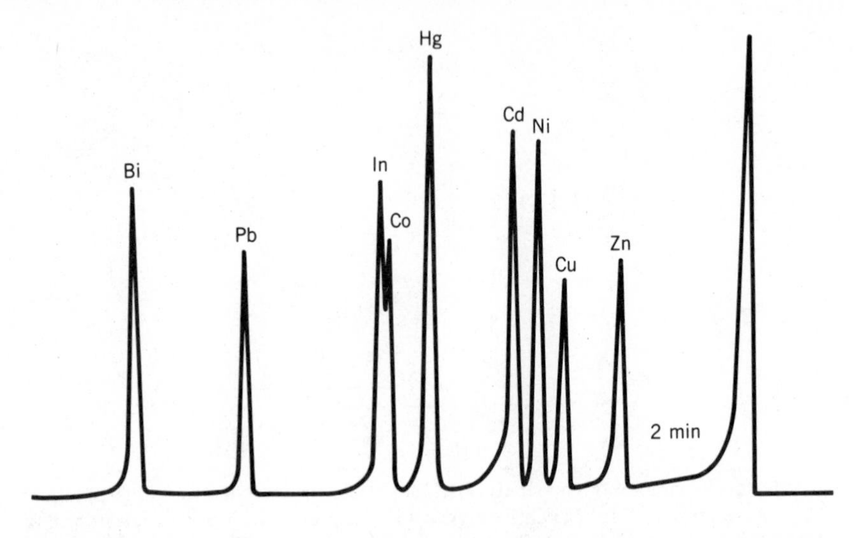

Figure 5.21. Separation of zinc, copper(II), nickel, cadmium, mercury, cobalt(II), indium, lead, and bismuth di(trifluoroethyl)dithiocarbamates. Column 90 cm × 2 mm 3% OV25 on Chromosorb W HP, 100–120 mesh. Temperature program 120 to 210°C at 2°C/min (Reference 199). Reprinted with permission from A. Tavlaridis and R. Neeb. *Zeitschrift für Analytische Chemie,* **292,** 199 (1978). Copyright 1978 Springer-Verlag.

is shown in Fig. 5.21, which illustrates a temperature programed separation from 120 to 210°C (199).

Lithium bis(trifluoroethyl)dithiocarbamate has been suggested as a reagent for the preparation of these chelates since it can be prepared in high purity and yield (200). Sucre and Jennings (160a) reported effective capillary separation of these complexes of nickel and cobalt(III) on 5-meter fused silica columns with a 150–190°C temperature program. It appears likely that further refinement of high resolution columns will broaden the application of these versatile complexes for analytical GC.

5.6.4 Dialkyldithiophosphinates

The analytical gas chromatographic investigation of dialkyldithiophosphinate (dithiophosphate) complexes has paralleled that of the dialkyldithiocarbamates. Cardwell and McDonagh (201) reported the packed column GC separation of zinc, nickel, palladium, and platinum chelates of 0,0′dialkyldithiophosphinates. Kleinmann and Neeb (202) carried out a parallel study for zinc, cadmium, and lead complexes. Typical packed

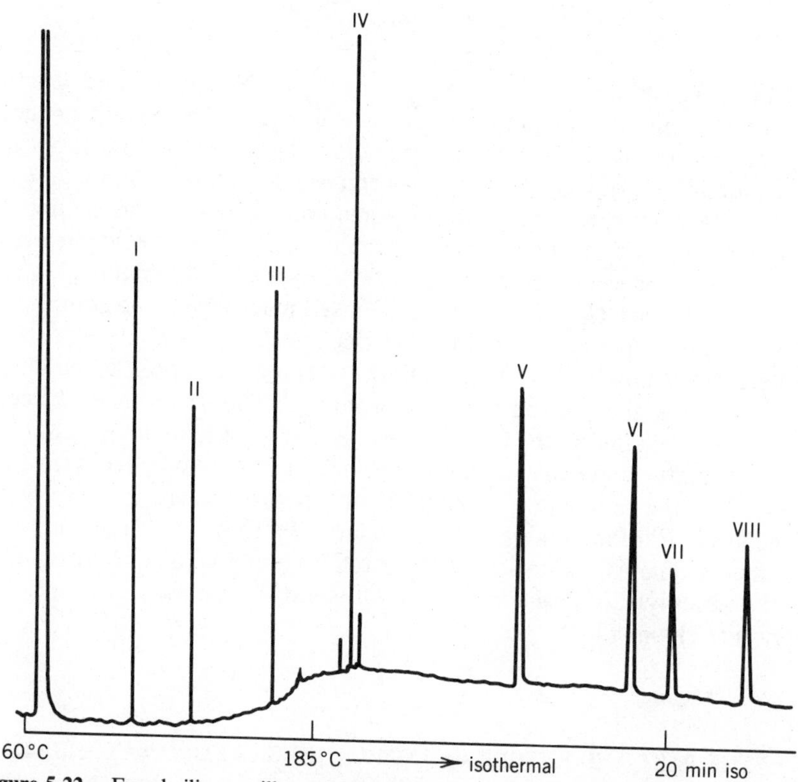

Figure 5.22. Fused silica capillary chromatogram of alkanes and metaldiisopropyl-dithiophosphinates. Column 25 m × 0.3 mm. OV 1. Hydrogen carrier gas at 57 cm/sec. Temperature program 60–185°C. FID detection. Peak identities: I—n-C_{12}, II—n-C_{14}, III—n-C_{17}, IV—n-C_{20}, V—Ni chelate, VI—n-C_{24}, VII—Pd chelate, VIII—Pt chelate (Reference 204). Reprinted with permission from P. J. Marriott and T. J. Cardwell, *Journal of Chromatography*, **234**, 157 (1982). Copyright 1982 Elsevier Science Publishers.

column temperatures were around 200°C. Fiegler et al. developed a procedure for the quantitation of cadmium in water, at concentrations in the range of 0.03–0.3 ppm (203). These complexes are suitable for selective detection by flame photometry with monitoring of either the S_2 or HPO emission modes. Sensitivities are found to be somewhat superior to flame ionization detection. An example of a parallel FID capillary column separation of alkanes and di-isopropyldithiophosphinates is shown in Fig. 5.22 (204), where the excellent peak shape for nickel, palladium, and platinum complexes is seen.

5.6.5 Metalloporphyrins

One of the most important demonstrations of the expanded range of sample applications for inorganic GC brought about by the advent of inert high resolution fused silica capillary columns has been the first reported conventional GC of transition metal porphyrin complexes. Although earlier there was reported separation of some metallo-porphyrins by hyperpressure supercritical fluid chromatography (205), their low vapor pressures was considered to preclude their GC. However, Marriott et al. have achieved capillary GC elution of these closed macrocylic ring complexes (206). Kovats indices in the range 5200–5600 necessitated the use of short (6-meter) columns. In Fig. 5.23 is shown a temperature programed chromatogram of *n*-alkanes and metallo-porphyrins. The first four peaks correspond to *n*-alkanes in the C_{40} to C_{50} range and the remainder are copper, nickel, vanadyl, and cobalt aetioporphyrin I, and octaethyl-porphyrin chelates. Only the nickel and vanadyl complexes of the latter group were unresolved. Elution of all peaks was in an isothermal 300°C region of the chromatogram. An extension of this area involved similar capillary GC of bis(trimethylsiloxy)silicon IV derivatives of porphyrins having polar ester side-chains (207).

5.7 THE PAST AND FUTURE OF INORGANIC GAS CHROMATOGRAPHY

The range of application of inorganic GC described in this chapter serves to emphasize that the scope of this technique is much broader than is often realized. While the history of inorganic GC has followed that of the discipline in general, the developments of recent years have served to widen greatly its areas of application. The most important advances that have impacted on inorganic GC may be noted thus:

1. The development of very inert silica capillary columns with very low levels of reactive residual metal impurities; the absence of interactive stationary phase supports plays a major role in quantitative elution of inorganics.
2. The parallel development of capillary columns with higher temperature limits.
3. The rapid development of specific element atomic spectral detection, which enables high sensitivity detection of inorganic species even in complex unresolved eluent profiles.

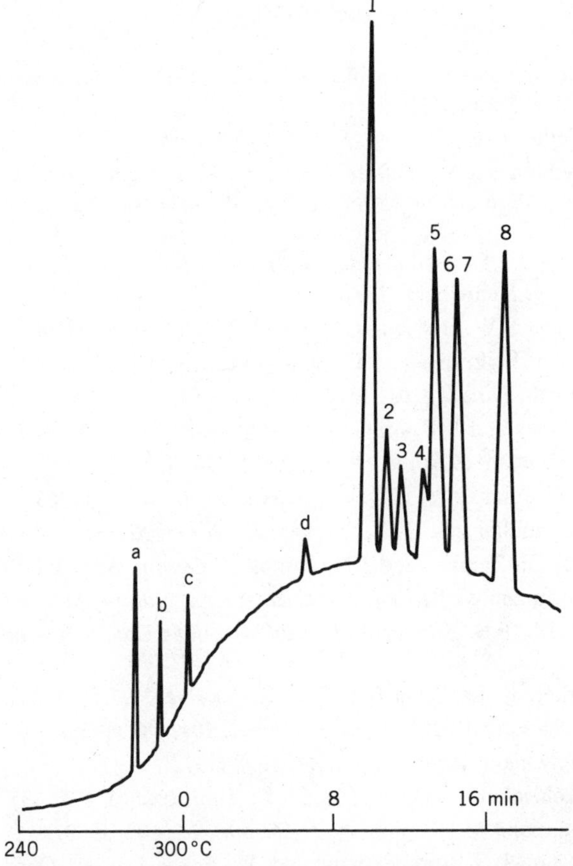

Figure 5.23. Fused silica capillary chromatogram of alkanes and metalloporphyrins. Column 6 meter × 0.3 mm. OV 1. Temperature program. 60–300°C at 7°C/min. Peak identities: a—n-C_{40}, b—n-C_{42}, c—n-C_{44}, d—n-C_{50}; 1—Cu Aetio, 2—Ni Aetio, 3—VO Aetio, 4—Co Aetio, 5—Cu OEP, 6—Ni OEP, 7—VO OEP, 8—Co OEP (Reference 206). Reprinted with permission from P. J. Marriott, J. P. Gill and G. Eglinton *Journal of Chromatography,* **236,** 395 (1982) Copyright 1982 Elsevier Science Publishers.

4. A probable major factor will also be the application of supercritical fluid chromatography, which will allow the separation of macrocyclic inorganic compounds.

It is clear that inorganic GC has much yet to offer the analytical chemist.

REFERENCES

1. A.V. Kiselev and Y.I. Yashin, *Gas Adsorption Chromatography,* Plenum Press, New York, 1969.
2. O.L. Hollis, *J. Chromatog. Sci.,* **11,** 335 (1973).
3. G. Guiochon and C. Pommier, *Gas Chromatography in Inorganics and Organometallics,* Ann Arbor Science Publishers, Ann Arbor, Michigan, 1973.
4. G. Schwedt, *Chromatographic Methods in Inorganic Analysis,* D.A. Hüthig Verlag, Heidelberg, 1981.
5. J. King and S.W. Benson, *J. Chem. Phys.,* **44,** 1007 (1966).
6. R. Aubeau, L. Leroy, and L. Champeix, *J. Chromatog.,* **19,** 245 (1965).
7. J.E. Purcell, *Nature,* **201,** 1321 (1964).
8. C.T. Hodges and R.F. Matson, *Anal. Chem.,* **37,** 1065 (1965).
9. G.M. Neumann, *Z. Anal. Chem.,* **244,** 302 (1969).
10. O.L. Hollis and W.V. Hayes, *J. Gas Chrom.,* **4,** 235 (1966).
11. E.L. Obermuller and G.O. Charlier, *J. Chrom. Sci.,* **7,** 580 (1969).
12. W. Gates, P. Zambri, and J.N. Armor, *J. Chrom. Sci.,* **19,** 183 (1981).
13. R.L. Hoffmann, G.R. List, and C.D. Evans, *Nature,* **211,** 965 (1966).
14. A.G. Datar, P.S. Ramanathan, and M. Snkar Das, *J. Chrom.,* **106,** 428 (1975).
15. R. Ellerker, F. Lax, and H.J. Dee, *Water Res.,* **1,** 243 (1967).
16. R.V. Holland and P.W. Board, *Analyst.,* **101,** 887 (1976).
17. K.L. McDonald, *Anal. Chem.,* **44,** 1298 (1972).
18. H.W. Kohlschutter and W. Hoppe, *Z. Anal. Chem.,* **197,** 133 (1963).
19. G. Manara and M. Taramasso, *J. Chrom.,* **65,** 109 (1973).
20. A.G. Datar, P.S. Ramanathan, and M. Snkar Das, *J. Chrom.,* **93,** 217 (1974).
21. O. Pitak, *Chromatographia,* **2,** 462 (1969).
22. A. Steffen and K. Bachman, *Talanta,* **25,** 551 (1978).
23. W.A. Robbins and J.A. Caruso, *J. Chrom. Sci.,* **17,** 360 (1979).
24. G. Rambeau, *Chim. Anal.,* **52,** 774 (1970).
25. R.P. Hirschmann and T.L. Mariani, *J. Chrom.,* **34,** 78 (1968).
26. M.L. Conti and M. Lesimple, *J. Chrom.,* **29,** 32 (1967).
27. H.W. Durbeck and R. Niehaus, *Chromatographia,* **11,** 14 (1978).
28. J.S. Fritz and R.C. Chang, *Anal. Chem.,* **46,** 938 (1974).
29. R.D. Kadeg and G.D. Christian, *Anal. Chim. Acta,* **88,** 117 (1977).
30. T.L.C. deDouza and S.P. Bhatia, *Anal. Chem.,* **48,** 2234 (1978).
31. J. Koch and K. Figge, *J. Chromatog,* **109,** 89 (1975).
32. G. Parissakis, D. Vrandit-Piscon, and J. Kohtoyannakos, *Z. Anal. Chem.,* **254,** 188 (1971).

33. O. Pitak, *Chromatographia,* **2,** 304 (1969).
34. J.J. Kaufman, J.E. Todd, and W.S. Koski, *Anal. Chem.,* **29,** 1032 (1957).
35. H.W. Myers and R.F. Putnam, *Anal. Chem.,* **34,** 664 (1962).
36. E.J. Sowinski and I.H. Suffet, *Anal. Chem.,* **44,** 2237 (1972).
37. E.J. Sowinski and I.H. Suffet, *Anal. Chem.,* **46,** 1218 (1974).
38. P.L. Timms, C.C. Simpson, and C.S.G. Phillips, *J. Chem. Soc.,* 1467 (1964).
39. F. Feher and H. Strock, *Naturwissenschaften,* **50,** 570 (1963).
40. C.S.G. Phillips and P.L. Timms, *Anal. Chem.,* **35,** 505 (1963).
41. F.H. Pollard, G. Nickless, and D.J. Cooke, *J. Chrom.,* **13,** 48 (1964).
42. C.J. Soderquist and D.G. Crosby, *Anal. Chem.,* **50,** 1435 (1978).
43. A.D. Zorin, G.G. Devyatykh, V.Y. Dudurov, and A.M. Amelschenko, *Russ. J. Inorg. Chem.,* **9,** 1364 (1964).
44. G.G. Devyatykh, A.D. Zorin, A.M. Amelschenko, S.B. Lyakhmanov, and A.E. Ezhelevc. *Doklady Chemistry,* **156,** 594 (1964).
45. R.D. Kadeg and G.D. Christian, *Anal. Chim. Acta,* **88,** 117 (1977).
46. A. Vinsjansen and K.E. Thrane, *Analyst,* **103,** 1195 (1978).
47. T. Dumas and E.J. Bond, *J. Chrom.,* **206,** 384 (1981).
48. W.B. Robbins and J.A. Caruso, *J. Chrom. Sci.,* **17,** 360 (1979).
49. M.H. Hahn, K.J. Mulligan, M.E. Jackson, and J.A. Caruso, *Anal. Chim. Acta,* **118,** 115 (1980).
50. P.R. Gifford and A. Bruckenstein, *Anal. Chem.,* **52,** 1028 (1980).
51. Y. Talmi and D.T. Bostick, *Anal. Chem.,* **47,** 2145 (1975).
52. Y. Talmi and V.E. Norvell, *Anal. Chem.,* **47,** 1510 (1975).
53. H. Freiser, *Anal. Chem.,* **31,** 1440 (1959).
54. S.T. Sie, J.P.A. Bleuner, and G.W.A. Rijnders, *Sep. Sci.,* **1,** 41 (1966).
55. R.S. Juvet and F.M. Wachi, *Anal. Chem.,* **32,** 290 (1960).
56. C. Pommier, C. Eon, H. Fould, and G. Guiochon, *Bull. Soc. Chim. Fr.,* 1401 (1969).
57. I. Tohyama and K. Otazai. *Z. Anal. Chem.,* **262,** 346 (1972).
58. F. Dienstbach and K. Bächmann, *Anal. Chem.,* **52,** 620 (1980).
59. G. Michael, U. Danne and G. Fisher, *J. Chrom.,* **118,** 104 (1976).
60. T.R. Crompton, *Gas Chromatography of Organometallic Compounds,* Plenum Press, New York, 1982.
61. L.F. Green, L.J. Schmaush, and J.C. Worman, *Anal. Chem.,* **36,** 1512 (1964).
62. S. Tsalas and K. Bächmann, *Talanta,* **27,** 201 (1980).
63. R.S. Juvet and R.L. Fisher, *Anal. Chem.,* **38,** 1860 (1966).
64. O. Pitak, *Chromatographia,* **3,** 29 (1970).
65. A.G. Hamlin, G. Iveson, and T.R. Phillips, *Anal. Chem.,* **35,** 2037 (1963).

66. A. Steffen and K. Bächmann, *Talanta,* **25,** 51 (1978).
67. J. Rudolph, K. Bächmann, A. Steffen, and S. Tsalas, *Microchem. Acta,* 471 (1978).
68. G. Schomberg, R. Koster, and D. Henneberg, *Z. Anal. Chem.,* **170,** 285 (1959).
69. R. Koster and G. Bruno, *Annalen der Chemie,* **629,** 89 (1960).
70. G. Schomberg, *Gas Chromatography 1962,* M. Van Swaay (Ed.), Butterworths, London, 1962, p. 292.
71. L.J. Kuhns, R.S. Braman, and J.E. Graham, *Anal. Chem.,* **34,** 1700 (1962).
72. J. Stuchlik and V. Packova, *J. Chrom.,* **174,** 224 (1979).
73. Y.N. Bortnikov, E.N. Vyankin, E.N. Gladyshev, and V.S. Andreevichev, *Zavodsk. Lab.,* **35,** 1445 (1969).
74. P. Longi and R. Mazzochi, *Chimica Ind. Milano,* **48,** 718 (1966).
75. K.K. Fukin, V.G. Rezchikov, T.S. Kuznetsova, and I.A. Frolov, *Zavodsk Lab.,* **39,** 993 (1973).
76. F.H. Pollard, G. Nickless, and P.C. Uden, *J. Chrom.,* **19,** 28 (1965).
77. S.A. Estes, C.A. Poirier, P.C. Uden, and R.M. Barnes, *J. Chrom.* **196,** 265 (1980).
78. J.A. Semlyen and C.S.G. Phillips, *J. Chem.,* **18,** 1 (1965).
79. L.B. Peetre, O. Ellren, and B.E.F. Smith, *J. Chrom.,* **88,** 295 (1974).
80. L.B. Peetre and B.E.F. Smith, *J. Chrom.,* **90,** 41 (1977).
81. T. Oiwa, M. Sato, Y. Miyakawa, and I. Miyazaki. *J. Chem. Soc., Japan, Pure Chem. Sect.,* **84,** 409 (1963).
82. K.R. Burson and C.T. Kenner, *Anal. Chem.,* **41,** 820 (1969).
83. J. Frank and M. Wurst, *Coll-Czech. Chem. Commun.,* **25,** 701 (1960).
84. M. Wurst and J. Churacek, *Coll-Czech. Chem. Commun.,* **36,** 3497 (1971).
85. J.A. Semlyen, G.R. Walker, R.E. Blofield, and C.S.G. Phillips, *J. Chem. Soc.,* 4948 (1964).
86. G.N. Bortnikov, A.V. Kiselev, N.S. Vyazankin, and Ya. I. Yashin, *Chromatographia,* **4,** 14 (1971).
87. F.H. Pollard, G. Nickless, and D.J. Cooke, *J. Chrom.,* **17,** 472 (1965).
88. R.C. Putnam and H. Pu, *J. Gas Chrom.,* **3,** 289 (1965).
89. W.R. Schulz and D.J. Leroy, *Can. J. Chem.,* **42,** 2480 (1964).
90. J. Tadmore, *Anal. Chem.,* **36,** 1565 (1964).
91. K. Hoppner, U.U. Prosch, and H. Wieglab, *Z. Chem.,* **4,** 31 (1964).
92. C.G. Flinn and W.A. Aue, *J. Chrom.,* **186,** 29a (1979).
93. S. Kapila and C.R. Vogt, *J. Chrom. Sci.,* **181,** 144 (1980).
94. D.R. Hansen, T.J. Gilfoil, and H.H. Hill, Jr., *Anal. Chem.,* **53,** 857 (1981).
95. J.A. Rodriguez-Vazquez, *Talanta,* **25,** 299 (1978).
96. P.T.S. Wong, Y.K. Chan, and P.L. Luxon, *Nature* (London), **253,** 263 (1975).

97. U. Schmidt and F. Huber, *Nature* (London), **259,** 159 (1976).
98. E.W. Abel, G. Nickless, and F.H. Pollard, *Proc. Chem. Soc.,* 288 (1960).
99. W.W. Parker and R.L. Hudson, *Anal. Chem.,* **35,** 1344 (1962).
100. J.E. Lovelock and A. Zlatkis, *Anal. Chem.,* **33,** 1958 (1961).
101. W.A. Aue and H.H. Hill Jr., *J. Chrom.,* **74,** 319 (1972).
102. M.D. DuPuis and H.H. Hill Jr., *Anal. Chem.,* **51,** 292 (1979).
103. P.M. Mutsaars and J.E. Van Steen, *J. Inst. Petrol.,* **58,** 102 (1972).
104. P.R. Ballinger and I.M. Whittemore, *Proc. Amer. Chem. Soc.,* Div. of Petroleum Chemistry, Atlantic City, New Jersey, **13,** 133 (1968).
105. Y.K. Chau, P.T.S. Wong, and P.D. Goulden, *Anal. Chim. Acta,* **85,** 421 (1976).
106. Y.K. Chau, P.T.S. Wong, G.A. Bengert, and O. Kramar, *Anal. Chem.,* **51,** 186 (1979).
107. B.Y.T. Radzuik, J.C. Van Loon, and Y.K. Chau, *Anal. Chim. Acta,* **255,** 105 (1979).
108. J.W. Robinson, E.L. Diesel, J.P. Goodbreed, R. Bliss, and R. Marshall, *Anal. Chim. Acta,* **92,** 321 (1977).
109. D.C. Reamer, W.H. Zoller, and T.C. O'Haver, *Anal. Chem.,* **50,** 1449 (1978).
110. A.J. McCormack, S.C. Tong, and W.D. Cooke, *Anal. Chem.,* **37,** 1470 (1965).
111. C.I.M. Beenakker, *Spectrochem. Acta,* **31B,** 483 (1976).
112. B.D. Quimby, P.C. Uden, and R.M. Barnes, *Anal. Chem.,* **50,** 2112 (1978).
113. S.A. Estes, P.C. Uden, and R.M. Barnes, *J. Chrom.,* **239,** 181 (1982).
114. S.A. Estes, P.C. Uden, and R.M. Barnes, *Anal. Chem.,* **53,** 1336 (1981).
115. S.A. Estes, P.C. Uden, and R.M. Barnes, *Anal. Chem.,* **54,** 2402 (1982).
116. B.J. Gudzinowicz and H.F. Martin, *Anal. Chem.,* **34,** 648 (1962).
117. G.E. Parris, W.R. Blair, and F.E. Brinkman, *Anal. Chem.,* **49,** 378 (1977).
118. G. Schwedt and H.A. Ruessel, *Chromatographia,* **5,** 242 (1972).
119. B.J. Gudzinowicz and J.L. Driscoll, *J. Gas Chrom.,* **1,** 25 (1963).
120. C.S. Evans and C.M. Johnson, *J. Chrom.,* **21,** 202 (1966).
121. Y. Talmi and W.W. Audren, *Anal. Chem.,* **46,** 2122 (1974).
122. S.A. Estes, P.C. Uden, and R.M. Barnes, *Anal. Chem.,* **53,** 1829 (1981).
123. G.N. Bortnikov, M.N. Bochkarev, N.S. Vyazankin, S.K.R. Ratushnaya, and Y.I. Yoshin, *Izv. Akad. Nank SSSR. Ser. Khim,* **4,** 851 (1971).
124. G.L. Baughman, M.H. Carter, N.L. Wolf, and R.G. Zepp, *J. Chrom.,* **76,** 471 (1973).
125. C.A. Bache and D.J. Lisk, *Anal. Chem.,* **43,** 950 (1971).
126. T.H. Risby and Y. Talmi, *C.R.C. Critical Reviews in Analytical Chemistry,* **14,** No. 3, 231 (1983).

127. J.E. Longbottom, *Anal. Chem.,* **44,** 1111 (1972).
128. G.W. Rice, J.J. Richard, A.P. D'Silva, and V.A. Fassel, *Anal. Chem.,* **53,** 1519 (1981).
129. M. Shaviat, *J. Chrom. Sci.,* **17,** 527 (1979).
130. G. Westoo, *Acta Chim. Scand.,* **22,** 2277 (1968).
131. R. Belcher, J.A. Rodriguez-Vazquez, P.C. Uden, and W.I. Stephen, *Chromatographia,* **9,** 201 (1976).
132. J.A. Rodriguez-Vazquez, *Talanta,* **25,** 299 (1978).
133. G. Schwedt and H.A. Russel, *Z. Anal. Chem.,* **264,** 301 (1973).
134. V. Luckow and H.A. Russel, *J. Chrom.,* **150,** 187 (1978).
135. P. Zarnegar and P. Mushak, *Anal. Chim. Acta,* **69,** 389 (1974).
136. P. Jones and G. Nickless, *J. Chrom.,* **76,** 285 (1973).
137. P. Mushak, F.E. Tibbetts III, P. Zarnegar, and G.B. Fisher, *J. Chrom.,* **87,** 215 (1973).
138. P. Jones and G. Nickless, *J. Chrom.,* **89,** 201 (1974).
139. A.L. Smith (Ed.), *Analysis of Silicones,* Wiley-Interscience, New York, 1974.
140. L.M. Brown and K.S. Mazdiyasni, *Anal. Chem.,* **41,** 1243 (1969).
141. R.S. Barratt, R. Belcher, W.I. Stephen, and P.C. Uden, *Anal. Chim. Acta,* **57,** 447 (1971).
142. T.J. Cardwell and M.R.L. Carter, *J. Chrom.,* **140,** 93 (1977).
143. C. Pommier and G. Guiochon, *J. Chrom. Sci.,* **8,** 486 (1970).
144. H. Veening, N.J. Graves, D.B. Clark, and B.R. Willeford, *Anal. Chem.,* **42,** 1655 (1969).
145. C. Segard, P.P. Rogues, C. Pommier, and G. Guiochon, *Anal. Chem.,* **43,** 1146 (1971).
146. W.J.A. VandenHeuvel, J.S. Keller, H. Veening, and B.R. Willeford, *Analyt. Letters,* **3,** 279 (1970).
147. C. Segard, C. Pommier, B.P. Rogues, and G. Guiochon, *J. Organomet. Chem.,* **77,** 49 (1974).
148. P.C. Uden, R.M. Barnes, and F.P. DiSanzo, *Anal. Chem.,* **50,** 852 (1978).
149. M.D. DuPuis and H.H. Hill, *Anal. Chem.,* **51,** 292 (1979).
150. S.A. Estes, P.C. Uden, M.D. Rausch, and R.M. Barnes, *J. High Res. Chrom. and Chrom. Comm.,* **3,** 471 (1980).
151. E.J. Forbes, M.K. Sultan, and P.C. Uden, *Analyt. Letters,* **5,** 927 (1972).
152. P.C. Uden, D.E. Henderson, F.P. DiSanzo, R.J. Lloyd, and T.P. Tetu, *J. Chrom.* **196,** 403 (1980).
153. K. Tanikawa and K. Arakawa, *Chem. Pharm. Bull* (Tokyo), **13,** 926 (1965).
154. O.E. Ayers, T.C. Smith, J.D. Burnett, and B.W. Ponder, *Anal. Chem.,* **38,** 1606 (1966).
155. M. Lederer, *Nature* **176,** 462 (1955).

156. W.J. Biermann and H. Gesser, *Anal. Chem.*, **32,** 1525 (1960).
157. R.W. Moshier and R.E. Sievers, *Gas Chromatography of Metal Chelates,* Pergamon, Oxford, London, 1965.
158. R.E. Sievers, B.W. Ponder, M.L. Morris, and R.W. Moshier, *Inorg. Chem.,* **2,** 693 (1963).
159. D.N. Sokolov, A.V. Davydov, S.Y. Prokofyev, S.S. Travnikov, E.V. Fedoseev, and B.F. Myasoedov, *J. Chrom.,* **155,** 241 (1978).
160. T. Fujinaga, T. Kuwamoto, and S. Murai, *Talanta,* **18,** 429 (1971).
160a. L. Sucre and W. Jennings, *J. High Res. Chrom. and Chrom. Commun.,* **3,** 452 (1980).
161. P.C. Uden and D.E. Henderson, *Analyst,* **102,** 889 (1977).
162. J.A. Rodriguez-Vazquez, *Anal. Chim. Acta.,* **73,** 1 (1974).
163. W.D. Ross, J.L. Pyle, and R.E. Sievers, *Env. Sci. Tech.,* **11,** 469 (1977).
164. M.L. Morris, R.W. Moshier, and R.E. Sievers, *Inorg. Chem.,* **2,** 411 (1963).
165. C.A. Burgett and J.S. Fritz, *J. Chromatog.,* **77,** 265 (1973).
166. C.S. Springer, Jr., D.W. Meek, and R.E. Sievers, *Inorg. Chem.,* **6,** 1105 (1967).
167. R. Fontaine, B. Santoni, C. Pommier, and G. Guiochon, *Chromatographia,* **3,** 532 (1970).
168. R. Belcher, J.R. Majer, W.I. Stephen, I.J. Thomson, and P.C. Uden, *Anal. Chim. Acta,* **50,** 423 (1970).
169. R. Belcher, J.R. Majer, R. Perry, and W.I. Stephen, *Anal. Chim. Acta,* **45,** 305 (1970).
170. R. Belcher, C.R. Cranley, J.R. Majer, W.I. Stephen, and P.C. Uden, *Anal. Chim. Acta,* **60,** 109 (1972).
171. R. Belcher, J.R. Majer, R. Perry, and W.I. Stephen, *J. Inorg. Nucl. Chem.,* **31,** 47 (1969).
172. R. Belcher, W.I. Stephen, I.J. Thomson, and P.C. Uden, *J. Inorg. Nucl. Chem.,* **33,** 1851 (1971).
173. R. Belcher, W.I. Stephen, I.J. Thomson, and P.C. Uden, *J. Inorg. Nucl. Chem.,* **34,** 1017 (1972).
174. P.C. Uden, K.A. Nonnemaker, and W.E. Geiger, *Inorg. Nucl. Chem. Letters,* **14,** 161 (1978).
175. R. Belcher, R.J. Martin, W.I. Stephen, D.E. Henderson, A. Kamalizad, and P.C. Uden, *Anal. Chem.,* **45,** 1197 (1973).
176. R. Belcher, K. Blessel, T.J. Cardwell, M. Pravica, W.I. Stephen, and P.C. Uden, *J. Inorg. Nucl. Chem.,* **35,** 1127 (1973).
177. P.C. Uden and D.E. Henderson, *J. Chrom.,* **99,** 309 (1974).
178. S. Dilli and E. Patsalides, *J. Chrom.,* **130,** 251 (1977).
179. P.J. Clark, Ph.D. Dissertation, University of Massachusetts, 1977.
180. P.C. Uden, D.E. Henderson, and A. Kamalizad, *J. Chrom. Sci.,* **12,** 591 (1974).

181. P.C. Uden and K. Blessel, *Inorg. Chem.*, **12,** 352 (1973).
182. R. Belcher, A. Khalique, and W.I. Stephen, *Anal. Chim. Acta,* **100,** 503 (1978).
183. W.I. Stephen, *Proc. Soc. Anal. Chem.*, **9,** 137 (1972).
184. S. Dilli and A.M. Maitra, *J. Chrom.*, **254,** 133 (1983).
185. S. Dilli and E. Patsalides, *J. Chrom.*, **134,** 477 (1977).
186. S. Dilli, A.M. Maitra, and E. Patsalides, *Inorg. Chem.*, **21,** 2832 (1982).
187. E. Patsalides, B.J. Stevenson, and S. Dilli, *J. Chrom.*, **173,** 321 (1979).
188. L.N. Bazhenova, K.I. Pashkevich, V.E. Kirichenko, and A. Ya Alzikovich, *Zh. Anal. Khim.*, **36,** 410 (1981).
189. G.D. Thorn and R.A. Ludwig, *The Dithiocarbamates and Related Compounds,* Elsevier, Amsterdam, 1962.
190. J. Krupcik, J. Garaj, S. Holotik, D. Oktavec, and M. Kosik, *J. Chrom.*, **112,** 189 (1975).
191. J. Masaryk, J. Krupcik, J. Garaj, and M. Kosik, *J. Chrom.*, **115,** 256 (1975).
192. T.J. Cardwell, D.J. DeSarro, and P.C. Uden, *Anal. Chim. Acta,* **85,** 415 (1976).
193. A. Radecki, J. Halkiewicz, J. Grzybowski, and H. Lamparaczyk, *J. Chrom.,* **151,** 259 (1978).
194. A. Radecki and J. Halkiewicz, *J. Chrom.*, **187,** 363 (1980).
195. M.L. Riekkola, *Finn. Chem. Lett.*, **3,** 83 (1980).
196. M.L. Riekkola, O. Makitie, and M. Sundberg, *Kem. Kemi.*, **6,** 525 (1979).
197. R.A. Zabairova, G.N. Bortnikov, F.A. Gorina, and K.M. Samarin, *Zavod. Lab.*, **47**(3), 22 (1981).
198. A. Tavlaridis and R. Neeb, *Z. Anal. Chem.*, **282,** 17 (1976).
199. A. Tavlaridis and R. Neeb, *Z. Anal. Chem.*, **292,** 199 (1978).
200. L. Sucre and W. Jennings, *Anal. Letters,* **13**(A6), 497 (1980).
201. T.J. Cardwell and P.S. McDonagh, *Inorg. Nucl. Chem. Letters,* **10,** 283 (1974).
202. A. Kleinmann and R. Neeb, *Z. Anal. Chem.*, **285,** 107 (1977).
203. B. Fiegler, V. Gemmer-Colos, A. Dencks, and R. Neeb, *Talanta,* **26,** 761 (1979).
204. P.J. Marriott and T.J. Cardwell, *J. Chrom.*, **234,** 157 (1982).
205. N.M. Karayannis and A.H. Corwin, *J. Chrom.*, **47,** 247 (1970).
206. P.J. Marriott, J.P. Gill, and G. Eglinton, *J. Chrom.*, **236,** 395 (1982).
207. P.J. Marriott and G. Eglinton, *J. Chrom.*, **249,** 311 (1982).

CHAPTER

6

HIGH-PERFORMANCE LIQUID CHROMATOGRAPHY OF INORGANICS AND ORGANOMETALLICS

JOHN C. MacDONALD

Department of Chemistry
Fairfield University
Fairfield, Connecticut

6.1 INTRODUCTION

The separation of inorganics and organometallics by gas chromatography (GC) is limited by compound vapor pressures and temperature instabilities. Although liquid chromatography (LC) has been used extensively in the form of thin-layer and classical ion exchange and overcomes these limitations, column chromatography with recently developed column packings, instrumentations, and detectors yields new chromatographic capabilities. Ion chromatography (Chapter 9) is such an example.

The literature of these LC separations of inorganics and organometallics up to 1970 is presented in the book by Michal (1) with over 700 references. The literature between 1970 and 1979 is surveyed in the book by Schwedt (2) with over 450 literature references to liquid chromatography. Most of that more recent research into these separations by LC involves high-performance liquid chromatography (HPLC).

Schwedt (3) also reviewed the literature in 1979 and presented 63 references. More recently, Willeford and Veening (4) reviewed 91 literature references to HPLC of organometallic and metal coordination compounds. Reviews appear also in the Japanese (5) and Chinese (6) literature. Research into these HPLC separations of inorganics and organometallics is the object of doctoral disserations (7–9), which, by their nature, also review the literature and thus serve as excellent sources.

This chapter presents some recent research in the HPLC of inorganics and organometallics. The presentation is not exhaustive but selective. The intent is to demonstrate how HPLC is available to solve problems of inorganic separations.

Table 6.1 Organometallics Separated by SEC

Compound	MW	Retention Volume (ml)
$Ni[P(O—p—tolyl)_3]_4$	1467	17.6
$Ni[P(O—C_6H_5)_3]_4$	1299	19.6
$Ni[P(O—CH_2CH_3)_3]_4$	723	20.6
$Ni[P(O—CH_3)_3]_4$	555	25.1
$P(O—p—tolyl)_3$	352	26.0
$P(O—C_6H_5)_3$	310	26.8
$P(C_6H_5)_2CH_3$	210	31.7
$P(C_6H_5)(OCH_3)_2$	170	32.3

Adapted from Yau et al. (11).

6.2 SIZE-EXCLUSION CHROMATOGRAPHY (SEC)

SEC utilizes a column to separate chemical species according to molecular size. The column contains particles of a rigid packing with pores. The size of the pores determines the molecular size range over which a column is useful. To give some insight into the relationship of pore size and molecular size, Vanacek and Regnier (10) find that the size exclusion effect is eliminated when pore size is 2–10 times the size of the molecule. Chemical species of sizes greater than the pore size are not retained. The smaller a species is the longer it is retained by the column and separation by size occurs.

Prior to the development of rigid particles, SEC was limited to low pressures and relatively slow separations. The development of rigid packings now allows higher pressures and more rapid separations. The field of modern SEC is presented in a monograph by Yau, Kirkland, and Bly (11).

Most of the work done in SEC involves organics, but applications to inorganic and organometallic species are appearing. Modern SEC is of value, for example, in separating the labile nickel complexes of the phosphorous esters in Table 6.1 (12). The column was 120 × 0.8 cm μ-Styragel, 100 Å, with dry tetrahydrofuran as eluent at a flow rate of 3.4 ml/min. Note that SEC is a method not only for separating complexes for analyses but also for purifying complexes. The reactants in Table 6.1 are of lower molecular weight and elute after the synthesized, higher molecular weight complexes.

Kirkland (13) has extended SEC from molecules to molecular clusters of inorganic colloids in the 1–50 μm range. The SEC column was <10 μm porous silica microspheres. Aluminosilicate sols were separated in a few minutes and SEC is suggested to be a useful technique for characterizing inorganic colloids.

An interesting application of SEC is to the separation of free Cd (II) and the complex with fulvic acid (FA), which is the putative soil organic acid for metal ion transport in the environment (14). Cadmium is of increasing concern as a heavy metal environmental pollutant and the understanding of cadmium transport requires knowledge of the equilibrium constant for

$$\text{Cd (II)} + \text{H}_x\text{FA} \rightleftharpoons \text{CdFA} + x\text{H}^+$$

The free and complexed Cd (II) are separated by two 25-cm HPLC columns of Sephadex G-10, a cross-linked dextran gel of 40–120 μm bead diameter. The mobile phase was distilled deionized water. This gel is used to fractionate species of molecular weight less than 700. Detection was by flame atomic fluorescence. The larger Cd-fulvic acid complex is unretained and elutes before hydrated Cd (II) with baseline separation. As with the phosphorous esters above, SEC is a viable method not only for separating these complexes for analysis but also for purification.

Many scientists still think of SEC as a slow separation method exclusively for polymers. The method is now rapid and applicable to much smaller molecules; with 100 Å columns it is a viable choice for heretofore intractable separation problems of inorganic chromatography.

6.3 REVERSED-PHASE HIGH-PERFORMANCE LIQUID CHROMATOGRAPHY (RPHPLC)

Most of the recent research in RPHPLC of inorganic and organometallics is conveniently characterized here by the detector being used.

6.3.1 Atomic Absorption (AA)

The peak storage sampling method has been used for determining the five lead species, tetramethyl-, trimethylethyl-, dimethyldiethyl-, methyltriethyl-, and tetraethyllead, by graphite furnace atomic absorption after separation with a Lichrosorb 10 μm-C-18 ODS column (15). The method is also applicable to organo tin compounds but both protocols are complicated by the need to digest the separated compounds with methanolic iodine to minimize errors in the atomic absorption analysis caused by incomplete decomposition in the flame and metal carbide formation. The more desirable detection is the continuous monitoring of the column ef-

Table 6.2 Compounds Separated on Lichrosorb C_{18}

Compound	Formula
Tetramethyllead	$(CH_3)_4Pb$
Trimethylethyllead	$(CH_3)_3(C_2H_5)Pb$
Dimethyldiethyllead	$(CH_3)_2(C_2H_5)_2Pb$
Methyltriethyllead	$CH_3(C_2H_5)_3Pb$
Tetraethyllead	$(C_2H_5)_4Pb$

Adapted from Messman and Rains (19).

fluent. This use of atomic absorption on-line in HPLC is included in the review by Van Loon in 1979 (16) and again in 1981 (17). The value of separating before using AA in metal speciation is clear. Without separation, only total content of a metal is determinable by AA. With separation, different compounds or different oxidation states of the same metal can be monitored.

The interfacing of column effluent and an AA spectrometer was done in 1973 (18) and used later by Messman and Rains (19). The tetraalkyllead compounds in Table 6.2 were separated (19) on the reversed-phase HPLC column C_{18} μBondapak, 300 × 3.9 mm i.d. The mobile phase was 70% acetonitrile and 30% water. Flow rate was 3.0 ml/min at ambient temperature. A short piece of small-diameter polyethylene tubing connected the column outlet and the atomic absorption nebulizer. The aspiration rate was adjusted to be slightly less than the column flow rate to avoid a post-column reduced pressure. Under these conditions baseline separation of the five tetraalkyllead compounds occurs in less than four minutes. Lead measurements were made at the 283.3-nm line using a 0.7-μm spectral band pass.

The superiority of AA detection over UV detection at 254 nm thus was shown, since gasoline samples do have components that mask the tetraalkyllead compounds. These interferences are coeluting unsaturated and aromatic hydrocarbons.

It has been found that the nebulization of sample solutions can be a critical problem in atomic absorption analyses. To minimize this, a nebulization technique based on a porous glass frit has been investigated (20). The device produces a smaller droplet size distribution than pneumatic nebulizers and has a higher sample transport efficiency. The mean droplet size is approximately 0.1 μm and up to 94% of the sample is converted to usable aerosol. Although a significant limitation may be a slow equilibration time, a sample delivery rate of 30 μl/min is used in continuous sampling mode. This is certainly not inconsistent with some use in HPLC.

6.3.2 Inductively Coupled Plasma (ICP)

ICP first appeared in the early 1960s (21–24) and became commercially available in 1974. The technique is replacing atomic absorption spectroscopy as the method of choice for metal analysis. Recent advances (25,26) have eliminated matrix and other problems, such as background emissions, previously present in ICP. The technique has been coupled to liquid chromatographs (27–29) and is being used for trace analysis and speciation of real world samples by Krull and his co-workers (30). That work used ion pair reversed-phase HPLC with a tetraalkylammonium salt as the ion pair reagent to separate arsenite, dimethyl arsenate and arsenate. The ICP detector gave calibration curves after hydride formation that were linear over several orders of magnitude with detection limits at 100–200 parts per billion or less. Separation of mercury cations employed an alkyl sulfonate anion as the ion pair reagent and ICP, after hydride formation, gave detection limits of 50–100 ppb.

To improve upon their detection limits Krull and his co-workers use direct current plasma spectroscopy (31). That work couples the best available separation technique and the technique of plasma spectroscopy to determine Cr(III) and Cr(VI) in real environmental samples.

The separation again used paired-ion reverse-phase HPLC. The HPLC conditions of mobile phase, column selection, flow rates, and capacity factors first were reported (32) and then applied (31) to ocean waters and other environmental samples. Calibration curves for the separated Cr(III) and Cr(VI) are linear over at least three orders of magnitude with a detection limit of 10 ppb. The authors found this chromatographic method of chromium ion speciation and plasma detection to be reproducible, valid, accurate, precise, and reliable for environmental samples. They suggest further that the method has, perhaps, the lowest detection limits ever reported for the analysis *and speciation* of chromium ions. The focus, however, should be on the use of chromatography to separate, and not upon detection limits solely. For example, in their work in interfacing HPLC and ICP, Hausler and Taylor (33) denote that detection limits in ICP vary with solvent and temperature of the detector chamber as in Table 6.3.

6.3.3 Visible Detection

The sixth paper of Schwedt in his series on the application of HPLC in inorganic analysis appeared in 1982 (34). PAN, 1-(2-pyridylazo)-naphthol,

Table 6.3 Detection Limits (ppb) as a Function of ICP Solvent

Element	Toluene without Cooling	Toluene with Cooling	Water without Cooling
Ag	318	25	7
Al	54,200	137	1,045
Ba	1,270	3	1
Cd	55	5	3
Cu	243	17	5
Fe	2,470	69	6
Si	3,050	197	27

Adapted from Hausler and Taylor (33).

is an excellent spectrophotometric reagent for many metals at low concentrations. The possibilities for separation of metal chelates of PAN by reversed-phase HPLC were investigated systematically. Only complexes of copper, nickel, and cobalt are separable. Other complexes show instability in the reverse-phase system. The LiChrosorbs RP-2, RP-8, and RP-18 were investigated and the best conditions for separations are a column of RP-2 and a mobile phase of acetonitrile/water/citrate buffer pH 5 (80:18:2), 0.01 M NH_4SCN. Flow rates are at 1.0 ml/min and detector wavelength setting is 565 nm. Baseline separations occur. The order of elution is copper (3.5 min), nickel (4.5 min), and cobalt (9.5 min). This method of separating complexes of PAN, and likely other organic spectrophotometric reagents, seems of less analytical utility than the method of Chapter 9, wherein ion chromatography is used to separate after which the metal complexes are formed for detection.

6.3.4 Ultraviolet Detection

In addition to those detection methods in which nonabsorbing species are converted to chromophores by complexing, ion pairing or more drastic chemical reactions, many methods that measure direct absorption of light by inorganic eluates exist. Method development is obviously indicated when the species absorbs in the visible region but light absorption may be overlooked if the absorption is in the ultraviolet, as in Table 9.6. Although desirable, it is not necessary for the peak absorption to be measured; the tail of an absorption peak at less than 190 nm may give sufficient energy for measurement in modern detectors set at wavelength higher than 190 nm. Measurements at wavelengths lower than 190 nm are limited by atmospheric absorption.

The recent paper by Skelly (35) is an example of direct use of UV detection in inorganic chromatography. Several reversed-phase columns were studied for separation of inorganic anions. From the point of retentiveness for IO_3^-, Br^-, NO_3^-, I^-, he found LiChrosorb RP-18 > LiChrosorb RP-8 > Partisil ODS-2 > Partisil C-8 > Partisil ODS-3 > Partisil ODS. The more lightly loaded columns give shorter retention times and sharper peaks. The eluent was 0.01 *M* octylamine in water, pH 6.0–6.5 at a flow rate of 2 ml/min. UV detection was at 205 nm. Under these conditions, the LiChrosorbs give poorer resolution and the retention times for Partisil-ODS are too short for practical applications.

The experiments done indicated that the separation involved an ion-exchange separation rather than ion pairing. The octylamine, however, must be in the eluent continuously if the method is to work.

Although direct UV detection does require that the eluate absorb, the method of detection has the advantage that large concentrations of non-absorbing species, such as chloride and sulfate, do not interfere. Thus, the method then was applied to analysis in sulfuric acid, brine, and silage.

6.3.5 Flame Photometric Detection (FPD)

The principle of flame photometric detection of nonmetals such as phosphorus and sulfur was originally described by Brody and Chaney (36) and later used in HPLC (37,38). In that use for phosphorus (38), the phosphorus-to-carbon selectivity ratio is 28,000, the limit of detection is 0.010 mg P/L, and a linear dynamic range of 50,000 is obtained for direct sample detection. Although phosphoric acids were used in that work, the detector can be used for nonionic phosphorus-containing materials and the detection limit improved by a factor of 20.

Later, McGuffin and Novotony (39) constructed a more sensitive phosphorus flame photometric detector for microcolumn liquid chromatography. The column, 0.5 mm i.d. × 10 m, was packed with silica or alumina particles of uniform size and operated typically at a flow rate of about 1 μl/min. The total column effluent is aspirated directly into a hydrogen-nitrogen-air flame and phosphorus emission monitored with a photomultiplier tube after wavelength selection by an interference filter at 530 nm. The method was applied to nanogram amounts of phosphorus in pesticides and to dimethylthiophosphinate derivatives of deoxycorticosterone, estradiol, cyclohexylamine, and phloroglucinol. The detection limit obtained with this system was 7×10^{-11} g/s phosphorus. Note that the method was applied not only to phosphorus-containing compounds directly but also to compounds that can form phosphorus-containing derivatives such as those in Table 6.1.

6.3.6 Amperometric Detection

Because of cost, selectivity, and sensitivity, electrochemical (EC) detection is becoming the method of choice in HPLC, but only if the electrochemistry is opportune. Although the exhaustive review of 61 papers of HPLC in inorganic analysis by Schwedt (3) has two references to coulometry (40,41), one to conductimetry (42), and one to potentiometry (43), there are no references to amperometric detection. Until very recently, amperometric detection in HPLC was limited to organic compounds, for which an excellent introductory and review paper is available (44). The technique of amperometric detection has been since extended to HPLC of organometallic compounds.

Dithiocarbamates form metal complexes that are used in separations by solvent extraction and in spectrophotometric analyses. Recently, these metal complexes were separated by reversed-phase HPLC (45) and shown to have electrochemical properties suitable for analysis (46). These properties then were used to simultaneously determine Cu(II), Ni(II), Co(II), Cr(III), and Cr(VI) as dithiocarbamates after HPLC separations (47).

Although both the oxidation and reduction reactions of these organometallics are well defined, the ubiqitous presence of reducible oxygen dissolved in the polar solvent eluting from the C-18 HPLC column results in the oxidative process being more desirable (47). Indeed, that research of Bond and Wallace demonstrates the necessity of the use of HPLC with amperometric detection to investigate and understand the electrochemistry, if spurious results are to be avoided. For example, in the case of nickel, the electrochemistry at gold electrodes is complicated by the presence of excess ligand, while on glassy carbon electrodes excess ligand has no effect on the oxidative analysis; in the case of cobalt, the glassy carbon electrode shows kinetics slower than at the gold electrode, but gives better results because of lower background signal; in the case of Cr(III), similar results were obtained with all electrodes: glassy carbon, gold, and platinum; in the case of Cr(VI), different results were obtained with the three electrodes. Only after suitable investigation by knowledgeable scientists of solvent, electrodes, and cyclic voltammetry can a correct HPLC-EC protocol be stated. If the investigator is not versed in electroanalytical chemistry and has no access to such expertise, incorrect analyses may result. The insightful investigator will understand the electrochemistry and be able to select potential for selectivity of detection and for minimizing error due to overlapping peaks.

An excellent example of where understanding of electrochemistry resulted in a new HPLC method for organometallics is in cancer chemotherapy (48). Cisplatin, mitomycin C, and mitoxanthrone are undergoing clinical use for the treatment of a variety of human cancers. The three drugs are separable on octadecylsilane columns. The two organics are detected easily by oxidation but cisplatin (dichlorodiammineplatinum (II)) can be oxidized only at potentials too anodic for practical use. The oxidation potential, however, is shifted by 0.10 *M* chloride to 0.80 V versus the saturated calomel electrode and quantitation is possible. These electrode potential shifts by solvent changes are well known to electroanalytical chemists. Use of such shifts will be of great value as electrochemical detection of column effluents of inorganic chromatography increases.

6.3.7 Differential Pulse Detection (DPD)

The best approach for determining easily oxidized or reduced eluates is simple amperometry where a constant potential is applied to the detecting electrode and the current monitored. For compounds that require relatively large anodic (oxidizing) or cathodic (reducing) potentials, the selectivity of this simple method is frequently inadequate.

MacCrehan (49) has improved selectivity in such cases by employing DPD, differential pulse detection (50,51). In this detection method, a constant potential called the base potential is applied to the detector. The base potential defines the background current and is selected to be just below the half-wave potential of the eluate. A potential pulse is periodically superimposed upon the base potential and selectivity is controlled by the magnitude of the pulse. Typical pulse heights are 25–100 mv and pulse time is 0.5 sec. To minimize interferences from capacitive currents, the faradaic components are measured after capacitive current decay. The electronic measurement time is less than in simple amperometry, where the current is monitored continuously. The faradaic current is measured for only 16 msec of the 500-msec cycle (49). The signal-to-noise ratio is thus less than in simple amperometry, and the limit of detection is typically less.

MacCrehan used DPD to determine methylmercury, tri-*n*-butyl tin, triethyl tin, and triphenyl tin. Columns were Spherisorb ODS (5 μm) or Whatman PXS cation exchange (10 μm). Solvents were 40–60% methanol in 0.1 mol/L sodium acetate buffer, pH 5.2.

MacCrehan developed a good separation for Hg^{2+}, CH_3Hg^+, $CH_3CH_2Hg^+$, and $C_6H_5Hg^+$ on the bonded-phase C_{18} column in less than 15 minutes. DPD gave a detection limit of about 2 ppb for CH_3Hg^+ and response was linear over 4 orders of magnitude.

Pertechnate TcO_4^- is the starting material for many diagnostic radiopharmaceuticals (52). At a reactor site ^{99}Mo as $^{99}MoO_4^{2-}$ is adsorbed on alumina and sent to a medical site. The decay scheme

$$^{99}\text{Mo} \xrightarrow[66\text{h}]{\beta^-} {}^{99}m\,\text{Tc} \xrightarrow[6\text{h}]{} {}^{99}\text{Tc}$$

generates radioactive ^{99}m TcO_4^- and stable $^{99}TcO_4^-$ in the column. The monovalent pertechnates are eluted with saline and divalent molybdate remains on the alumina. The pertechnate is then reduced and converted to the desired radiopharmaceutical. Although radioactive ^{99}m Tc is easily measured, quality control requires that total TcO_4^- be determinable. Concentrations of the eluate are 5×10^{-8} *M* to 5×10^{-6} *M* and liquid chromatography/electrochemistry has now been used for determining these low concentrations (53). The column was 5-μm NH_2-bonded Spherisorb, 25 cm × 4.6 mm i.d. The mobile phase was 0.01 *M*, pH 5.7 phosphate at flow rate of 1.0 ml/min. Cyclic voltammetry was used to investigate the electrochemistry of TcO_4^- and then differential pulse, normal pulse, and sampled DC investigated to determine the optimum detection technique. Although normal pulse gave greatest current, signal to noise is about equal for the three techniques. Sampled DC at mercury, however, was used because of a more desirable baseline. In the reductive mode, though, deoxygenating of the solvent is necessary. Linearity exists between 2.1×10^{-8} *M* to 1.0×10^{-4} *M* TcO_4^- with corresponding precisions of 8.3% and 0.2%. This application demonstrates the capabilities of reductive LCEC for determining *ionic* inorganic species eluting from the liquid chromatographic column.

6.4 ION-PAIR CHROMATOGRAPHY (IPC)

Although ion chromatography (IC, Chapter 9) has recently become the method of choice for determining anions and, less frequently, cations, some do not consider IC a high-performance system when compared to modern HPLC (54).

Those workers (54) have combined the principles of paired-ion chromatography and the performance of modern 5- and 10-μm C_{18} phases bonded to silica. The ion-pairing reagents in Table 6.4 were investigated. At high carbon number, the reagents are effectively permanently bound to the HPLC column, which acts then as a high-performance ion-exchange column. For example, using a 30-cm Varian 10-μm C_{18} column coated

Table 6.4 Ion-Pairing Reagents

Name	Structure	No. of Carbon Atoms
Cation exchange reagents		
l-hexanesulfonate	$CH_3(CH_2)_5SO_3^-$	C_6
l-octanesulfonate	$CH_3(CH_2)_7SO_3^-$	C_8
l-dodecylsulfate	$CH_3(CH_2)_{11}SO_4^-$	C_{12}
l-eicosylsulfate	$CH_3(CH_2)_{19}SO_4^-$	C_{20}
Anion exchange reagents		
tetraethylammonium-salt	$(CH_3CH_2)_4N^+$	C_8
tetrabutylammonium-salt	$(CH_3(CH_2)_3)_4N^+$	C_{16}
trioctylmethylammonium-salt	$CH_3(CH_3(CH_2)_7)_3N^+$	C_{25}
tetraoctylammonium-salt	$(CH_3(CH_2)_7)_4N^+$	C_{32}
tridodecylmethyl-ammonium salt	$CH_3(CH_3(CH_2)_{11})_3N^+$	C_{37}

Adapted from Cassidy and Elchuk (54).

with $C_{20}H_{41}SO_4Na$, baseline separation of Cu(II), Co(II), and Mn(II) occurred in 200 seconds at a flow rate of 2 ml/min of the eluent 0.075 mol/L tartrate, pH 3.4. Similar column efficiencies were found for anion IPC/HPLC columns. Using a 25-cm Whatman C_{18} column and coating with tridodecylmethylammonium iodide, baseline separation of IO_3^-, $S_2O_3^{2-}$, NO_2^-, NO_3^-, and I^- occurred in 400 seconds and the height equivalent of theoretical plate for each anion was 0.07, 0.07, 0.12, 0.08, and 0.14 mm, respectively.

The eluted metal ions were detected by absorption spectrophotometry at 530–540 nm after post-column reaction with 4-(2-pyridylazo)-resorcinol (PAR). Anion detection was by ultraviolet absorption spectrophotometry at 215 nm and by conductivity to detect such anions as Cl^-, F^-, SO_4^{2-}, and PO_4^{3-}, which do not absorb light sufficiently at 215 nm. The technique was used to quantitate NO_3^- and NO_2^- in process media such as 20% hydrogen peroxide, 6 molar ammonia as well as in deionized water. Cassidy and Elchuk (54) state that problems previously noted for conventional ion chromatography (55) were not observed by them.

O'Laughlin and Hanson (56) report the HPLC separation of iron(II), nickel(II), and ruthenium (II) as the 1,10-phenanthroline complexes. The μ-Bondapak/C18 and μ-Bondapak-CN columns were used. The cationic chelates, $Fe(Phen)_3^{2+,3+}$, $Ni(Phen)_3^{2+}$ and $Ru(Phen)_3^{2+}$ form ion pairs with alkylsulfonic acids. Separations of these ion pairs by reversed-phase HPLC are dependent on the reagents, pH, methanol/water, or acetonitrile/water ratios. The optimum composition of the mobile phase was 20%

of the organic component. At lower organic ratios, resolution increased but retention volumes are too high. The authors confirm a previous report (57) that plate counts and resolution similar to organic HPLC separations are obtained by forming ion pairs of inorganic complexes.

6.5 CROWN ETHERS

The geometry of large cyclic ethers is such that the molecules appear as and are called "crown ethers," which can form stable complexes with metal cations (58). These complexes have been used in solvent extraction, in ion-selective electrodes, and most recently in HPLC (59) as detailed here.

Silica gel (Wako-gel LC-10H, 10 μm, irregular) was reacted with (3-aminopropyl) triethoxysilane. The NH_2-modified silica gel was reacted with methacrylic anhydride to form a vinyl-modified silica gel, which then was co-polymerized with a vinyl crown ether. Two columns, 4 mm i.d., 30 cm, were studied: A. poly (benzo-15-crown-5)-modified silica and B. bis (benzo-15-crown-5)-modified silica.

The columns were investigated using alkali and alkaline earth metal ions. The research demonstrated the differences between ion exchange resin columns and these crown ether columns.

With cation exchange resins having sulfonic acid or carboxylic acid as exchange sites, retention is typically dependent on mass/charge ratios:

$$Li^+ < Na^+ < K^+ < Rb^+ < Cs^+; Mg^{2+} < Ca^{2+} < Sr^{2+} < Ba^{2+}; Na^+ < Ca^{2+} < Al^{3+}$$

With crown ethers, retention is dependent upon the relative size of the cation and the ring of crown ether. The ring in this research was $(\text{—C—O—C—})_5$ and the orders of elution were $Li^+ < Na^+ < Cs^+ < Rb^+ < K^+$ and $Mg^{2+} < Ca^{2+} < Sr^{2+} < Ba^{2+}$. Flow rates were 1.0 ml/min and pressure drops were 40 to 140 kg/cm^2. The effect of solute-solvent interactions upon the separations also was investigated by changing solvents (water/methanol ratios) and counterions. Decreasing water content increases retention time, but differently for each cation. For example, in one experiment when water content was decreased to 50% by using methanol, the retention times for Li^+, Na^+, Cs^+, Rb^+, and K^+ were 1.11, 1.63, 1.85, 4.09, and 6.64 times higher. When counterions are compared, in retention KCl < KBr < KI.

The authors (59) do denote that the theoretical plate numbers usually characteristic of HPLC could not be obtained. Separations of five ions require at least fifteen minutes. Attempts at improved efficiencies will involve use of regular rather than irregular silica gel, endcapping of unreacted silanol, and improved packing methods.

6.6 CONCLUSIONS

The use of HPLC for inorganic chromatographic analysis is now a standard procedure. The high resolution available for organic analyses is equally applicable to inorganic analyses. The method is particularly valuable for speciation of inorganic ions, whether they be free, complexed, or in compounds. The recent coupling of various detectors and chemical reactors in chromatographic effluents has hastened this inorganic use of HPLC.

The choice in inorganic column liquid chromatography is now between reverse phase and ion chromatography. Reverse-phase columns presently have more resolution than do ion-exchange resin columns. Since charged species can be ion paired for HPLC separations that have high resolution, the technique is now a routine method.

REFERENCES

1. J. Michal, *Inorganic Chromatographic Analysis,* Van Nostrand, New York, 1973.
2. G. Schwedt, *Chromatographic Methods in Inorganic Analysis,* Huthig Verlag, Heidelberg, New York, 1981.
3. G. Schwedt, *Chromatographia,* **12,** 613, 1979.
4. B.R. Willeford and H. Veening, *J. Chromatogr.,* **251,** 61 (1982).
5. K. Saitoh, *Bunseki,* **1979,** 548; *Chem. Abstr.,* **91,** 221722.
6. W. T. Kau, *Fen Hsi Hua Hsueh,* **8,** 474 (1980); *Chem. Abstr.,* **94,** 184834.
7. Z. Iskandarani, *Diss. Abstr. Int. B.,* **43,** (4), 1088 (1982).
8. W.A. MacCrehan, *Diss. Abstr. Int. B.,* **39,** (7), 3283 (1979).
9. D.W. Hausler, *Diss. Abstr. Int. B.,* **43,** (3) (1981).
10. G. Vanacek and F.E. Regnier, *Anal. Biochem.,* **121,** 156 (1982).
11. W.W. Yau, J.J. Kirkland and D.D. Bly, *Modern Size Exclusion Liquid Chromatography,* Wiley, New York, 1979.
12. C.A. Tolman and P.E. Antle, *J. Organomet. Chem.,* **159,** C5 (1978).
13. J.J. Kirkland, *J. Chromatogr.,* **185,** 273 (1979).
14. W.J. Taraszewski, M.R. Pitluck, D.T. Haworth, and B.D. Pollard, 1983 Abstracts, Pittsburgh Conference, Atlantic City, New Jersey, Paper No. 851.

15. T.M. Vickrey, H.E. Howell, G.V. Harrison, and G.J. Ramelow, *Anal. Chem.,* **52,** 1743 (1980).
16. J.C. Van Loon, *Anal. Chem.,* **51,** 1139A (1979).
17. J.C. Van Loon, *Am. Lab.* (Fairfield, CT), **13,** (5), 47 (1981).
18. S.E. Manahan and D.R. Jones, *Anal. Lett.,* **6,** 245 (1973).
19. J.D. Messman and T.C. Rains, *Anal. Chem.,* **53,** 1632 (1982).
20. L.R. Layman and F.E. Lichte, *Anal. Chem.,* **54,** 638 (1982).
21. T.B. Reed, *J. Appl. Phys.,* **32,** 821, 2534 (1961).
22. T.B. Reed, *Int. Sci. Technol.,* June, 42 (1962).
23. S. Greenfield, I.L. Jones, and C.T. Berry, *Analyst,* **89,** 713 (1964).
24. R. Wendt and V.A. Fassel, *Anal. Chem.,* **37,** 920 (1965).
25. H. Linn, *ICP Information Newsletter,* **2,** (2), 51 (1976).
26. J.C. MacDonald, *Amer. Lab.* (Fairchild, Ct), **15,** (9), 90 (1983).
27. D. Bushee, I.S. Krull, R.N. Savage, and S.B. Smith Jr., *J. Liquid Chrom.,* **5,** (3), 463 (1982).
28. D. Bushee, I.S. Krull, R.N. Savage, and S.B. Smith Jr., *J. Liquid Chrom.,* **5** (4), 693 (1982).
29. I.S. Krull, D. Bushee, R.N. Savage, R.C. Schleicher, and S.B. Smith Jr., *Anal. Letters,* **15** (A3), 267 (1982).
30. D. Bushee, I.S. Krull, S.B. Smith Jr., and P.R. Demko, 1983 Abstracts, Pittsburgh Conference, Atlantic City, New Jersey, Paper No. 397.
31. I.S. Krull, K.W. Panaro, and L.L. Gershman, 1983 Abstracts, Pittsburgh Conference, Atlantic City, New Jersey, paper No. 557A.
32. I.S. Krull, D. Bushee, R.N. Savage, R.G. Schleicher, and S.B. Smith Jr., *Anal. Lett.,* **15** (A3), 267 (1982).
33. D. Hausler and L.T. Taylor, *Anal. Chem.,* **53,** 1223 (1981).
34. G. Schwedt and R. Budde, *Chromatographia,* **15,** 527 (1982).
35. N.E. Skelly, *Anal. Chem.,* **54,** 712 (1982).
36. S.S. Brody and J.E. Chaney, *J. Gas Chromatogr.,* **4,** 42 (1966).
37. T.L. Chester, *Anal. Chem.,* **52,** 638 (1980).
38. T.L. Chester, *Anal. Chem.,* **52,** 1621 (1980).
39. V.L. McGuffin and M. Novotny, *Anal. Chem.,* **53,** 946 (1981).
40. L.R. Taylor and D.C. Johnson, *Anal. Chem.,* **46,** 262 (1974).
41. R.J. Davenport and D.C. Johnson, *Anal. Chem.,* **46,** 1971 (1974).
42. H. Small, T.S. Stevens, and W.C. Bauman, *Anal. Chem.,* **47,** 1801 (1975).
43. M.C. Francks and D.L. Pullen, *Analyst,* **99,** 503 (1974).
44. P.T. Kissinger, *Anal. Chem.,* **49,** 447A (1977).
45. A.M. Bond and G.G. Wallace, *Anal. Chem.,* **53,** 1209 (1981).
46. D. Coucouvanis, *Prog. Inorg. Chem.,* **26,** 301 (1979).
47. A.M. Bond and G.G. Wallace, *Anal. Chem.,* **54,** 1706 (1982).

48. R.P. Baldwin, W.N. Richmond, S. Houpt, K. Korfhage, T. Woodcock and K.A. Lalley, 1983 Pittsburgh Conference, Atlantic City, New Jersey, Abstract No. 006.
49. W.A. MacCrehan, *Anal. Chem.,* **53,** 74 (1981).
50. W.A. MacCrehan and R.A. Durst, *Anal. Chem.,* **50,** 2108 (1978).
51. D.G. Schwartzfager, *Anal. Chem.,* **48,** 2189 (1976).
52. G.P. Saha, *Fundamentals of Nuclear Pharmacy,* Springer-Verlag, New York, (1979).
53. J. Lewis, J.P. Fodda, E. Deutsch, and W.R. Heineman, *Anal. Chem.,* **55,** 708 (1983).
54. R.M. Cassidy and S. Elchuk, *Anal. Chem.,* **54,** 1558 (1982).
55. W.F. Koch, *Anal. Chem.,* **51,** 1571 (1979).
56. J.W. O'Laughlin and R.S. Hanson, *Anal. Chem.,* **52,** 2263 (1980).
57. S.J. Valenty and P.E. Behnken, *Anal. Chem.,* **50,** 834 (1978).
58. C.J. Pedersen, *J. Am. Chem. Soc.,* **89,** 7017 (1967).
59. M. Nakajima, K. Kimura, and T. Shono, *Anal Chem.,* **55,** 463 (1983).

CHAPTER

7

THIN-LAYER CHROMATOGRAPHY OF INORGANIC IONS AND COMPOUNDS

JOSEPH SHERMA

Department of Chemistry
Lafayette College
Easton, Pennsylvania

7.1 INTRODUCTION

Thin-layer chromatography (TLC) of inorganic ions and compounds involves techniques that are very similar to those used earlier in paper chromatography (1,2). In general, TLC is faster than paper chromatography, resolution is often superior, and the choice of sorbents is wider. The usual sorbent for the TLC of inorganics is silica gel, with some use of cellulose, alumina, and other miscellaneous layers. The sorbent can be used untreated or impregnated with various reagents to improve resolutions and/or detection.

Silica gel G has been applied most often, but more recently organic-bound "hard layers" of silica gel have found increased favor. Much of the work has been empirical in nature because of the uncertain purity of the silica gel and the variable effectiveness of attempts to purify the layers (3). In addition, the mechanism governing inorganic TLC on silica gel is not totally clear. Under some conditions, silica gel can act as a cation exchanger (4,5), and the selectivity of the exchanger and the resultant migration of solutes will be influenced by the nature of the mobile phase and the charge on the ions during chromatography (6). For example, in HCl media, certain ions such as Co and Cu form negative chloride complexes that may alter R_F values, compared to systems containing $HClO_4$ in which cations would be present. In addition to ion exchange, partition and adsorption mechanisms may be operative, depending upon the solutes and the mobile phase.

The presence of ions in a sample can cause zones with distorted shapes (e.g., tailing or streaking) or even multiple zonation, as well as interference with detection reagents. This can occur when two inorganic solutes are in close proximity on the chromatogram, or for organic and inorganic ions. To preclude the possibility of interaction, it is best to arrange conditions so that the cations and anions migrate independently on the layer. For example, if weak organic acids are chromatographed in a mobile phase containing a sufficiently strong, volatile acid, the undissociated acid will be formed; the original cation will then migrate in the dissociated form.

Theoretical aspects of inorganic TLC have been studied. Takitani (7), who developed a scheme for separating 20 common metal ions with three mobile phases, studied the relation between R_F and solvent composition, as well as the separation of different valency states of the same elements (As, Sn, Fe, Sb, and Hg). He also investigated the influence of relative humidity on R_F values and the effect of various anions on the TLC of cations (8). Seiler (5) has published a systematic study of the behavior of Co^{2+}, Cu^{2+}, Fe^{2+}, and Fe^{3+} in order to assess the factors involved in their separation on silica gel layers. A variety of solvents were evaluated, consisting of mixtures of either HCl or $HClO_4$ with lower alcohols, ketones, and cyclic ethers.

The following sections will review procedures for and applications of TLC as applied to the separation of cations, anions, and inorganic compounds, complexes, and derivatives.

7.2 SAMPLE PREPARATION

Methods for preparation of samples are similar for TLC and paper chromatography. Standard solutions of the common metal cations are prepared by dissolving reagent grade nitrates or chlorides in water, 0.1 *M* HNO_3, or 0.1 *M* HCl to a final metal concentration of 0.05–0.2 *M*. Some ions require the presence of larger amounts of acid or other reagents to stabilize the solutions. For example, Sn(II), Sn(IV), Sb(III), and As(III) are prepared by dissolving chlorides in 4–6 *M* HCl; oxides of the rare earths and related compounds are dissolved in 0.5–6 *M* HNO_3; Mo(VI) is prepared as a solution of molybdic acid made slightly basic with NaOH; and Ti(IV) is often used as a solution of $TiCl_4$ in 0.2 *M* H_2SO_4 and 0.3% hydrogen peroxide (9–12). From 2 to about 25 μl of test solution is spotted at the origin, the exact amount depending upon the ultimate sensitivity of the detection method. Overloading should be avoided, as it may cause tailing.

For systematic analysis of cation unknowns, samples are prepared by procedures generally employed in classical inorganic qualitative analysis. In one method, a small amount (about 0.2 g) of an unknown sample is boiled with 2 ml of 2 *N* HNO_3, adding a few drops of H_2O_2 if dichromate,

chromate, or permanganate ions are present. The solution is centrifuged after cooling, if necessary, and the clear supernate is diluted with water until the acid concentration is about 0.1 *N* (13).

Anions such as tartrate, oxalate, formate, etc., may interfere with the chromatographic analysis of cations and can be removed by carbonate fusion (NH_4^+ may be lost as well), exchange for acetate on a column of resin (1,14), or precipitation (15).

Organic or biological samples are generally ashed before chromatographic analysis. For example, in the determination of Fe, Mn, Zn, and Cu in plant material on cellulose with direct reflectance densitometry, the sample is dry-ashed to remove organic matter, and the elements of interest are separated from phosphates and other cations on an ion-exchange column (16).

Anion samples are prepared in a manner similar to that of cations. Soluble samples are dissolved in a small volume of water or dilute acid or alkali and spotted directly. Samples are concentrated by precipitating the anions and redissolving in a smaller volume, or the precipitates can be applied directly to the layer (17). Interfering cations are removed by ion exchange (18). Biological materials are usually fused with carbonate or ashed (19).

A solvent extraction system was applied prior to TLC for the individual determination of numerous cations and anions in solutions of alloys. Steel, ceramics, rocks, dust, and many other types of inorganic samples have also been analyzed by TLC after appropriate sample-preparation procedures (20). Trace analysis of toxic elements in organic, biological, and environmental samples by TLC in combination with extraction as metal dithizonates has proven to be particularly valuable (21).

A detailed description of a scheme for separation of a sample into analytical groups for total analysis of cations and anions was given by Seiler (22). The groups obtained include anions, basic thioacetamide precipitate (copper group), acidic thioacetamide precipitate, ammonium carbonate precipitate, and the alkali group. Cellulose has been used for the comprehensive TLC separation of main and subgroup elements, including U, Ce, and Th (23).

Complexes of metals are sometimes formed prior to TLC to improve separations compared to the simple ions. For example, diethyl dithiocarbamate (24), dithizonate (25), acetylacetone (26), and EDTA (27) complexes have been used. More often, complexes formed in situ by the presence of a complexing agent in the mobile phase (28) or layer (29) are the basis of the TLC resolution.

7.3 THIN LAYERS

The first application of inorganic TLC was reported by Meinhard and Hall in 1949 (30). Starch-bound layers of alumina plus Celite were used by these workers for the radial separation of zinc and iron. Since this historic work, layers of silica gel G (gypsum binder) and S (starch binder) have been mainly used for inorganic separations, wtih some use of cellulose and other assorted sorbents. The layer has been impregnated with complexing agents or liquid ion exchangers in some cases to improve separations. Layers are usually prepared on glass and are typically 20 cm × 20 cm × 250 μm in size.

Since silica gel contains metallic impurities (e.g., Na^+, Mg^{2+}, Ca^{2+}, Fe^{3+}), washing with acid and water is usually carried out to effect their removal. Calcium sulfate would also be removed from silica gel G by this treatment, so this must be replaced or substituted with starch if this binder would be preferable. Specially purified silica gel powder (e.g., silica gel HR) is available commercially and may be suitable for use without further purification. instructions for purifying silica gel for inorganic TLC were given by Seiler and Rothweiler (29).

Commercially precoated "hard" silica gel layers, containing various organic binders in place of gypsum or starch, are becoming increasingly popular compared to "homemade" layers. Layers prepared from specially washed silica (silica gel HR) should always be purchased for inorganic work and will often be directly usable. If additional purification is found to be necessary, a predevelopment with the chosen mobile phase may be advantageous.

Mobile phases will usually not be directly transferable from paper chromatography to silica gel TLC, because of the ion-exchange properties of the silica gel. Mobile phases from paper chromatography may be directly applicable to cellulose TLC (31,32), but this may not always be true because of differences in the fibrous or crystalline structure of the cellulose comprising the papers and layers.

Commercial high-performance silica gel plates (10 × 10 cm), containing a dense layer of small-particle silica gel, have not yet been used to any great extent for inorganic TLC, although a few applications have been reported (33). However, HP layers should provide the potential for rapid, high-efficiency, high-sensitivity separations of inorganic ions and compounds (34).

Aluminum oxide (alumina), organic ion-exchange resin (Amberlite, Dowex) (35–37), inorganic ion exchanger (38), and cellulose ion exchanger (e.g., sulfoethyl, ECTEOLA, carboxymethylcellulose, and cellulose phosphate) layers (39–41) have all been used for inorganic TLC. Cellulose and silica gel have been impreganted with a complexing agent such as EDTA (29) or liquid ion exchangers [e.g., tri-*n*-octylamine, bis(2-ethylhexyl) phosphate, or Amberlite LA-1] to improve resolution of certain mixtures (42–48). A double-layer plate consisting of ion-exchange resin and cellulose impregnated with dimethylglyoxime has been used for separation of Fe^{3+}, Ni^{2+}, and Co^{2+} (49).

Whatman LKD channeled silica gel layers contain an inert preadsorbent zone that facilitates sample application. Different volumes of standards and crude samples are applied to the individual channels, and the material moves with the solvent front to the preadsorbent-silica gel interface, where the solutes are uniformly concentrated into tight bands. Separation occurs as development continues into the silica gel area. Little work has been reported for inorganic TLC on these layers, but they should be ideal for those systems in which the inorganic compounds of interest are not retained on the preadsorbent.

C_{18} and C_8 bonded silica gel layers were used for the separation of metal tetraphenylporphyrin chelates. Mutual separations of the Mg^{2+}, Ni^{2+}, Cu^{2+}, Zn^{2+}, Cd^{2+}, Mn (III), and Cd^{2+} chelates were accomplished by two-dimensional TLC with acetone-propylene carbonate (2:8) and acetone as the two solvents (50).

Eight novel bonded-phase silica gel substrates were evaluated for the separation of Fe, Cu, Ni, and Zn with 29 mobile phases. Detections were made by solutions of 1% diphenylcarbazide in ethanol and 1% $K_4Fe(CN)_6$ in water. The best separation was obtained with an immobilized trifluoroacetylacetone substrate and 10% trifluoroacetylacetone in acetone mobile phase (51).

The technology of TLC sorbents has been discussed by Beesley (52).

7.4 APPLICATION OF SAMPLES

Samples are applied in TLC with a variety of manual and automated devices that have been previously described (53). For convenience, the point of sample application (and also the development finish line) can be marked lightly with a pencil, if the surface of the layer is not disturbed. This is easier to accomplish on precoated organic-bound hard silica gel layers than on soft silica gel G or cellulose.

Application of sample solutions and standards to be used for comparison is most often made using a calibrated syringe or micropipet, such as Drummond Microcaps (Fig. 7.1). To sharpen resolution, initial zone sizes should be kept as small as possible by application of low microliter volumes. Larger volumes can be applied to a given origin in small portions

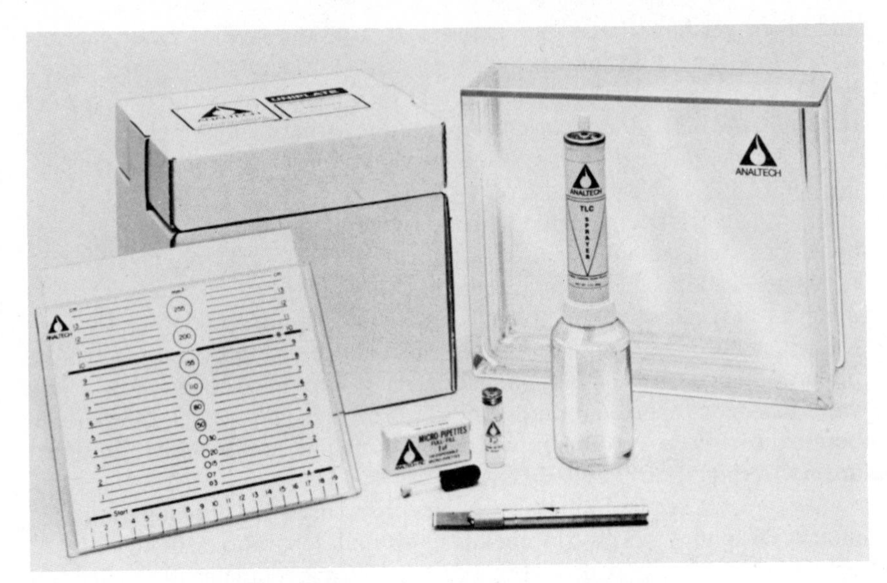

Figure 7.1. Basic equipment for TLC, including precoated thin layers (in box), rectangular developing chamber, spotting guide, detection reagent sprayer, disposable micropipets, and adsorbent scraper. (Photograph courtesy of Analtech, Newark, Delaware.)

with drying of the solvent in between individual aliquots. The weight applied should be as low as possible consistent with the sensitivity of the detection method, in order to avoid overloading. Because of the difficulty in drying aqueous solutions, addition of a portion of an organic solvent to the sample solutions, if solubility considerations allow, will speed sample application.

Improved resolution is often achieved if the samples are applied as very thin, short bands or streaks rather than individual round spots. With practice, streaking can be accomplished manually, using conventional syringes or micropipets; or small dots can be placed side by side to simulate a streak. Streaking can also be done with automated devices, such as the Camag Linomat. Devices that apply streaks across a 20×20 cm plate for preparative TLC are also available commercially.

For in situ quantitation, uniformity of the initial zone sizes of samples and standards is especially critical. Use of automated sample applicators or Whatman preadsorbent plates is recommended for accurate and precise quantitative TLC.

Effects of sample application on TLC separations have been reviewed by Felton (54), including spotting, streaking, and preadsorbent plates.

The solution from which sample is applied and the form of cations can be critical in obtaining good separations. For example, the separation of UO_2^{2+} from other metal ions obtained in ethyl acetate-water saturated ether-tri-*n*-butylphosphate mobile phase (55,56) required adjustment of the sample solution to 4.7 *N* in HNO_3. This resulted in formation of a uranyl nitrate-(tri-*n*-butylphosphate) complex that migrated readily in the mobile phase, while the other cations remained at or near the origin. Alkaline earths were applied to silica gel S layers in the form of the acetates, and 0.001 ml of acetic acid was added to the sample spots to avoid streaking (22).

7.5 MOBILE PHASES (SOLVENT SYSTEMS)

The solvent systems for inorganic TLC are mostly extensions of those used earlier for similar substances in PC, with modifications to suit the type of sorbent being used. For ionic species, a mixture of organic solvents together with some aqueous acid, base, or buffer is typical. For nonionics, water-containing and anhydrous organic solvents are typically used. Some unusual mobile phases have also been reported, such as a molten eutectic mixture of $Li(NO_3)$ and $K(NO_3)$ at 270°C to separate Ag^+, Tl^+, Pb^{2+}, Hg_2^{2+}, and Hg^{2+} (57).

A search of the literature will usually reveal a layer-mobile phase system that has been found suitable for the mixture of interest. Or, a mobile phase for a similar mixture can be modified to provide the required separation. Sources of information on numerous TLC systems and separations are given in references (22,58–60).

Typical mobile phases for different types of thin layers are shown in Table 7.1. Numerous other solvent systems are reported below in the section on applications.

7.6 DEVELOPMENT

Techniques for development of layers have been documented in standard texts on TLC (53,64). A single one-dimensional ascending development in a vapor saturated rectangular glass tank (Figure 7.1) is usually carried out for inorganic TLC, although multiple (65), two-dimensional (65,66), and radial (67) development have also been utilized, as well as low-volume sandwich chambers (68). Since many mobile phases for inorganic separations are partially or totally aqueous, development times may be relatively lengthy.

Table 7.1 **Typical Mobile Phases for Different Types of Thin Layers**[a]

Layer	Mobile Phase	Ions Separated
Silica gel G	Butanol-1.5 N HCl-2,5-hexanedione (100:20:0.5)	Hydrogen sulfide group
Silica gel G	Acetone-conc. HCl-2,5-hexanedione (100:1:0.5)	Ammonium sulfide group
Silica gel G	Ethyl acetate (H_2O sat.)-tributyl phosphate (50:2)	U, Ga, Al
Silica gel G	Ethanol-acetic acid (100:1)	Alkali metals
Silica gel G	Acetone-butanol-conc. NH_4OH—H_2O (65:20:10:5)	Halogens
Silica gel G	Methanol-conc. NH_4OH—10% trichloroacetic acid-H_2O (50:15:5:30)	Phosphates
Dowex 1-cellulose (1:1)	1 M aqueous $NaNO_3$	Halogens
Cellulose	Halogen acid-alcohol mixtures	Groups IA, IIA, IIIB, IVB, VB, VIB, and transition metal cations
Cellulose	Butanol-water-HCl (8:1:1)	Fe, Al, Ga, Ti, In
Cellulose	Acetic acid-pyridine-conc. HCl (80:6:20)	Ammonium sulfide group
Ammonium phosphomolybdate	0.1–5 M NH_4NO_3	Alkali metals
DEAE cellulose	Sodium azide-HCl mixtures	Cd, Cu, Hg
Amberlite CG400 and CG120	Various concentrations of HCl and HNO_3	Pb, Bi, Sn, Sb, Cu, Cr, Hg

[a] From Kirchner (1978); Bobbitt (1963); Randerath (1963); and Gagliardi and Brodar (1970).

7.7 DETECTION

The detection methods for inorganics are similar to those in paper chromatography, but corrosive reagents that attack paper can, if necessary, be used on layers of inorganic adsorbents such as silica gel or alumina. Typical chromogenic and fluorogenic spray reagents for cations include potassium iodide (0.2%, aqueous), hydrogen sulfide (saturated aqueous solution), ammonium sulfide (0.2 *N*, aqueous), quercetin (0.1%, alcoholic), 1-(2-pyridylazo)-2-naphthol (PAN) (0.2%, methanolic), oxine (8-hydroxyquinoline) (1%, methanolic, view under visible and UV light), and sodium rhodizonate (0.5%, aqueous). Reaction with dithizone to produce colored dithizonate chelates of many metals is particularly suitable if quantitative spectrometric analysis (in situ or after elution) is to be carried out. Anions are detected with bromocresol purple (0.1%, alcoholic), 1% ammoniacal silver nitrate + 0.1% alcoholic fluorescein/UV light, zirconium alizarin lake (0.1% in HCl solution), and ammonium molybdate (1%, aqueous) followed by $SnCl_2$ (1%) and HCl (10%). Details of the reagent preparation and conditions for these and many other detection methods are given in reference (58). Typical detection limits range from 10 ng to several μg. The Weisz ring oven technique has been combined with TLC for determination of various metals (69).

If chromogenic or fluorogenic reagents are not suitable, labeled species can be detected using radioactivity scanners and detectors with counting equipment. This is an especially valuable method for detection and quantitation of very low levels of inorganic material (70–72).

7.8 IDENTIFICATION AND QUANTITATION

Identification of an inorganic substance is made by comparing R_F values of samples and standards chromatographed on the same layer. A coincidence of colors with one or more selective chromogenic reagents is added confirmatory evidence. It is not always true that ions chromatographed alone and in mixtures will have the same R_F values, as is usually the case for low concentrations or organic compounds. For example, R_F values for Rb and Cs individually and in an alkali metal ion mixture were 0.5/0.27 and 0.38/0.15, respectively (73). This must be considered when identifying inorganic substances based on R_F values.

Quantitation is obtained by visual comparison of zone sizes and color intensities between samples and bracketing standards on the same plate. Errors are $\sim\pm 30\%$ by this procedure. Planimetric evaluation of spot areas, which are proportional to concentration, can reduce the errors to

$\sim\pm10\%$. K^+ and Mg^{2+} have been determined on silica gel G using ethanol-methanol (1:1) + 1% acetic acid as the mobile phase, acid violet 6BN for detection, and the square root of spot area for quantitation (74).

More accuracy and precision are obtained if zones of samples and standards are scraped and collected and the analytes eluted for determination by some microanalytical procedure (e.g., electroanalytical, spectrometric, or radiochemical). Procedural details have been described earlier (60).

The most convenient approach to quantitation in TLC is by in situ measurement of the absorption of visible or ultraviolet light or of fluorescence (70). Transmission or reflectance scanning can be used for this photometric evaluation. An entire book has been devoted to the principles, practice, and applications of scanning densitometric quantitation of thin-layer chromatograms (75), which results in errors and standard deviations of $\sim\pm1$–5% for inorganic TLC (22).

Application of samples and standards in small, uniform zones is probably the most critical step in densitometry. Manual and automatic sample application to TLC, HPTLC, and preadsorbent plates for quantitative TLC has been discussed by Touchstone and co-workers in two papers (76,77). These papers illustrate the advantages of preadsorbent plates, application of thin lines or streaks rather than spots for optimum resolution, and the importance of using automatic spotting if volumes greater than 15 μl are applied to nonpreadsorbent plates.

An example of an application of densitometry in inorganic quantitative TLC is the determination of boron in water and soil samples at ppb levels (78). Soil samples were dried at 200°C for 2 hours, 5 g was sieved, and 5 ml of deionized water and 0.5 ml of 4 *N* HCl were added. After Vortex mixing and filtration, the extraction was repeated. The combined extracts were concentrated, acidified to pH 2, and filtered through a 40 μ Millipore filter. Water samples were concentrated from 25 to 0.5 ml and acidified to pH 1. Ten to 25 μl aliquots of the samples were applied to an MN 300 cellulose layer using an automatic syringe-type spotter. The layer was developed with butanone-water-ethylene glycol (85:13:2) to provide an R_F of 0.43 for boron. After drying, the chromatogram was sprayed heavily with azomethine H reagent (0.1 g per 10 ml of 1% ascorbic acid). After one hour, the boron zone was scanned at 400 nm. Calibration curves were linear for 50–450 ng of boron standards. Boron-free (soft) glassware was used throughout the procedure. Figure 7.2 shows the calibration curve for boron and Fig. 7.3 shows typical scans of boron zones.

Inorganic anions, in mixtures with organic compounds, have also been quantitated by densitometry. Separations of NO_2^-, IO_4^-, ClO_3^-, BrO_3^-, $Cr_2O_7^{2-}$, MnO_4^-, SeO_3^{2-}, and other anions were obtained on silica gel

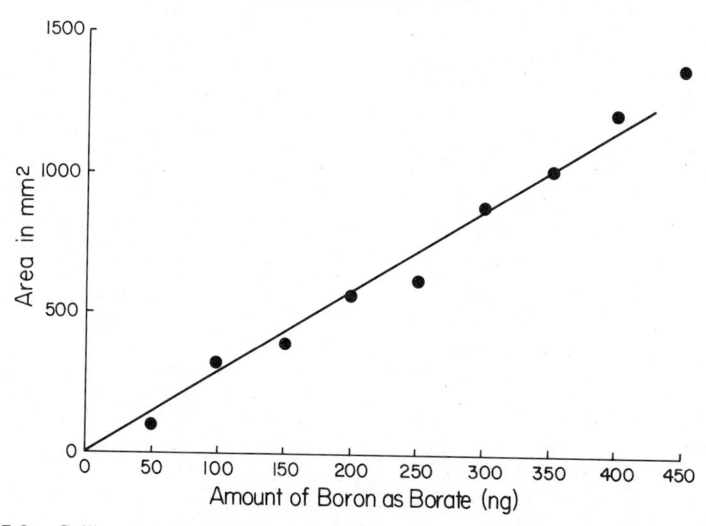

Figure 7.2. Calibration curve for 50 to 450 ng of boron (boric acid). After Touchstone et al., Reference 78.

layers developed with propanol-ammonium hydroxide (2:1). Detection as blue spots was made by spraying with 2% diphenylamine in sulfuric acid. Calibration curves were linear for 1–5 μg of anion, and detection limits were ~100 ng. The relative standard deviations were 6% and 8%, respectively, for determining NO_3^- and $Fe(CN)_6^-$ in molasses (79). Polyphosphates in foods were determined by densitometry after separation by TLC (80).

Additional approaches to quantitation of inorganic substances include in situ electrometric measurements, radiometry, and X-ray fluorescence. Several quantitative TLC evaluation methods, such as determination of spot size, in situ measurement of absorption in transmitted light, and radiometric measurement of radioactive isotopes, have been compared for practicality and results (81). A large number of quantitative applications have been tabulated in a recent book by Schwedt (82), including detection limits, errors, and precision.

7.9 APPLICATIONS OF TLC

The following subsections present a selective review of published references on the TLC of inorganic cations, anions, and compounds. The references were chosen to include important species and a variety of chromatographic systems. In all cases, the substances studied, the layer, the

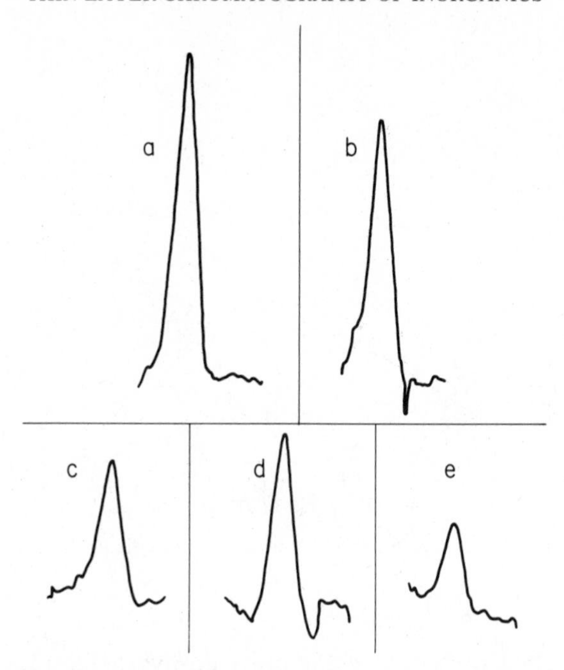

Figure 7.3. Densitometer scans of (*a*) 100 ng of boron (borate) standard; (*b*) tap water from rural town (109 ng); (*c*) river water at high tide (63 ng); *d*) creek water above the tide line (92 ng); (*e*) creek at low tide (38 ng). The scans are from chromatograms of extracts of water from the stated sources. After Touchstone et al., Reference 78.

mobile phase (in v/v proportions, unless otherwise noted), and the literature source are given. In many cases, detection spray reagents and representative, approximate R_F values, or migration sequences are also included. The references are arranged according to the type of ions or compounds chromatographed.

Applications of TLC to inorganic substances have been made for the purpose of analysis of certain pairs or groups (e.g., alkali metals, halides), or for so-called total analysis of complex mixtures. As an example of the latter, 20 common metals were analyzed on silica gel S with three developing solvents: acetone-3 *N* HCl (99:1) for separation of Ni, Co, Cu, Fe, Pb, Mn, Cr, and As; methanol-butanol-35% HCl (8:1:1) for Ba, Sr, Ca, Mg, Al, NH_4, Na, K, and Li; and butanol-benzene-1 *N* HNO_3-1 *N* HCl (50:46:2.6:1.4) for Sb, As, Cu, Cd, Sn, Bi, Zn, and Hg. The separation of some elements with different valence states was also examined. TLC detection limits were found to be 10 to 100 times more sensitive than those with paper chromatography (7,8). Other examples of total cation analysis include the studies of Zetlmeisl and Haworth (21 cations on cellulose) (32); Kaushik and Johri (27 cations on silica gel G) (83); Husain and Eivazi (44 cations on the inorganic exchanger stannic acetate) (84); and Hashmi et al. (39 cations on silica gel and alumina, after solvent extraction) (67).

Kawanabe et al. (85) applied TLC on purified silica gel S to the total analysis of anions. Group A, containing SCN^-, I^-, Cl^-, $Fe(CN)_6^{3-}$, $Fe(CN)_6^{4-}$, ClO_3^-, BrO_3^-, IO_3^-, and NO_3^-, was separated with acetone-water (10:1). Group B, containing F^-, NO_2^-, $S_2O_3^{2-}$, SO_4^{2-}, CrO_4^{2-}, PO_4^{3-}, AsO_4^{3-}, and AsO_3^{3-}, was chromatographed in methanol-butanol-water (3:1:1). Group C, containing CrO_4^{2-} and BO_2^{2-}, was developed with butanol saturated with 2 *N* nitric acid.

References (20) and (21) contain tabulations of applications of TLC and HPTLC to the analysis of a variety of inorganic, organic, biological, and environmental samples. Reference (53) should also be consulted for references on numerous applications of inorganic analysis.

7.9.1 Cations

(a) Copper (hydrogen sulfide) subgroup (analytical group IIa); MN silica gel S-HR; *n*-butanol-1.5 N HCl-hexanedione (acetonylacetone) (100:20:0.5); 2% KI, NH_3 vapor, H_2S gas; R_F values: Cu^{2+} 0.15, Pb^{2+} 0.33, Cd^{2+} 0.50, Pb^{2+} 0.64, Hg^{2+} 0.90 (3).

(b) $(NH_4)_2S$ group (analytical groups IIIa and IIIb); MN silica gel S-HR; acetone-conc. HCl-acetonylacetone (100:1:0.5); NH_3 vapor, 0.5% oxine in 60% ethanol, 366 nm UV light; R_F values: Ni^{2+} 0.20, Al^{3+} 0.30, Cr^{3+} 0.38, Mn^{3+} 0.60, Co^{2+} 0.70, Zn^{2+} and Fe^{3+} 0.90 (5).

(c) Ammonium carbonate group (alkaline earths; analytical group IV); MN silica gel S-HR; ethanol-*n*-propanol-acetic acid-acetonylacetone-water (37.5:37.5:5:1:20); 1.5% aqueous violuric acid (acid violet); R_F values: Ba^{2+} 0.56, Sr^{2+} 0.64, Ca^{2+} 0.70 (22).

(d) Alkaline earths; cellulose MN 300 HR; dioxane-conc. HCl-H_2O (58:12:30) and methanol-conc. HCl-H_2O (73:12:15 or 8:1:1); conc. NH_4OH followed by 2% oxine in ethanol and viewing under 366 nm UV light; migration sequence: $Be^{2+} > Mg^{2+} > Ca^{2+} > Sr^{2+} > Ba^{2+} > Ra^{2+}$ (62,86).

(e) Alkali metals (analytical group V); MN silica gel S-HR; absolute ethanol-acetic acid (100:2); 1.5% violuric acid; R_F values: K^+ 0.10, Na^+ 0.40, Li^+ 0.50 (29).

(f) UO_2^{2+}, Fe, Co, Cu, Ni, Al, Th; MN silica gel S-HR; ethyl acetate-water saturated ether-tri-*n*-butylphosphate (50:50:2); 0.25% PAN in ethanol; UO_2^{2+} migrates while the other ions remain at or near the origin (3,55).

(g) Ga, Al; MN silica gel S-HR; acetone-conc. HCl (100:0.5); 0.5% oxine in 60% ethanol, NH_3 vapor, 366 nm UV light; Ga^{3+} migrates ahead of Al^{3+} (56).

(h) Mn, Fe, Co, Ni, Cr, As, Sb, Th, U, V, Se, Mo; silica gel G impregnated with NTA; 5% aqueous NH_3-ethanol-acetone-acetic acid (80:120:4:1.5); 0.5% dithizone in chloroform; representative R_F values: Fe^{2+} 0.40, Ni^{2+} 0.75, As (III) 0.66, Th(IV) 0.32, U(V) 0.06, V(V) 0.22, Se(VI) 0.83 (87).

(i) Toxic metals; cellulose MN 300; (A) acetone-4 N HCl (70:30), (B) acetone-25% HNO_3 (70:30); 0.5% aqueous Na_2S or saturated alizarin in 96% ethanol, exposure to NH_3 vapors, plus other general and specific reagents; R_F values in mobile phase (B): Ag^+ 0.30, As^{3+} 0.40, Ba^{2+} 0.10, Be^{2+} 0.60, Bi^{3+} 0.90, Cd^{2+} 0.40, Ce^{4+} 0.10, Co^{2+} 0.30, Cu^{2+} 0.50, Hg^{2+} 0.90, Mn^{2+} 0.50, Ni^{2+} 0.30, Pb^{2+} 0.30, SeO_3^{2-} 0.50, Sb^{3+} 0.90, TeO_3^- 0.20, Tl^+ 0.20, UO_2^{2+} 0.90, Zn^{2+} 0.30 (88).

(j) Hydrogen sulfide group; cellulose MN 300; *n*-butanol saturated with 3 *N* HCl; aqueous 0.2% KI, hydrogen sulfide water, 0.2 *N* $(NH_4)_2S$, alcoholic 0.1% quercetin followed by NH_3 vapor; migration sequence: $Ag^+ < Cu^{2+} < Pb^{2+} < Sb^{5+} < Bi^{3+} < As^{5+} < Cd^{2+} < Sn^{4+} < Hg^{2+}$ (89).

(k) Ammonium sulfide group; cellulose MN 300; acetic acid-pyridine-conc. HCl. (80:6:20); 0.2% methanolic PAN followed by NH_3 vapor, and 1% methanolic oxine followed by viewing under 366 nm UV light; migration sequence: $Cr^{3+} < Al^{3+} < Ni^{2+} < Mn^{2+} < Co^{2+} < Zn^{2+} < Fe^{3+}$ (89).

(l) Analytical group IIa; (A) cellulose and (B) silica gel; (A) *n*-butanol-$CHCl_3$ – 6 *M* HCl (8:1:1), (B) *n*-butanol-1.5 *M* HCl-methyl ethyl ketone (75:15:2); 0.5–1% aqueous Na_2S and 0.05% dithizone in $CHCl_3$; R_F values: (A) Pb^{2+} 0, Cu^{2+} 0.06, Bi^{3+} 0.38, Cd^{2+} 0.61, Hg^{2+} 0.71; (B) Pb^{2+} 0.38, Cu^{2+} 0.23, Bi^{3+} 0.60, Cd^{2+} 0.66, Hg^{2+} 0.86 (90).

(m) Hg, Cd, Ni, Cu, Fe, Pb, Sb, Ag; acid washed alumina; isopropanol-1 *M* HCl + 0.5 *M* Br^- (1:1); dimethylglyoxime and $(NH_4)_2S$; migration sequence: $Hg^{2+} > Cd^{2+} > Ni^{2+} = Cu^{2+} > Fe^{3+} > Pb^{2+} > Sb^{3+} > Ag^+$ (91).

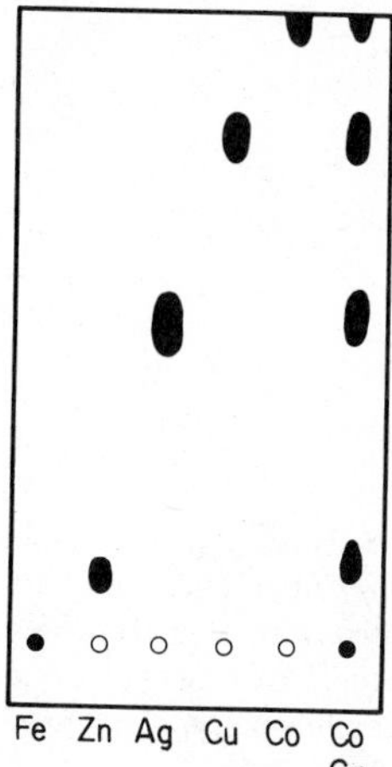

Figure 7.4. Separation of a number of elements on TBP Corvic layers by development with 4 *N* hydrochloric acid. After Pierce and Flint, Reference 92.

(n) Rare earths; Corvic (vinyl chloride-vinyl acetate copolymer) coated with bis(2-ethylhexyl)phosphoric acid; 0.25 or 0.80 *M* HCl; 1% oxine in ethanol-H_2O (1:1) followed by NH_3 vapor; migration sequence: La > Pr > Sm; Gd > Tb > Er (42).

(o) Ni, Cu, Ag, Zn, Fe; Corvic (vinyl chloride-vinyl acetate polymer) coated with tri-*n*-butyl phosphate; 4 *N* HCl; oxine, diphenylthiocarbazone, or radioactivity; migration sequence: $Co^{2+} > Cu^{2+} > Ag^{+} > Zn^{2+} > Fe^{3+}$ (Fig. 7.4) (92).

(p) Ni, Co, Mn, Zn, Al, Fe, Cu, Cd, Pb, Bi, Hg, Sb; silica gel impregnated with Amberlite LA-1 liquid anion exchanger; 6–10 *N* HCl; H_2S, dithiooxamide, or oxine; migration sequence: $Ni^{2+} > Cu^{2+} > Co^{2+} > Fe^{3+}$ in 8.5 N HCl; $Pb^{2+} > Bi^{3+} > Hg^{2+}$ in 6 *N* HCl (93).

(q) Nineteen assorted metal ions; zirconium phosphate; 0.5 *M* HCl; PAN, dithizone; R_F values: Sn^{2+}, Pb^{2+}, Fe^{3+} 0; La^{3+} 0.10; Ga^{3+} 0.09; Y^{3+} 0.15; In^{3+} 0.20; Cs^{+} 0.10; Co^{2+}, Cu^{2+} 0.55; Cd^{2+} 0.53; Fe^{2+} 0.56; Ca^{2+} 0.42; Zn^{2+} 0.52; Ni^{2+} 0.50; Ba^{2+} 0.63; Na^{+} 0.55; Sr^{2+} 0.60; Hg^{2+} 0.73 (94).

(r) Forty one assorted metal ions; alginic acid; 0.01–1 *M* solutions of HCl, HNO_3, $HClO_4$, H_3PO_4, CH_3COOH, and $(COOH)_2$; 12 detection reagents; typical R_F values in 0.05 *M* $HClO_4$: Pb^{2+} 0, Tl^{2+} 0.12, Hg^{2+} 0.63, Cu^{2+} 0.16, As^{3+} 0.50, Fe^{3+} 0.08, Zn^{2+} 0.30, Ba^{2+} 0.05, Mg^{2+} 0.42 (95).

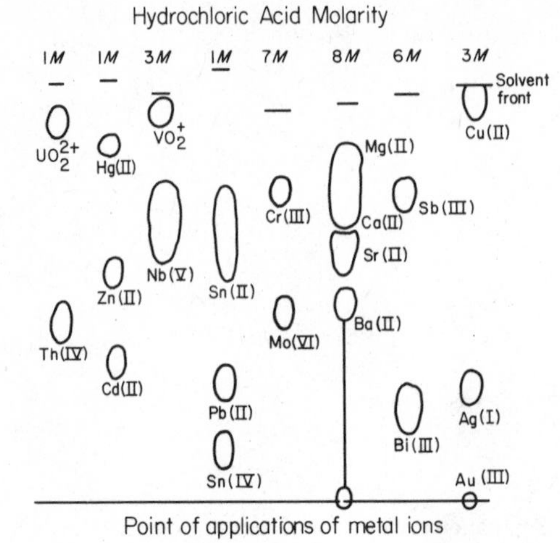

Figure 7.5. Chromatograms of some metal ions chromatographed on cellulose impregnated with Primene JM-T hydrochloride (0.3 M) using hydrochloric acid as the mobile phase. After Graham et al., Reference 46.

(s) Analytical group IIb; silica gel; (A) *n*-butanol-1 *M* tartaric acid-3 *M* HCl (10 : 1 : 1), (B) *n*-butanol-1 *M* HCl (20 : 1); 0.05% dithizone in $CHCl_3$; R_F values: (A) As^{3+} 0.55, Sb^{3+} 0.64, Sn^{2+} 0.77, (B) As^{3+} 0.38, Sb^{3+} 0.75, Sn^{2+} 0.87 (90).

(t) Analytical group IIIb; cellulose MN 300 HR; acetone-conc. HCl-H_2O (75 : 13 : 12); 1% alizarin in ethanol, followed by NH_3 vapor; R_F values: Ni^{2+} 0.10, Mn^{2+} 0.32, Co^{2+} 0.64, Zn^{2+} 0.96 (90).

(u) Analytical group V; silica gel; (A) 0.1 *M* I_2 in nitromethane-benzene (2 : 3), (B) 0.01 *M* I_2 in nitrobenzene; bromocresol green; R_F values: (A) K^+ 0.68, Na^+ 0.32, NH_4^+ 0.52, Li^+ 0.18, (B) K^+ 0.36, Na^+ 0.18, NH_4^+ 0.24, Li^+ 0.06 (90).

(v) Alkali, alkaline earth, transition, lanthanide, and actinide metals; cellulose + 0.3 *M* Primene JM-T-HCl (liquid C_{18}-C_{24} trialkyl methylamine anion exchanger); 1–9 *M* HCl; PAN followed by NH_3 vapor, UV light (Fig. 7.5) (46).

(w) Fe, Hg, Sb, Tl valance states; silica gel G; (A) *n*-butanol-3 *N* HCl-2% tartaric acid (4 : 1 : 1), (B) *t*-butanol-2 *N* HCl (4 : 1), (C) *t*-butanol-acetic acid (2 : 1), (D) isobutanol-glacial acetic acid (4 : 1); potassium thiocarbonate reagent; R_F values: (A) Hg^+ 0.95, Hg^{2+} 0; (B) Sb^{3+} 0.25, Sb^{5+} 0; (C) Tl^+ 0.15, Tl^{3+} 0.92; (D) Fe^{2+} 0, Fe^{3+} 0.85 (96).

(x) Sixty metal ions; cellulose; tartaric acid (0.1–0.5 mol/dm^3)-NH_3 (0–1.3 mol/dm^3)-ethanol (5–80%); separations reported include: Ta, Nb; Nb; V; Sn^{2+}, Ge, Pb; Li, Rb, Cs; Ce^{4+}, Ce^{3+}; Cr, W, Mo; Se, Te; Au, Cu, Ag; Sm, Ce^{3+}; Pr, Ce^{3+} (Fig. 7.6) (97).

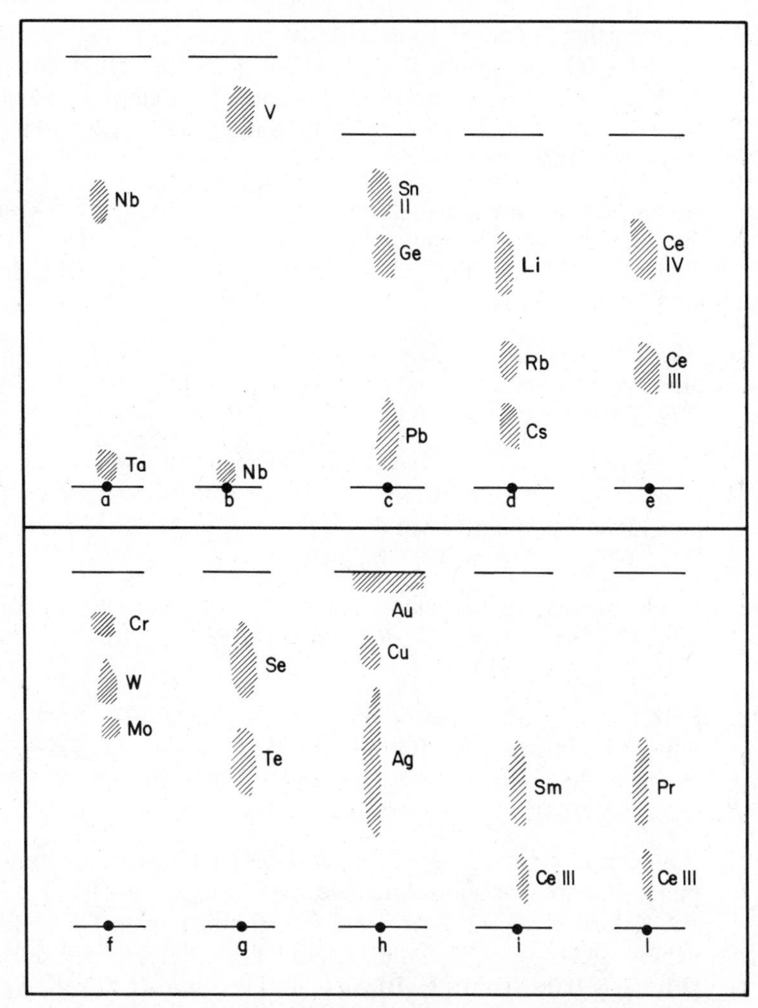

Figure 7.6. Separation of inorganic ions on cellulose layers in tartaric acid-water-ethanol-NH_4OH media. Eluents: 0.2 mol/dm^3 tartaric acid: (*a*) 5% ethanol, (*b*) 20% ethanol, (*c*, *d*, *e*) 80% ethanol; 0.1 mol/dm^3 tartaric acid: (*f*, *g*) 60% ethanol, (*h*) 80% ethanol; 0.1 mol/dm^3 tartaric acid-0.3 mol/dm^3 NH_4OH: (*i*, *l*) 40% ethanol. After Baffi et al., Reference 97.

(y) ^{235}U fission products; silica gel treated with EDTA; 0.0025 *M* EDTA-methanol (92:8); radioactivity; R_F values: ^{141}Ce 0.34, ^{113}Sn 0.69, ^{115}Cd 0.85. The following mixture was also separated using a combination of two dimensional and stepwise development: ^{140}Ba, ^{140}La, ^{95}Zr, ^{181}Hf, ^{95}Nb, ^{113}Sn, and ^{113}In (65).

(z) Rare earths; silica gel containing 10 mg of starch and 100–150 mg NH_4NO_3 per gram; 0.11 *M* HCN in methyl ethyl ketone; 0.1% arsenazo I in 35% urotropine in aqueous ethanol; R_F values: La 0.08, Ce, 0.24, Pr 0.31, Nd 0.38, Sm 0.52, Eu 0.56, Gd 0.60, Y 0.65, Er 0.68 (98).

(aa) Alkali and alkaline earth metals; silica gel containing 5% ammonium phosphododecamolybdate; 0.01 *N* HCl; R_F values: Na^+ 0.85, K^+ 0.52, Rb^+ 0.21, Cs^+ 0.05, Ba^{2+} 0.55, La^{3+} 0.12, Sr^{2+} 0.58, Y^{3+} 0.03 (99).

(bb) Alkali metals; silica gel G, prewashed with 0.05 *M* HCl to remove Na^+; phenol saturated with 6 *N* HCl; R_F values: Na^+ 0.0, K^+ 0.12, Rb 0.23, Cs 0.43 (100).

(cc) Noble metals; silica gel; (A) *t*-butanol-acetic acid (6:1), and (B) *t*-butanol-acetic acid-HCl (20:3:1); potassium thiocarbonate and benzidine; R_F values: (A) Pd^{2+} 0.58, Rh^{3+} 0.33, Ru^{3+} 0.77; (B) Os^{8+} 0.08, Ir^{4+} 0.68, Pt^{4+} 0.93 (101).

(dd) Noble metals; diethylaminocellulose; 5 *M* LiCl; Br_2 and 2% $SnCl_2$-2% KI in 3 *M* HCl; R_F sequence: $Rh^{3+} > Ru^{3+} > Pd^{2+} > Pt^{4+} > Au^{3+}$ (102).

(ee) Rare earths; silica gel; diethyl ether-THF-*bis*-(2-ethylhexylphosphate)-nitric acid (100:15:1:3.5); R_F sequence: Lu > Yb > Tm > Er > Ho > Dy > Tb > Gd > Eu > Sm > Nd > Pr > Ce > La (103).

(ff) Eighteen inorganic ions; silica gel impregnated with diethyltriamine; ethanol-acetone-acetic acid, (A), (7:5:2), or (B) (4:5:2); 1% $K_4Fe(CN)_6$ in 2% HCl and 0.5% dithizone in $CHCl_3$; R_F values: (A) Co^{2+} 0.54, Zn 0.38, Ni^{2+} 0.42, Fe^{2+} 0.71, Cu 0.22, U 0.77, V 0.26, Th 0.13; (B) As 0.28, Pb^{2+} 0.61, Cd 0.56, Hg^{2+} 0.81, Ag 0.40, Zr 0.0, Sb 0.33, Ti^{4+} 0.16, Se^{4+} 0.21, Sn^{4+} 0.06 (specified oxidation states shown) (104).

(gg) Fifty-eight metal ions; cellulose phosphate; 0.025 *M* H_2SO_4-1.0 *M* $(NH_4)_2SO_4$. See Fig. 7.7 for representative separations (105).

(hh) Forty four metal ions; ECTEOLA-cellulose; methanol-6 *M* HCl (95:5); R_F values: Be^{2+} 0.80, Al^{3+} 0.70, Sc^{3+} 0.60, V^{4+} 0.60, Mn^{2+} 0.70, Co^{2+} 0.60, Cu^{2+} 0.55, Ga^{3+} 0.30, Sr^{2+} 0.70, Nb^{5+} 0.20, Pd^{2+} 0.25, Cd^{2+} 0.35, Sb^{3+} 0.15, U^{6+} 0.30, Bi^{3+} 0.25, Hg^{2+} 0.30, Pt^{4+} 0.15, Re^{7+} 0.68, Yb^{3+} 0.62 (39).

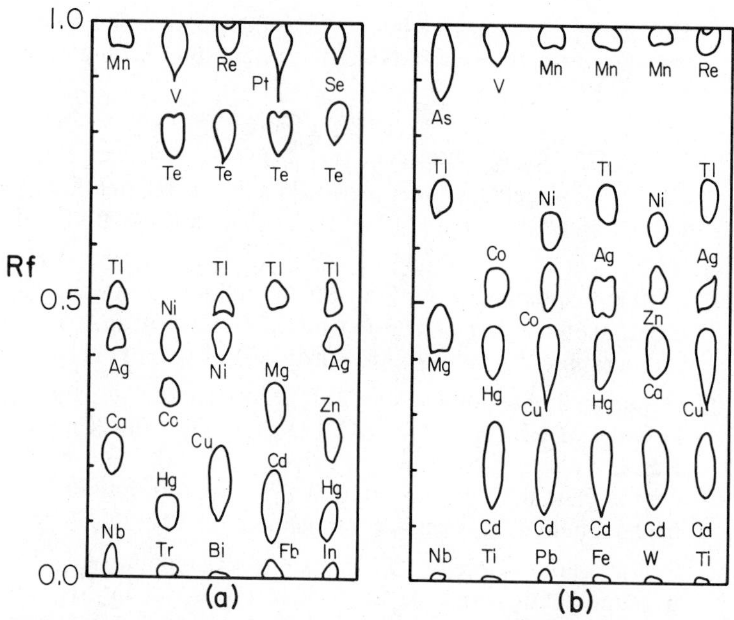

Figure 7.7. Multicomponent separations on P-cellulose layers. Mobile phases: (*a*) 0.025 *M* H_2SO_4-0.5 *M* $(NH_4)_2SO_4$; (*b*) 0.025 *M* H_2SO_4-1.0 *M* $(NH_4)_2SO_4$. After Shimizu et al., Reference 105.

7.9.2 Anions

(a) Halides; MN silica gel S-HR; acetone-*n*-butanol-conc. NH_4OH-H_2O (13:4:2:1); 0.1% bromocresol purple and NH_4OH in ethanol, 1% ammoniacal $AgNO_3$ + 1% ethanolic fluorescein, and 0.1% zirconium-alizarin lake in conc. HCl; R_F values; F^- 0.10, Cl^- 0.30, Br^- 0.50, I^- 0.86 (106).

(b) Phosphates; MN silica gel S-HR; methanol-conc. NH_4OH-10% trichloroacetic acid-H_2O (10:3:1:6); 1% aqueous ammonium molybdate followed by 1% $SnCl_2$ in 10% HCl; migration sequence: $H_2PO_2^- > H_2PO_3^- > H_2PO_4^- > H_2P_2O_7^{2-}$ (107).

(c) Sulfates and polythionates; MN silica gel S-HR; (A) methanol-*n*-propanol-conc. NH_4OH-H_2O (10:10:1:2) and (B) methanol-dioxane-conc. NH_4OH-H_2O (3:6:1:1); 0.1 *M* $AgNO_3$ + 2 *N* NH_4OH or 0.1% aqueous bromocresol green; migration sequence: (A) $S_2O_8^{2-} > SO_3^{2-} > S_2O_3^{2-} > SO_4^{2-}$ and (B) $S_5O_6^{2-} > S_4O_6^{2-} > S_3O_6^{2-} > S_2O_6^{2-} > S_2O_3^{2-}$ (108,109).

(d) Halogens; Dowex 1-cellulose MN 300G (1:1); 1 *M* aqueous $NaNO_3$; detection via radioactivity; migration sequence: $Cl^- > Br^- > I^-$ (110).

(e) (A) SCN^-, I^-, Cl^-, $Fe(CN)_6^{3-}$, $Fe(CN)_6^{4-}$, ClO_3^-, BrO_3^-; (B) F^-, NO_2^-, $S_2O_3^{2-}$, SO_4^{2-}, CrO_4^{2-}, PO_4^{3-}, AsO_4^{3-}; (C) $C_2O_4^{2-}$, BO_2^-; silica gel; (A) acetone-water (10:1), (B) methanol-*n*-butanol-water (3:1:1), and (C) *n*-butanol saturated with 2 *N* HNO_3 (85).

(f) Orthophosphate and pyrophosphate; cellulose; dioxane-H_2O-trichloroacetic acid-NH_4OH (65 ml:27.5 ml; 5 g; 0.25 ml); autoradiography; R_F values: 0.83 (ortho), 0.61 (pyro) (111).

(g) Condensed phosphates; starch; trichloroacetic acid-isopropanol-H_2O-0.1 *M* EDTA-25% NH_4OH (5 g:80 ml:39 ml:1 ml:0.3 ml); Hanes and Isherwood reagent; R_F values: monophosphate 0.70, diphosphate 0.22, triphosphate 0.13, tetraphosphate 0.10, cyclotriphosphate 0.32, cyclotetraphosphate 0.18 (112).

(h) Sulfur anions; Gelman ITLC-SA and-SG media; (A) *n*-butanol-pyridine-acetic acid-H_2O (90:3:1:6), (B) *n*-propanol-H_2O (95:5); $AgNO_3$ reagent and radioactivity; R_F values: (A) thiosulfate 0.02, trithionate 0.26, tetrathionate 0.39, thiocyanate 0.62; (B) sulfate 0, thiosulfate 0.06, trithionate 0.54, tetrathionate 0.63, thiocyanate 0.71 (113).

(i) Halides; silica gel H and MN G-HR; (A) methyl ethyl ketone-ethanol-NH_3 (5:5:2), (B) *n*-butanol-benzylamine (9:1); R_F values: (A) Cl^- 0.55, Br^- 0.70, I^- 0.92, (B) F^- 0, Cl^- 0.55, Br^- 0.64, I^- 0.75 (90).

(j) Condensed phosphates; acid washed cellulose powder, S and S 142 dg; methanol + isopropanol-water (7:1) + trichloroacetic acid solution (125 g and 32 ml of 25% NH_4OH diluted to 1 L) + 96% acetic acid-H_2O (1:4) (75:20:25:6); detection by spraying with molybdate and sulfite reducing solutions in sequence; R_F values; mono 0.55, di 0.72, tri 0.58, tetra 0.45, penta 0.30, hexa 0.19, hepta 0.11, octa 0.06, trimeta 0.34, tetrameta 0.18, pentameta 0.12, hexameta 0.08, heptameta 0.05 (66).

(k) Sulfur anions; microcrystalline cellulose; *n*-propanol-1 *M* NH_4OH-acetone (15:10:1); R_F values: S^{2-} 0, SO_3^{2-} 0.70, SO_4^{2-} 0.28, $S_2O_3^{2-}$ 0.51 (114).

(l) Halates; cellulose; butanol-pyridine-1 *N* NH_4OH (2:1:2); R_F values: perbromate 0.67, bromate 0.27, periodate 0.10, iodate 0.07 (115).

(m) Oxyanions and esters; cellulose; 28% NH_4OH-acetone-*n*-butanol (6:13:3); fluorescent morin-Al complex; R_F values: PO_4^{3-} 0, $B_4O_7^{2-}$ 0.05, SO_4^{2-} 0.05, CrO_4^{2-} 0.06, $Cr_2O_7^{2-}$ 0.09, IO_4^- 0.12, IO_3^- 0.14, BO_3^{3-} 0.15, SiO_3^{2-} 0.16, ClO_2^- 0.35, BrO_3^- 0.43, NO_2^- 0.55, NO_3^- 0.60, ClO_3^- 0.63, ClO_4^- 0.80 (116).

(n) Monovalent anions, halo acids, halo oxyacids, pseudohalo acids; Sephadex LH gel; (A) 3 *N* lithium acetate, (B) ethanol-0.01 *N* HNO_3 (1:1); R_F sequence: (A) $Co(NH_3)_6^{3+} > Cl^- > Br^- > I^- > CNS^- > CNSe^-$, (B) $Co(NH_3)_6^{3+} > IO_4^- > IO_3^- > BrO_3^- > ClO_3^- > ClO_4^- > BrO_4^-$ (117).

7.9.3 Inorganic Compounds, Complexes, and Derivatives

(a) Ferrocene derivatives; silica gel G; benzene, benzene-ethanol (30:1 and 15:1), propylene glycol-methanol (1:1), and propylene glycol-chlorobenzene-methanol (1:1:1); bromine and $NaIO_4$ (118).

(b) Chromium and cobalt complexes of *o*-hydroxy-*o*′-carboxyazo dyes; aluminum oxide; methanol; natural color (119).

(c) Organotin stabilizers in polyvinyl chloride; silica gel G + "Komplexon III"; water-butanol-ethanol-acetic acid (10:5:5:0.15); diphenylthiocarbazone (120).

(d) Diethyl dithiothiocarbamate complexes; silica gel G; (A) xylene, (B) benzene; (A) freshly prepared $SnCl_2$, (B) 0.5% oxine in alcohol; R_F values of complexes: (A) Au^{3+} 0, Se^{4+} 0.33, Te^{4+} 0.50, Pt^{4+} 0.10, Pd^{2+} 0.20 (121); (B) Ga^{3+} 0.07, In^{3+} 0.64, Ti^{3+} 0.57, Zn^{2+} 0.78, Cd^{2+} 0.30 (122).

(e) Dithizonates; silica gel G; benzene-methylene chloride (5:1); detection by natural color; R_F values of complexes: Cd^{2+} 0.13, Bi^{3+} 0.37, Pb^{2+} 0.34, Cu^{2+} 0.48, Zn^{2+} 0.50, Hg^{2+} 0.58 (25). Dithizonate derivatives were used for isolating and identifying Ag, Cd, Co, Cu, Hg, Ni, Pb, and Zn in toxicological cases; TLC was on silica gel 60 layers with benzene mobile phase (123).

(f) EDTA complexes; silica gel H; H_2O-glycol monomethyl ether-methyl ethyl ketone-acetone-NH_4OH (40:20:20:20:0.15); natural color or 20% ammonium peroxydisulfate solution; R_F values of complexes: Co 0.86, Cu 0.79, Ni 0.69, Mn 0.71, Cr 0.44, Fe 0.17 (124).

(g) Metal complexes with morpholine-4-carbodithioate; silica gel G; acetone-benzene (1:4), $CHCl_3$, or *n*-propanol; natural color; R_F sequence of complexes in acetone-benzene (1:4): Pt > Fe > Cu > Co > Ru > Rh > Ir > Te (125).

(h) Mercury compounds; silica gel; isobutyl methyl ketone-benzene (3:1); radioactivity; R_F values: diphenylmercury 0.60; phenylmercuric chloride 0.40, mercuric chloride 0 (126).

(i) Tropolone chelates; silica gel G; (A) chloroform-benzene (3:1), (B) chloroform-butanol (5:2), (C) chloroform-pyridine-MIBK (15:2:5), (D) chloroform-MIBK (3:1), (E) pyridine; 0.05 *M* aqueous potassium thiocarbonate; R_F values for chelates: (A) Ru^{3+} 0.86, Rh^{3+} 0.38, Pd^{3+} 0.64, (B) Ti^{4+} 0.57, V^{5+} 0, Fe^{3+} 0.88, (C) Ni^{2+} 0.76, Co^{2+} 0.88, U^{6+} 0, (D) Cu^{2+} 0.75, Fe^{3+} 0.45, Cd^{2+} 0, (E) Pb^{2+} 0.76, U^{6+} 0.57, Fe^{3+} 0.92 (127).

REFERENCES

1. F.H. Polard and J.F.W. McOmie, *Chromatographic Methods of Inorganic Analysis—with Special Reference to Paper Chromatography*, Academic press, New York, 1958.
2. J. Sherma and G. Zweig, *Paper Chromatography*, Academic Press, 1971, pp. 500–538.
3. H. Seiler and M. Seiler, *Helv. Chim. Acta,* **43,** 1939 (1960).
4. F. Umland and K. Kirchner, *Z. Anorg. Allgem. Chem.,* **280,** 211 (1955).
5. H. Seiler, *Helv. Chim. Acta,* **45,** 381 (1962).
6. B. Sansoni, *Z. Naturforsch.,* **11b,** 117 (1956).
7. S. Takitani, *Bunseki Kagaku,* **12,** 1156 (1963).
8. S. Takitani, N. Fukuoka, Y. Iwasaki, and H. Hasegawa, *Bunseki Kagaku,* **13,** 469 (1964) and **14,** 652 (1965).
9. M. Zureshi and F. Khan, *J. Chromatogr.,* **34,** 222 (1968).
10. J.S. Fritz and J. Sherma, *J. Chromatogr.,* **25,** 153 (1966).
11. J. Sherma and A.D. Finck, *Anal. Chim. Acta,* **43,** 503 (1968).
12. J.S. Fritz and L.H. Dahmer, *Anal. Chem.,* **37,** 1272 (1965).
13. I.I.M. Elbeih and M.A. Abou-Elnaga, *Anal. Chim. Acta,* **17,** 397 (1957).
14. H. Seiler, E. Sorkin, and H. Erlenmayer, *Helv. Chim. Acta,* **35,** 120 (1952).
15. H.T. Gordon and C.A. Hewel, *Anal. Chem.,* **27,** 1471 (1955).
16. R.A. Webb, D.G. Hallas, and H.M. Stevens, *Analyst,* **94,** 794 (1969).
17. K. Yamaguchi, *J. Pharm. Soc. Japan,* **74,** 1276 (1954).
18. H.G. Linskens, *Papierchromatographie in der Botanik,* Springer, Berlin, 1955.

19. E. Maly, *J. Chromatogr.*, **19,** 206 (1965).
20. G. Schwedt, *Chromatographic Methods in Inorganic Analysis,* D.A. Hüthig, Verlag, Heidelberg, 1981, pp. 115–118.
21. G. Schwedt, *Chromatographic Methods in Inorganic Analysis,* Huthig, Verlag, Heidelberg, 1981, pp. 131–134.
22. H. Seiler, in E. Stahl (Ed.), *Thin Layer Chromatography—A Laboratory Handbook,* Springer-Verlag, New York, 1969, p. 837.
23. F.W.H.M. Merkus, *Progr. Separ. Purif.*, **3,** 233 (1970).
24. J. Rai and V.P. Kukerja, *Int. Symp. Chromatogr. Electrophor., Lect. Pap.*, 6th 1970, Ann Arbor Science Publishers, Ann Arbor, Michigan, 1971, p. 453.
25. M. Hranisavlejevic-Jakovljevic, I. Pejkovic-Tadic, and K. Jakovljevic, in G.B. Marini-Bettolo (Ed.), *Thin Layer Chromatography,* Elsevier, Amsterdam, 1964, p. 221.
26. Y. Tsunoda, T. Takeuchi, and Y. Yoshino, *Nippon Kagaku Zaschi,* **85,** 275 (1974).
27. J. Vanderdeelen, *J. Chromatogr.*, **39,** 521 (1969).
28. M.R. Verma and J. Rai, *Int. Symp. Chromatogr., Electrophor.*, 4th, 1966, Ann Arbor Science Publishers, Ann Arbor, Michigan, 1968, p. 544.
29. H. Seiler and W. Rothweiler, *Helv. Chim. Acta,* **44,** 941 (1961).
30. J.E. Meinhard and N.F. Hall, *Anal. Chem.*, **21,** 185 (1949).
31. F.D. Houghton, *J. Chromatogr.*, **24,** 494 (1966).
32. M.J. Zetlmeisl and D.T. Haworth, *J. Chromatogr.*, **30,** 637 (1967).
33. W. Funk, in W. Bertsch, S. Hara, R.E. Kaiser, and A. Zlatkis (Eds.), *Instrumental HPTLC,* D.A. Hüthig Verlag, New York, 1980.
34. D.C. Fenimore and C.M. Davis, *Anal. Chem.*, **53**(2), 252A (1981).
35. L. Lepri, P.G. Desideri, and R. Mascherini, *J. Chromatogr.*, **88,** 351 (1974).
36. R. Frache and A. Dadone, *Chromatographia,* **4,** 156 (1971); **6,** 475 (1973).
37. L. Lepri and P.G. Desideri, *J. Chromatogr.*, **84,** 155 (1973); L. Lepri, P.G. Desideri, V. Coas, and D. Cozzi, *J. Chromatogr.*, **47,** 442 (1970).
38. B.A. Zabin and C.B. Rollins, *J. Chromatogr.*, **14,** 534 (1964).
39. K. Oguma, *Chromatographia,* **8,** 669 (1975).
40. R. Kuroda, N. Kojima, and K. Oguma, *J. Chromatogr.*, **69,** 223 (1972).
41. T. Shimizu, Y. Kogure, H. Aria, and T. Suda, *Chromatographia,* **9,** 85 (1976).
42. T.B. Pierce and R.F. Flint, *Anal. Chim. Acta,* **31,** 595 (1964).
43. P. Markl and F. Hecht, *Mikrochim. Acta,* 1963, p. 970.
44. U.A.Th. Brinkman, G. de Vries, and E. van Dalen, *J. Chromatogr.*, **22,** 407 (1966); **25,** 447 (1966).
45. H.R. Leene, G. de Vries, and U.A.Th. Brinkman, *J. Chromatogr.*, **80,** 221 (1973).

46. R.J.T. Graham, L.S. Bark, and D.A. Tinsley, *J. Chromatogr.*, **35,** 416 (1968); **39,** 200 (1969).
47. U.A.Th. Brinkman and G. de Vries, *J. Chromatogr.*, **56,** 103 (1971).
48. U.A.Th. Brinkman, P.J.J. Steerenburg, and G. de Vries, *J. Chromatogr.*, **54,** 449 (1971).
49. J.A. Berger, C. Meyniel, J. Petit, and P. Blanquet, *Bull. Soc. Chim. Fr.*, 1963, 2662.
50. K. Saitoh, M. Kobayashi, and N. Suzuki, *Anal. Chem.*, **53,** 2309 (1981).
51. K.T. Den Bleyker and T.R. Sweet, *Chromatographia,* **13,** 114 (1980).
52. T.E. Beesley, in J.C. Touchstone and D. Rogers (Eds.), *Thin Layer Chromatography, Quantitative Environmental and Clinical Applications,* Wiley-Interscience, New York, 1980, p. 171.
53. B. Fried and J. Sherma, *Thin Layer Chromatography/Techniques and Applications,* Marcel Dekker, New York, 1982.
54. H.R. Felton, *Am. Lab.,* **12**(5), 105 (1980).
55. H. Seiler and M. Seiler, *Helv. Chim. Acta,* **48,** 117 (1965).
56. H. Seiler and M. Seiler, *Helv. Chim. Acta,* **44,** 939 (1961).
57. L.F. Druding, *Anal. Chem.* **35,** 1744 (1963).
58. G. Zweig and J. Sherma, *CRC Handbook of Chromatography,* Vols. 1 and 2, Academic Press, New York, 1972.
59. J.G. Kirchner, *Thin Layer Chromatography,* 2nd ed., Wiley-Interscience, New York, 1978.
60. J. Sherma and B. Fried, *Anal Chem.,* **56**(5), 48R (1984), and earlier reviews of TLC in the April Fundamental Reviews issues every even-numbered year.
61. J.M. Bobbitt, *Thin Layer Chromatography,* Reinhold, New York, 1963, p. 149.
62. K. Randerath, *Thin Layer Chromatography,* Academic press, New York, 1963, p. 226; 2nd edition, Verlag Chemie, Weinheim, Ger. (1975).
63. E. Gagliardi and B. Brodar, *Chromatographia,* **2,** 267 (1969); **3,** 7 and 320 (1970).
64. J.C. Touchstone and M.F. Dobbins, *Practice of Thin Layer Chromatography,* 2nd edition, Wiley-Interscience, New York, 1983.
65. G. Bottura, A. Breccia, F. Marchetti, and F. Spalletti, *Ric. Sci., Rend., Ser. A,* **6,** 373 (1964).
66. T. Rossel, *Z. Anal. Chem.,* **197,** 33 (1963).
67. H.M. Hashmi, M.A. Shahid, A.A. Ayaz, F.R. Chugtai, N. Hassan, and A.S. Adil, *Anal. Chem.,* **38,** 1554 (1966); *Talanta,* **12,** 713 (1965).
68. H. Hammerschmidt and M. Mueller, *Papier,* **17,** 448 (1963).
69. L.J. Ottendorfer, *Anal. Chim. Acta,* **33,** 115 (1963).
70. H. Seiler, *Helv. Chim. Acta,* **46,** 2629 (1963).

71. A. Moghissi, *J. Chromatogr.*, **13,** 542 (1964).
72. A. Breccia and F. Spalletti, *Nature,* **198,** 756 (1963).
73. L.F. Druding, *Anal. Chem.*, **35,** 1582 (1963).
74. S.J. Purdy and E.V. Truter, *Analyst,* **87,** 802 (1962).
75. J.C. Touchstone and J. Sherma, *Quantitative Thin Layer Chromatography,* Wiley-Interscience, New York, 1979.
76. J.C. Touchstone, R.E. Levitt, and G.J. Hansen, in J.C. Touchstone and D. Rogers (Eds.), *Thin Layer Chromatography, Quantitative Environmental and Clinical Applications,* Wiley-Interscience, New York, 1980, p. 36.
77. J.C. Touchstone and S.S. Levin, *J. Liq. Chromatogr.*, **3,** 1853 (1980).
78. J.C. Touchstone, M.F. Dobbins, M.L. Mallinger, and J. Strauss, in J.C. Touchstone and D. Rogers, Eds., *Thin Layer Chromatography, Quantitative Environmental and Clinical Applications,* Wiley-Interscience, New York, 1980, p. 151.
79. J. Franc and E. Kosivoka, *J. Chromatogr.*, **187,** 462 (1980).
80. M. Covello and O. Schettino, *Farmaco (Pavia), Ed. Prat.*, **20,** 396 (1965).
81. H. Seiler and M. Seiler, *Helv. Chim. Acta,* **50,** 2477 (1967).
82. G. Schwedt, *Chromatographic Methods in Inorganic Analysis,* D.A. Huthig, Verlag Heidelberg, 1981, pp. 80–88.
83. N.K. Kaushik and K.N. Johri, *Chromatographia,* **9,** 233 (1976).
84. S.W. Husain and F. Eivazi, *Chromatographia,* **8,** 277 (1975).
85. K. Kawanabe, S. Takitani, M. Miyazaki, and Z. Tamura, *Bunseki Kagaku,* **13,** 976 (1964).
86. E. Gagliardi and W. Likussar, *Mikrochim. Acta,* 1965, 765.
87. S.P. Srivastava and V.K. Gupta, *Chromatographia,* **12,** 496 (1979).
88. F.W.H.M. Merkus, *Pharm. Weekblaad,* **98,** 947 (1963).
89. E. Pfeil, A. Friedrich, and T. Wachsmann, *Z. Anal. Chem.*, **158,** 429 (1957).
90. J. Gasparic and J. Churacek, *Laboratory Handbook of Paper and Thin Layer Chromatography,* Halsted Press, New York, 1978, p. 315.
91. E.J. Goller, *J. Chem. Educ.*, **42,** 442 (1965).
92. T.B. Pierce and R.F. Flint, *J. Chromatogr.*, **24,** 141 (1966).
93. U.A.Th. Brinkman and G. de Vries, *J. Chromatogr.*, **18,** 142 (1965).
94. K.-H. König and K. Demel, *J. Chromatogr.*, **39,** 101 (1969).
95. D. Cozzi, P.G. Desideri, L. Lepri, and G. Ciantelli, *J. Chromatogr.*, **35,** 396 and 405 (1968).
96. K.N. Johri and H.C. Mehra, *Chromatographia,* **4,** 80 (1971).
97. F. Baffi, A. Dadone, and R. Frache, *Chromatographia,* **9,** 280 (1976).
98. K.S. Babayan and G.M. Varshal, *Zh. Anal. Khim.*, **28,** 921 (1973).
99. M. Lesigang-Buchtela, *Mikrochim. Acta,* 1969, 1027.
100. A.C. Handra and K.N. Johri, *Chromatographia,* **4,** 530 (1971).

101. K.N. Johri, B.S. Saxena, and H.C. Mehra, *Chromatographia,* **4,** 351 (1971).
102. K. Ishida and T. Saito, *J. Chromatogr.,* **152,** 191 (1978).
103. K. Jung, J. Maurer, J. Urlichs, and H. Specker, *Z. Anal. Chem.,* **291,** 328 (1978).
104. S.P. Srivastava, V.K. Dua, and K. Gupta, *Chromatographia,* **11,** 539 (1978).
105. T. Shimizu, K. Nakazawa, and T. Kikuchi, *Chromatographia,* **9,** 574 (1976).
106. H. Seiler and T. Kaffenberger, *Helv. Chim. Acta,* **44,** 1282 (1961).
107. H. Seiler, *Helv. Chim. Acta,* **44,** 1753 (1961).
108. H. Seiler and H. Erlenmayer, *Helv. Chim. Acta,* **47,** 264 (1964).
109. H. Seiler and M. Seiler, *Helv. Chim. Acta,* **48,** 117 (1965).
110. J.A. Berger, G. Meyniel, and J. Petit, *Compt. Rend.,* **255,** 1116 (1962); *J. Chromatogr.,* **29,** 190 (1967).
111. N.L. Clesceri, and G.F. Lee, *Anal. Chem.,* **36,** 2207 (1964).
112. V.D. Canic, M.N. Turcic, S.M. Petrovic, and S.E. Petrovic, *Anal. Chem.,* **37,** 1576 (1965).
113. D.P. Kelly, *J. Chromatogr.,* **51,** 343 (1970).
114. A.C. Handra and K.N. Johri, *Talanta,* **20,** 219 (1973).
115. M. Lederer and M. Sinibaldi, *J. Chromatogr.,* **60,** 275 (1971).
116. T. Okumura, *Talanta,* **26,** 171 (1979).
117. V. Di Gregorio and M. Sinibaldi, *J. Chromatogr.,* **129,** 407 (1976).
118. K. Schlögl, H. Pelousek, and A. Mohar, *Monatsch. Chem.,* **92,** 533 (1961).
119. G. Schetty and W. Kuster, *Helv. Chim. Acta,* **44,** 2193 (1961).
120. M. Türler and O. Högl, *Mitt. Gebiete u. Lebensm. Hyg.,* **52,** 123 (1961).
121. J. Rai and V.P. Kukerja, *Chromatographia,* **3,** 500 (1970).
122. J. Rai and V.P. Kukerja, *Chromatographia,* **3,** 499 (1970).
123. P. Baudot, J.L. Monal, M.H. Livertoux, and R. Truhaut, *Chromatographia,* **128,** 141 (1976).
124. J. Vanderdeelen, *J. Chromatogr.,* **39,** 521 (1969).
125. N. Singh, R. Kumar, and B.D. Kansal, *Chromatographia,* **11,** 408 (1978).
126. V. Luckow and H.A. Ruessel, *Chromatographia,* **9,** 578 (1976).
127. K.N. Johri and H.C. Mehra, *Sep. Sci.,* **11,** 171 (1976).

CHAPTER

8

ION EXCHANGE IN RADIOCHEMISTRY

JAMES W. MITCHELL

AT&T Bell Laboratories
Murray Hill, New Jersey

8.1 INTRODUCTION

Ion-exchange techniques find versatile and unique applications in radiochemistry, defined here as any operation in which radioactive isotopes are present and chemical parameters can be manipulated to influence the behavior of the radioisotope. Equally important applications exist for radiotracer determination, detection, or monitoring of the behavior of stable elements at ultralow concentrations, for investigations of naturally existing radioactive species, and for radionuclidic species being produced via nuclear transformation of stable ones. Radioisotope analyses, and the combined use of radiotracers and ion exchange to monitor the efficiency of chemical purification procedures, radiochemical separations in activation analysis, separation of radionuclides from nuclear wastes and cooling waters, and the production of high specific activity and medicinal isotopes are some important examples treated in this chapter. Specific attention is focused on the requirements for treatment of ultratrace (≤ 1 μg/ml) levels of the elements.

In radionuclear analysis, the use of separations techniques is perhaps the ultimate example of the significant enhancement of the scope of application of an analytical method resulting from the use of selective separations methods. Separation steps following nuclear irradiation (activation) of the sample provide the following advantages: Complex multielement-containing samples that would give rise to many superimposed gamma-ray isotopic peaks are simplified considerably by separating elements into groups. Thus, when obscured by massive background irradiation from matrix elements, the radioisotope of a single element of interest can be separated selectively to circumvent interferences. This selective separation has the benefit also of lowering the measurement limits due to the reduction in the background activity level. Separation methods in activation analysis have the added useful feature that the element to be determined need not be quantitatively recovered. Due to the above advantages resulting from separation techniques, activation analysis is ranked among the most widely used and dependable methods for quantitative determination of elements at the ultratrace level. Among the most widely used separation schemes amenable to use in radioactivation analysis, precipitation, solvent extraction, and ion exchange, the latter is a very prominently applied and important technique. Its application for multielement group separations and for single element sequential separations is discussed via examples.

The growth of nuclear technology has been accompanied by an increasing concern with minimizing environmental contamination. Environmental contamination problems due to radioisotopes occurring naturally. ^{14}C, ^{41}K, ^{238}U, ^{232}Th, ^{226}Ra, and a host of intermediate isotopes resulting from the natural decay of either ^{238}U or ^{232}Th, are virtually unimportant. Even with the decay of the naturally occurring radioactive series, and the production of a multitude of additional isotopes—^{234}Th, ^{234}Pa, ^{234}U, ^{230}Th, ^{222}Rn, ^{218}Po, ^{214}Pb, ^{218}At, ^{214}Bi, ^{214}Po, ^{210}Tl, ^{210}Pb, ^{210}Bi, ^{206}Tl, ^{210}Po, ^{228}Ra, ^{228}Ac, ^{226}Th, ^{224}Ra, ^{220}Rn, ^{216}Po, ^{212}Pb, ^{216}At, ^{212}Bi, ^{212}Po, and ^{208}Tl—environmental contamination is not a serious problem, since nature has generally distributed these species in minute amounts within the geosphere. Further proliferation of radioisotopes results from by-products of the fission of the natural and synthetically produced materials of ^{235}U, ^{233}U, and ^{239}Pu, and from the manufacture of radioisotopes via nuclear transformation. In these cases, highly radioactive species are produced. Developments in these areas have given birth to several fast-growing nuclear technologies and have caused increasing concern about worldwide nuclear pollution. Ion-exchange methods have not only played an important role in curtailing nuclear pollution of the environment but have been employed throughout many aspects of applied nuclear research, development, and technology. This section will not treat in detail the technological application of ion-exchange methods in uranium ore refining, disposal of by-product wastes, or nuclear materials reprocessing. Occasional highlights are provided, and more details are given where ion-exchange methods have been essential for the analysis or preconcentration of traces of naturally radioactive or nuclear transformed isotopes of environmental concern.

8.2 ANALYSIS WITH RADIOISOTOPES

8.2.1 Quantitative Separations of Ultratrace Elements

Separations of trace elements from others and from interfering matrix constituents are frequently necessary in determinations at the ultratrace level, ≤ 1 μg/g. Although ion-exchange techniques have proven to be extremely valuable for treatment of 0.01–1 mg amounts of cations and anions, applications for separations of traces in the μg range and below have been limited due to contamination problems, uncertainties concerning the irreversible exchange of extremely small amounts of ions onto the resin column, adsorption losses to column walls, and exchangeable impurities initially present on the resin. By using extreme precautions to minimize contamination problems (1,2) and radioisotopes with very high specific activities, it is possible to examine systematically the exchange behavior of ultratrace elements. In the example discussed, it was necessary to devise a highly efficient quantitative procedure for the separation of submicrogram amounts of iron from a phosphate-chloride medium resulting from hydrolysis of phosphorus oxychloride ($POCl_3$), a reagent for production of optical waveguide fibers (3). Iron is a detrimental impurity that renders the reagent useless for optical waveguide applications when it is present at the few ng/mL level. Determinations of iron in the reagent depended upon its separation from the phosphate-chloride medium generated via hydrolysis of phosphorus oxychloride.

Fe^{3+} as the chloride complex was removed from the hydrochloric-phosphoric acid medium by exchange onto an anion column (Amberlite IR 400) in the chloride form. Conditions for the retention of $FeCl_4^-$ while phosphate ions were eluted completely with 5.0 *M* HCl were established by isotope techniques using $^{32}PO_4^{3-}$ and $^{59}Fe^{3+}$. Elution volumes for removal of iron were also determined. Previous literature reports of successful elutions of iron with 0.1 *M* HCl, which works well for levels >0.1 mg, removed only 27.3 ± 0.1% of the iron from the exchange column when initial amounts applied were 20.0 μg or less. A successful quantitative ultratrace separation depended upon efficient recovery of the strongly retained iron. Quantitative elutions were eventually achieved with 0.15 *M* HCl − 1% hydrazine solutions. Reproducible recoveries of 10, 25, 50, and 125 ng amounts of iron were achieved as demonstrated by data in Table 8.1. Results are given for the retention of iron during elution of phosphate and for the final recovery of the phosphate-free iron from the column.

Some details of the execution of this investigation typify the extreme attention, precautions, and detailed considerations necessary to authenticate the usefulness of ion-exchange separations at the ultratrace level. Because of copious sources of environmental contamination by iron, all work at the ultratrace level was conducted within class 100 laminar flow hoods located inside of a class 1000 analytical clean room. Thoroughly cleaned and leached containers and fused silica resin columns were employed. Chemical reagents were produced in the purest possible forms. Hydrochloric acid, doubly subboiling distilled, was obtained from the

NBS. Analyses of similarly prepared HCl showed Fe at the 3 ng/ml level (4). Purified phosphoric acid was produced by hydrolyzing Ultrex (J. T. Baker) phosphorus pentoxide with demineralized doubly distilled water. A high-purity synthetically prepared hydrolyzed solution of $POCl_3$ was prepared by dissolving the appropriate quantity of P_2O_5, mixing with the required amount of pure HCl, and diluting to volume with pure water. This solution, which was 2.18 *M* and 6.72 *M* respectively in H_3PO_4 and HCl, contained a maximum of 0.7 ng/ml of iron. The 0.15 *M* HCl – 1% hydrazine solution for eluting ultratraces of iron from the column was tested for iron by a semiquantitative carbon-rod atomic absorption procedure. Total iron in the acids, hydrazine reagent, and in effluents of the latter from ion-exchange columns was below the detection limit of 5 ng/ml.

The radiolabeled iron tracer, ^{59}Fe, was obtained in a high specific activity grade, 15 Ci/g. It was necessary to preliminarily treat the tracer to remove a ^{60}Co impurity and to reduce the $^{59}Fe^{2+}$ level. Following oxidation of the tracer under controlled conditions with hydrogen peroxide purified by low temperature sublimation (5), the remaining $^{59}Fe^{2+}$ and ^{60}Co were removed via ion exchange on columns of Amberlite IR 400 in the Cl^- form. After sorption of the tracer onto the column from 5.0 *M* HCl, ^{60}Co and $^{59}Fe^{2+}$ were eluted from the column with 5.0 *M* HCl. The $^{59}Fe^{3+}$ tracer was then recovered by elution with 0.15 *M* HCl. Freshly oxidized tracer with negligible amounts of $^{59}Fe^{2+}$ was required for the ion-exchange separations experiments. During ion exchange Fe^{2+} and Fe^{3+} exhibit markedly different chemical behavior due to valence state differences. The level of iron carrier in the tracer solution was estimated from spot test comparisons with known aqueous solutions of iron. Less than 0.1 ng/ml of iron was added to any solution due to the addition of the tracer.

Purification of the ion-exchange resin before its use in preparing the $^{59}Fe^{3+}$ tracer and in subsequent separation experiments was accomplished in the following way. Precleaned fused silica columns were filled to a height of 10 cm with an aqueous slurry of Amberlite IR 400 (40–60 mesh) resin, a strongly basic quarternary ammonium anion exchange resin in the Cl^- form. A small pad of quartz wool ($\approx$1 cm, Spectrosil grade) was placed at the bottom and top of the resin column. Each resin bed was prepurified by passage of 10 column volumes of 1:1 HNO_3, followed by 5 column volumes of purified 5.0 *M* HCl. A third elution with 3 column volumes of 0.15 *M* HCl – 1% hydrazine eluate was followed by five column volumes of pure 5.0 *M* HCl. When this procedure was executed after deliberately exchanging $^{59}FeCl_4^-$ tracer onto the column, only background levels of gamma radiation from ^{59}Fe were detected using efficient

Table 8.1 Quantitative Separation of $^{59}Fe^{3+}$ from Phosphate-Chloride Medium

Amount of Fe (ng)[a]	125	50	25	10
% Retention[b]	99.5 ± 0.3	99.1 ± 0.4	97.6 ± 0.8	98.9 ± 1.0
% Recovery[c]	100	99.6 ± 0.9	97.1 ± 0.1	96.8 ± 1.0

[a] Mean of three separate samples for each level.
[b] $FeCl_4^-$ retained by column from medium 2.18 *M* in H_3PO_4 and 6.72 *M* in HCl, phosphate-containing anions eluted with 5.0 *M* HCl.
[c] Fe^{2+} recovered by eluting column with 0.15 *M* HCl–1% hydrazine.

well-type sodium iodide detectors. Thus it is unlikely that influential levels of exchangeable iron contamination resides on ion exchange columns prepurified as described. This procedure was found to be the ultimate one for removal of exchangeable iron from an anion exchange resin to be used in the Cl^- form.

Throughout the entire investigation, precautions were specifically taken to minimize the introduction of iron contamination during studies with submicrogram amounts. By using microgram amounts of radiolabeled ^{59}Fe and $H_3^{32}PO_4$, conditions were first established for the elution of phosphate with 5.0 *M* HCl while retaining iron on the column, and for the subsequent recovery of iron by elution with 0.15 *M* HCl − 1% hydrazine using microgram amounts of radiolabeled ^{59}Fe and $H_3^{32}PO_4$. The submicrogram separations of iron were then executed under contamination control conditions with the results shown in Table 8.1. Iron separated from a processed sample and the associated blank can be quantitatively measured by an exceptionally sensitive laser intracavity absorption technique (6).

Assessments of the utility of ion-exchange separations at the ultratrace level can usually only be made with radioisotope studies. The characteristic radiations accompanying the decay of the isotope permit its identification. Such radiation can be readily detected with extremely high sensitivity, thus the physical and chemical behavior of minute quantities of ions and chemical species can be monitored through complex chemical environments such as ion exchange resins. Contamination conrol methods must be practiced to ensure that actual levels of the element under study are not orders of magnitude greater than the investigator had assumed. Depending upon the specific activity of the radioisotope, conditions can be optimized for the ion-exchange separation, concentration, and recovery of ng levels of cations and anions. Sixty-nine of the elements in the periodic table have isotopes with suitable properties for analytical investigations. Those available commercially are given in Fig. 8.1.

COMMERCIALLY AVAILABLE RADIONUCLIDES

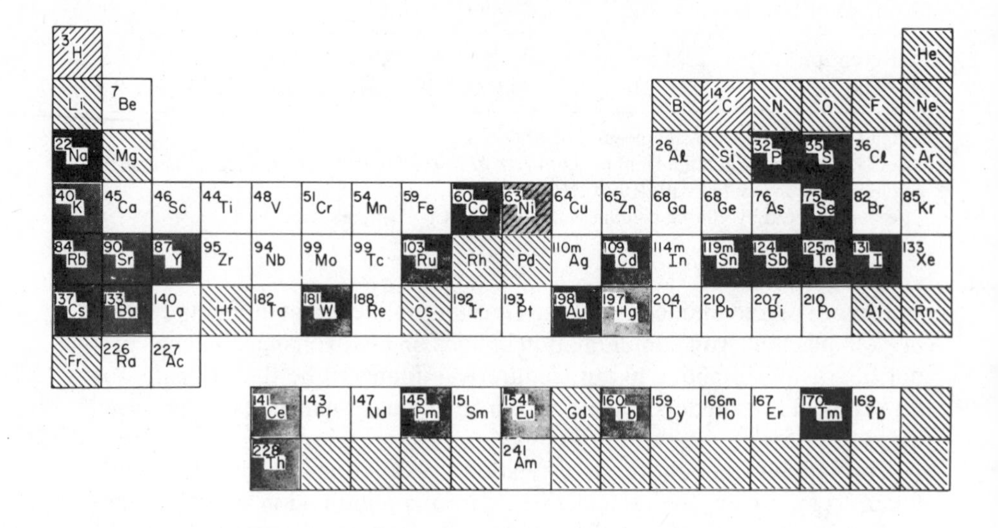

β-EMITTER ONLY

ADDITIONAL RADIOISOTOPES AVAILABLE

NOT AVAILABLE OR NO ISOTOPE

Figure 8.1. Commercially available radionuclides.

8.2.2 Substoichiometric Radioisotope Dilution

Quantitative analysis by isotope dilution depends upon the reduction in the specific activity of a radioisotope when it is mixed with nonactive atoms of the same chemical form. The specific activity of a tracer solution (S^*) is defined as the ratio of the total activity (A^*) to the total quantity of isotope (W^*) in the solution. The specific activity of the tracer represented by

$$S^* = \frac{A^*}{W^*} \tag{1}$$

changes to

$$S = \frac{A^*}{W^* + W} \tag{2}$$

when a solution containing the amount W of the nonactive material is added. By combining eqs. 1 and 2 the expression

$$W = W^* \left(\frac{S^*}{S} - 1 \right) \tag{3}$$

is obtained. The weight W of an unknown amount of the nonradioactive element in a sample can be quantitatively measured by isolating a known quantity of the element from the mixture of the tracer and the unknown sample, and measuring the resulting activity. Since quantitative results depend only on a measurement of the specific activity, it is not necessary to isolate quantitatively or recover the entire quantity of the substance from the sample solution.

The greatest advantage of analysis by isotope dilution is this ability to provide quantitative results without quantitatively recovering the entire amount of the material initially contained in the sample. However, the necessity to determine the specific activity of the sample requires a measurement of the weight of the isolated component. Usually this measurement is accomplished by weighing a precipitated compound; however, other analytical methods can be employed. This requirement has primarily restricted the use of isotope dilution methods to the determination of milligram or larger quantities of constituents in samples. As other more convenient analytical methods were developed, the use of isotope dilution procedures rapidly declined.

Ruzicka and Stary (7) devised a new procedure for isotope dilution analysis that has extended the application of the method into the submicrogram region. A discussion of this method is given. Upon substitution into Equation (3) for the specific activity of the standard radioisotope and sample solutions one obtains

$$W = W^* \left(\frac{a_1^*/w_1^*}{a_2/w_2} - 1 \right) \tag{4}$$

where a_1^* and a_2 are the measured relative activities emitted by the quantities, w_1^* and w_2 of the element isolated from the solutions containing the standard and sample, respectively. If the same weight of material is isolated from the standard and sample solutions, eq. 4 reduces to

$$W = W^* \left(\frac{a_1^*}{a_2} - 1 \right) \tag{5}$$

Equation 5 shows that the amount of material in a sample can be obtained from a simple measurement of the relative activities of equal quantities of material isolated from the standard and sample solutions. The factors that limit the sensitivity now become the detection limit for the measurement of the activity and the smallest quantity of the material that can be reproducibly isolated from the solutions.

A substoichiometric amount of a multidentate chelating agent such as EDTA can be used to isolate a quantity of cation by forming an anionic chelate complex. Upon passage of the solution through a cation-exchange resin in the sodium form, the excess cation is retained on the column and the metal complex is eluted. Appropriate conditions for substoichiometric isolations can be chosen by examining stability and equilibrium constants of EDTA complexes.

A general procedure for trace analysis by isotope dilution and substoichiometric ion exchange separation is outlined below:

1. Add a known quantity of the standard radioisotope to the sample and dissolve the mixture under conditions that prevent contamination.
2. Chemically treat, in the same manner as the sample solution, a blank solution containing the same amount of radioisotope as used in step (1).
3. Adjust pH and other conditions appropriately and add substoichiometric amounts of EDTA to complex equal amounts of the element in both solutions.
4. Remove the excess cation by ion exchange and recover the anionic EDTA fractions.
5. Measure the relative activities of the isolated fractions under identical counting conditions.

The method is simple in principle. However, to obtain reliable data a number of details must receive appropriate attention. In practice the chemist must eliminate or minimize all possible sources of contamination in order to obtain maximum sensitivity. A few of the normal precautions are listed.

1. Everything will contaminate unless adequately cleaned and proven otherwise.
2. All reagents must be specially purified or purchased in ultrahigh purity. Demineralized water that is subsequently distilled in an all-quartz apparatus is used in the preparation of solutions, etc.

Table 8.2 Isotope Dilution Data for the Analysis of Iron

Grams of Fe			Activity in C/unit time	
Added	In tracer	Found	Tracer Solution	Sample Solution
1.39×10^{-6}	1.89×10^{-6}	1.34×10^{-6}	54660	32520
			55520	31960
2.79×10^{-7}	2.79×10^{-7}	2.77×10^{-7}	69160	38240
			71140	33520
			74240	33520
1.39×10^{-8}	1.53×10^{-7}	0.995×10^{-8}	40080	36530
			39570	37720
				37960

3. Substitute appropriately cleaned and leached plastic or Teflon vessels for glassware.
4. Eliminate contamination from airborne particulate matter by working in all-plastic, positive-pressure, clean hoods. Evaporations and dissolutions are done in Teflon or plastic chambers that are continuously purged with filtered nitrogen.
5. Standard radiotracer solutions are prepared immediately before use by spiking nonactive solutions of known concentration with the carrier-free isotope and diluting to the desired concentration. Direct counting or reversed isotope dilution analysis of the diluted solution gives a precise determination of the quantity of element in the standard.

By using the general procedure, attempts to quantitatively determine iron have been made at the author's laboratory. The data shown in Table 8.2 were obtained. The activity separated from solutions containing known quantities of carrier iron shows reasonable precision in most cases. Except for the set of results at the 10^{-8} g level, the measured iron concentration is in good agreement with the amount added. Many of the difficulties with the direct isotope dilution method were circumvented by using a technique employing a calibration curve. Standard curves for iron in the 10^{-7} and 10^{-8} g region are shown in Fig. 8.2. Although enough sensitivity exists for quantitative measurements at the 10^{-8} g level and below, usable quantitative results were obviated, presumably by blank values becoming significant in comparison to the added carrier levels being measured. Experience has shown that quantitative results via this

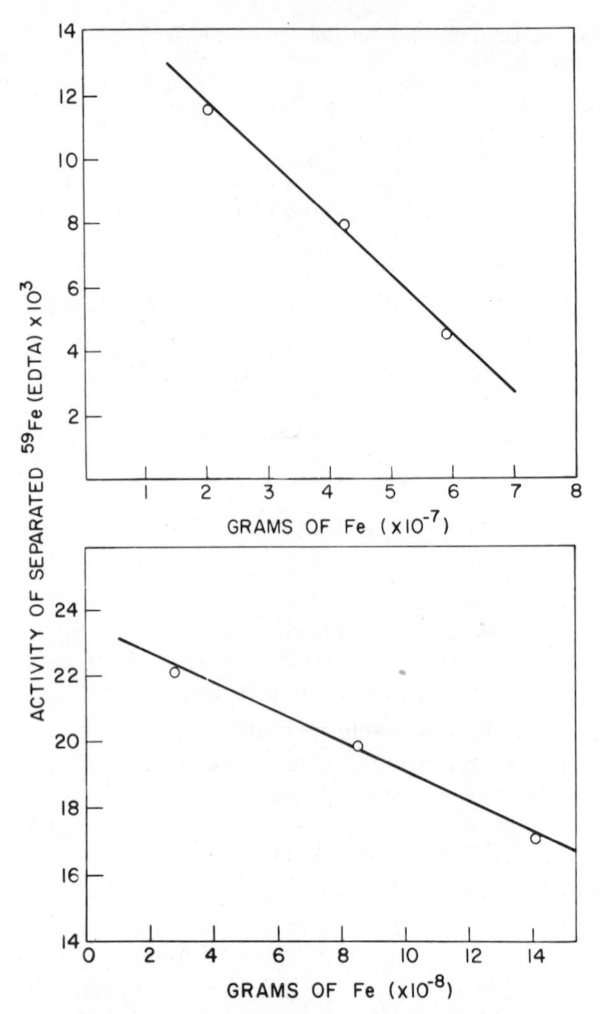

Figure 8.2. Calibration curves for substoichiometric radioisotope dilution ion exchange separation of EDTA complexes of Fe^{3+}.

method are to be attempted only if there are no other alternative procedures capable of providing the desired information.

Available radioisotopes and information on the EDTA complexes suggest that substoichiometric ion exchange radioisotope dilution methods could be developed for Cr^{3+}, Fe^{3+}, In^{3+}, Tm^{3+}, (R.E.), and Y^{3+} when present simultaneously with other cations. In reasonably pure monocomponent trace element samples such as radiotracer and synthetically prepared dilute standard solutions, methods for Th^{+4}, Bi^{3+}, Ni^{2+}, Pb^{2+}, and Cu^{2+} could be developed. Although some of these methods have been reported in Ruzicka and Stary's book (7), widespread use has not been adopted. The arduous and very difficult task of purifying EDTA (8) to obtain $<$ ng/ml levels of impurities, successfully executing quantitative chemical reactions of minute amounts of cations, preventing adsorption losses, minimizing blank contamination problems, and accurately determining the initial level of element in the radioisotope standard are formidable challenges to be overcome. Nevertheless, the potential for the extraordinary high sensitivity of the substoichiometric ion-exchange radioisotope dilution method, which is in principle a simple procedure, still perhaps merits further investigations. Several attempts by Western World analytical chemists to apply these methods have been unsuccessful.

8.2.3 Diagnostics of Ultrapurification Procedures

Ultrapure materials and chemicals with total elemental impurity content ≤ 1 μg/g are now required routinely, particularly in the telecommunications industry and for use in trace analysis. The purification of the material can be based on any process which separates the desired constituent from undesirable ones; or in the reverse situation, unwanted impurities are removed from the desired matrix component. Many physical or chemical methods can be applied to achieve desired purifications. When such methods are applied and material purity with respect to key elements is found by direct analytical determinations to be insufficient, it is necessary to scrutinize the purification process to determine whether contamination problems, or flaws in the execution or efficiency of the process are causative factors. Inherent deficiencies of the process or errors in its execution can be identified unequivocally by radioisotope techniques. Then direct, accurate analyses of processed materials can be used to pinpoint contamination problems.

For optical waveguide production, highly pure sodium and calcium carbonate raw materials were sought to meet projected specifications for Co, Cr, Cu, Fe, Mn, Ni, and V of 2, 20, 50, 20, 100, 20, and 100 ng/g, respectively. Since analysis of commercially available reagents showed them to be insufficiently pure (9), a purification process was needed to produce optical waveguide quality reagents.

The decontamination of sodium or calcium solutions with respect to transition elements can be effected by an ion exchange technique. It has already been shown earlier in this chapter that submicrogram quantities of trace elements can be effectively removed by ion-exchange techniques, provided extremely careful procedures are executed under stringent contamination control conditions. Such precautions are even more critical to the success of a purification process based on ion exchange. The scheme for production of high-purity sodium or calcium solutions, from which the solid carbonates could be produced, is shown in Fig. 8.3.

Reagent grade sodium nitrate was preliminarily purified by solvent extraction, and sodium then loaded onto a cation-exchange column in the NH_4^+ form. Any multivalent impurities escaping the extraction step and subsequently loaded onto the column with sodium would then be removed from the sodium-loaded column by conversion to the anionic complex during elution with EDTA. Purified sodium would be stripped from the column with high-purity ammonium carbonate, and optical waveguide quality sodium carbonate would be recovered eventually from the purified sodium solution. In principle, the proposed chemical purification scheme was sound. For practical use, it was necessary to determine exactly how effective the method would be for removing submicrogram quantities of the trace transition elements. Each step in the purification process was examined using high specific activity radioisotopes to pinpoint deficiencies with respect to removal of elements of interest. Microgram amounts of each purity were monitored through each step of the purification scheme.

The solvent extraction (Step 1) failed to remove some Cr^{3+} and VO_2^+, as shown by data in Table 8.3. Optimum organic-to-aqueous phase volume ratios and the required number of individual extractions to reduce impurities to the low ppm range can be calculated from the measured distribution ratios using the equation

$$W_n = W_0 \left(\frac{V_{aq}}{K_D V_{org} + V_{aq}} \right)^n \qquad (6)$$

where W_0 and W_n are the weights of cation initially present in the aqueous feed solution and the amount remaining after n extractions, respectively, K_D is the distribution ratio and V_{aq} and V_{org} are the volumes of the aqueous and organic phases.

Purification of the ion-exchange resin prior to its use was accomplished in a 2.5 × 75 cm polypropylene column provided with a polyethylene frit at the bottom to support the 50-cm column of resin. The column was also accommodated with a polyethylene stopcock to control the flow of eluate. The most effective method for purification of the ion-exchange resin prior to conversion to the NH_4^+ form was achieved by the following steps: (1) elution with 1.0 M HNO_3 and rinsing with pure H_2O; (2) elution with a mixture of 1:7 (1 M NaOH) – (30% H_2O_2); and (3) elution with hot (4 g/L) EDTA at pH 9.0. After each step the resin was rinsed with copious

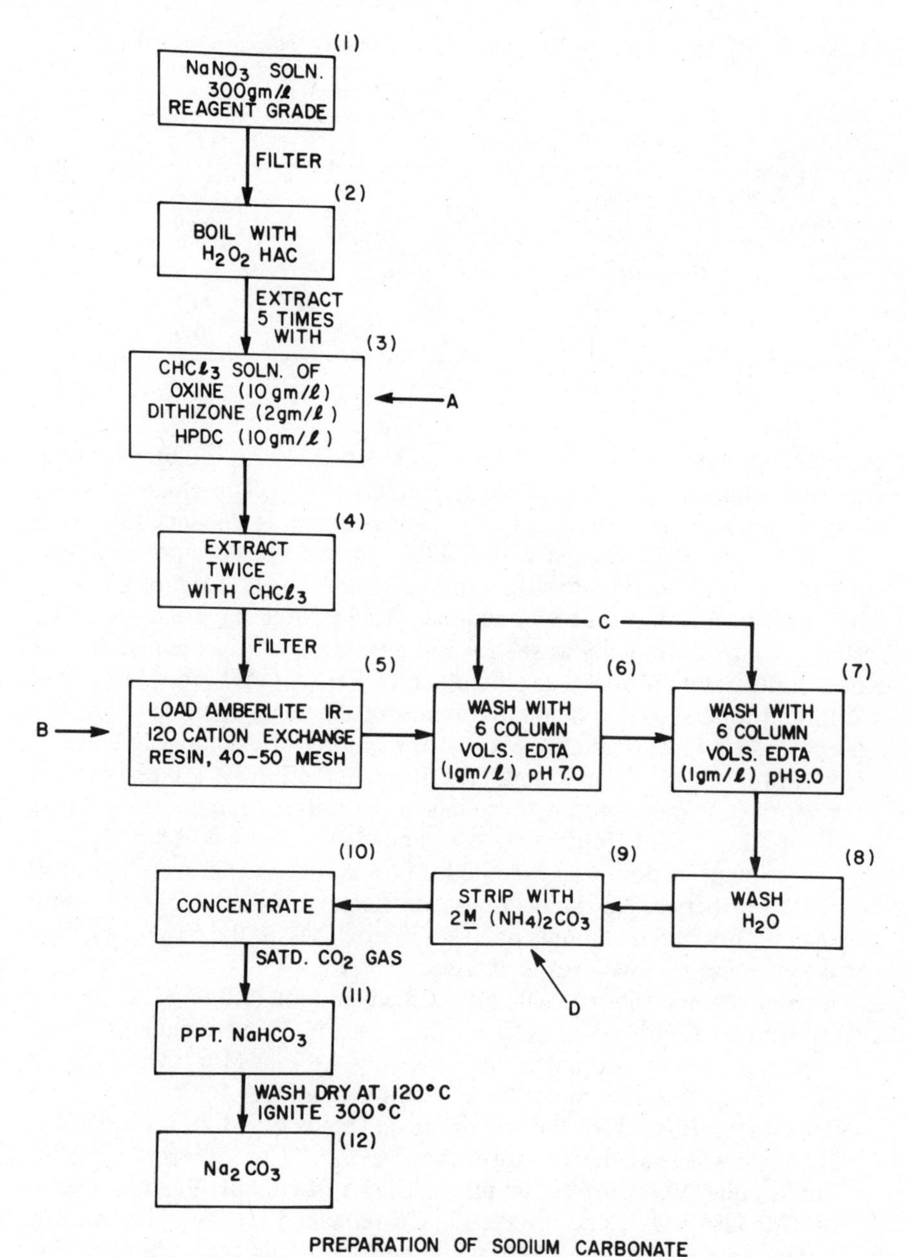

Figure 8.3. Solvent extraction ion-exchange purification of sodium solutions.

Table 8.3 Purifying Sodium Reagent Feed Solution By Solvent Extractions

	γ-Ray Activity (counts/min)				
Element	Aqueous	Organic	K_D	%E	pH
$^{51}Cr^{3+}$	26,581	103,393	3.87	79.5	8.2
$^{48}VO_2^+$	190,321	319,631	1.67	62.6	8.2
$^{54}Mn^{2+}$	35	152,164	4347	>99.9	7.9
$^{60}Co^{2+}$	19	100,160	5271	≥99.9	3.2
$^{59}Fe^{3+}$	77	27,295	354	99.7	3.2
$^{64}Cu^{2+}$	18	35,583	1977	≥99.9	3.2

amounts of high-purity water. The resin in the H^+ form is eluted with up to 50 column volumes of 1(1 M NaOH):7(30% H_2O_2). This eluate oxidizes several trace metals, Fe^{2+} and Co^{2+}, for example, to enhance their subsequent removal by elution with EDTA. Some Cr^{3+} is removed as the CrO_2^{2-} anion. Vanadium can be removed from the H^+ form of the resin by elution with dilute acid containing H_2O_2. After the basic peroxide eluted resin is thoroughly washed, it is then treated with 6 column volumes of a dilute solution of the disodium salt of EDTA (4 g/L adjusted to pH 7.0), and followed by 6 column volumes of the same solution at pH 9.0 preheated to 70°C. The efficiency of the elution of exchangeable Fe^{3+}, Co^{2+}, Mn^{2+}, Cr^{3+}, Cu^{2+}, and VO_2^+ from Amberlite IR 120 resin in the Na^+ form was monitored with radioisotopes. After complete sorption of 2 μg of ^{59}Fe, ^{60}Co, ^{54}Mn, ^{51}Cr, ^{64}Cu, and $^{48}VO_2^+$ labeled impurities, the purification procedure was executed. Mn, Co, Fe, and Cu were removed to 99.9% or better. No radioactivity was detected from isotopes of these elements. For initial amounts of $^{51}Cr^{3+}$ and $^{48}VO_2^+$ at the 1 μg level, 92% and 91% were removed respectively.

During loading (Step 5), 400 mL of feed solution (1.0 M in Na^+) with 0.005 μg/mL of Fe^{3+}, Mn^{2+}, Cr^{3+}, Cu^{2+}, and VO_2^+ were eluted through the column. Thirty percent of the Cr^{3+} and 4.8% of Mn^{2+} were loaded with sodium. To determine the efficiency of EDTA for removing exchanged impurities from the sodium resin (Steps 6 and 7), a prepurified NH_4^+ resin was loaded with 4 μg each of $^{59}Fe^{3+}$, $^{60}Co^{2+}$, $^{54}Mn^{2+}$, $^{51}Cr^{3+}$, $^{62}Cu^{2+}$, and $^{48}VO_2^+$. After 700 mL of EDTA (4 g/L, pH 9) were eluted, 76% Cr, 40% VO_2^+, and <0.0001% Cu remained on the resin. During stripping (Step 9), 2 μg of the trace impurities were partially and rapidly removed during elution with 6 column volumes of 0.9 M $(NH_4)_2CO_3$. Continuous tailing of Cr, Co, and Fe was then observed.

The purification process was thus found to have several deficiencies:

1. The extraction step failed to remove some Cr^{3+} and VO_2^+.
2. Any Cr^{3+} and Mn^{2+} escaping the extraction step would be loaded onto the resin with sodium.
3. Cr^{3+} was not removed from the sodium resin during elution with EDTA.
4. Any trace impurity retained by the Na^+ loaded column would be partially stripped during removal of Na^+ by elution with ammonium carbonate.

Investigations with radioisotopes showed that the following modifications were necessary. Following the extraction process (Step 1), 4 g/L of EDTA was added, the pH adjusted to 9.0, and the feed solution heated to 80°C before being loaded onto the resin while hot. Under these conditions no detectable Cr^{3+} was taken up by the resin. The most effective method for purification of the ion-exchange resin prior to conversion to the NH_4^+ was also disclosed by radioisotope studies and previously reported in this section.

The diagnostic examination of this ion-exchange purification process is a specific example of analytical information that would have been virtually impossible to obtain without the use of radioisotopes.

Reagents much purer than commercially available stocks have been produced via this process. However, the ultimate purification of these reagents has been based on mercury cathode electrolysis removal of trace transition element impurities (10). Limitations of the ion-exchange method were imposed by handling of large volumes of solutions. Routine operations including filtering, column adsorption, elution, and vessel transfers, as well as carbonaceous particulates from the ion-exchange resin by-products, were sources of contamination. Electrolysis in closed cells has eliminated most of the handling problems.

8.3 SEPARATIONS IN ACTIVATION ANALYSIS

8.3.1 Multielement Analysis

Multielemental analysis of many types of materials have been performed successfully by neutron activation in combination with ion-exchange separations of elements into small groups of components, which are compatible with counting simultaneously via high-resolution gamma-ray spectrometry. Often materials are sufficiently complex that analysis without chemical separations would be severely limited or totally obviated. In the

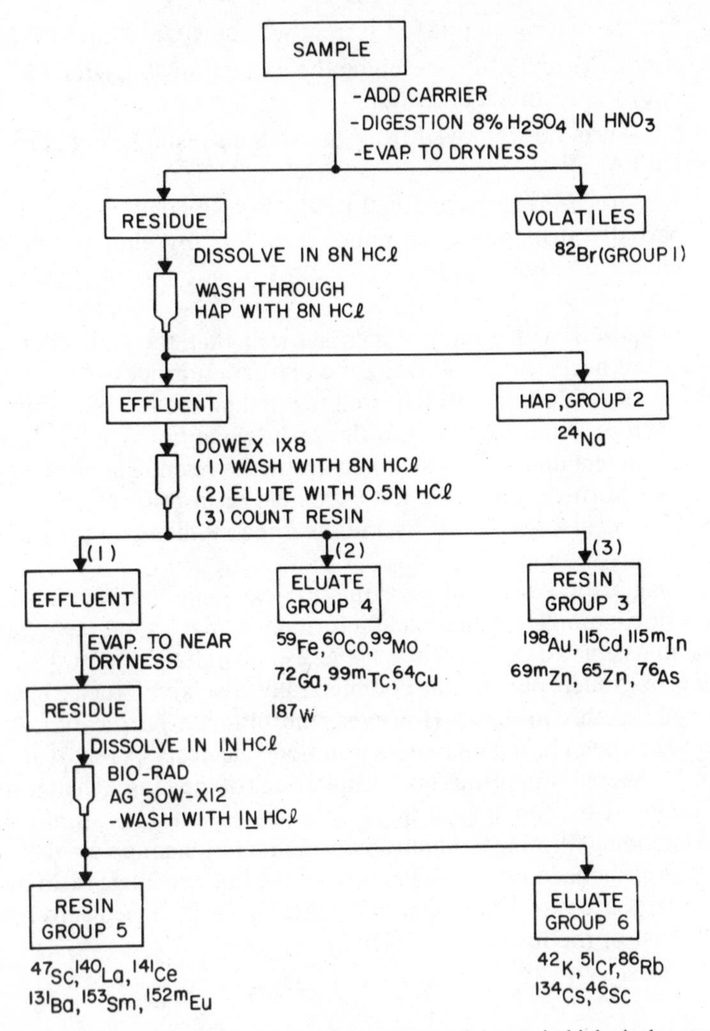

Figure 8.4. Radiochemical group separations of trace elements in biological materials.

case of biological materials, the trace elements Fe, Cu, Zn, Mn, B, Na, Co, Mo, and V are of interest in studies of plant life, while Fe, I, Cu, Zn, Mn, Co, Mo, Se, F, B, and Sr are essential elements for animal life. In addition to providing information on these elements, the analytical process must be able to cope with the presence of many other possible trace and major constituents. The diagram in Fig. 8.4 shows the ion exchange separation scheme used by Morrison and Potter (11) for radiochemical group separation of trace elements in freeze dried organ samples and NBS Standard Reference Material 1571, orchard leaves. After proper digestion of the irradiated sample in 8% H_2SO_4 in HNO_3, with precautions to trap any evolved volatiles, sodium-24 is removed by exchange onto hydrated antimony pentoxide (HAP) column. The effluent in 8 *N* HCl is taken up by a Dowex 1 × 8 anion-exchange resin which is subsequently washed several times with 8 *N* HCl. The effluent and washings are combined, evaporated, and saved for further processing. The resin is eluted with 0.5 *N* HCl to remove Group 4 elements. Direct counting of the resin provides information on Group 3 components. The concentrated effluent referred to above is processed further by dissolution into 1 *N* HCl and passage through a column of Bio-Rad AG50W-X12. Thorough washing of the column with 1 *N* HCl provides Group 6 elements, while the resin is counted directly to detect those of Group 5. With this rather straightforward adaption of well-known ion-exchange separations, 26 elements were quantitated in orchard leaves and 31 in heart and kidney samples. The excellent agreement of the quantitative results obtained for this large number of elements determined simultaneously is shown in Table 8.4. In analyses without ion exchange separations in which four separate irradiations of the sample were done, results were obtained only for Al, Cl, Mg, Mn, V, Sr, As, Br, Sb and Se.

As previously shown, samples containing decades of elements may be analyzed rather completely by neutron activation, provided group separations are made. Any of the existing ion-exchange schemes can be adopted for separating elements in the activated sample. It is only necessary that the analytical chemist properly chooses a valid separation scheme for the sample matrix to be analyzed. Fortunately, validation of the selected scheme and determinations of the efficiency of the separations can be assessed easily by spiking a sample with radioisotopes and carrier, and then executing the procedure to determine separation efficiency and yields.

8.3.2 Single-Element Determinations

Selective separations of matrix compoonents of a high-purity material are often prerequisites to the determination of a few trace constituents. The recent example of the analysis of high-purity niobium metal made possible by the use of ion-exchange techniques is described and further illustrates the unique quantitative activation analyses which would otherwise not be possible to execute in the absence of appropriate separation steps (12). Pure niobium, a scientifically and technologically important metal, has properties strongly dependent on its trace elemental content. Nuclear

Table 8.4 Results of Analysis of NBS Standard Reference Material 1571, Orchard Leaves

Element (ppm unless indicated)	Analysis 1	Analysis 2	NBS[a]
Al	460.	420.	. . .
As	10.	10.	14 ± 2
Au	0.0001	0.0001	—
Ba	50.	52.	—
Br	8.2	8.3	(10)
Ca	2.15%	2.04%	2.09% ± 0.03%
Cd	≤50.	≤50.	0.11 ± 0.02
Ce	1.	1.	—
Cl	790.	790.	(700)
Co	0.1	0.1	(0.2)
Cr	2.5	2.5	(2.3)
Cs	≤0.06	≤0.06	—
Cu	11.	9.	12 ± 1
Eu	0.3	0.3	—
Fe	290.	290.	300 ± 20
Ga	0.089	0.082	—
K	1.54%	1.47%	1.47 ± 0.03%
La	1.2	1.2	—
Mg	0.60%	0.59%	0.62% ± 0.02%
Mn	86.	86.	91 ± 4
Mo	≤5.	≤5.	—
Na	80.	74.	82 ± 6
Rb	11.	11.	12 ± 1
Sb	3.0	3.0	—
Sc	0.20	0.21	—
Se	0.08	—	0.08 ± 0.01
Sm	0.15	0.14	—
Sr	43.	37.	(37)
V	≤0.7	≤0.6	—
W	≤2.	≤2.	—
Zn	27.	23.	25 ± 3

[a] Numbers in parentheses are NBS noncertified values, Ref (11).

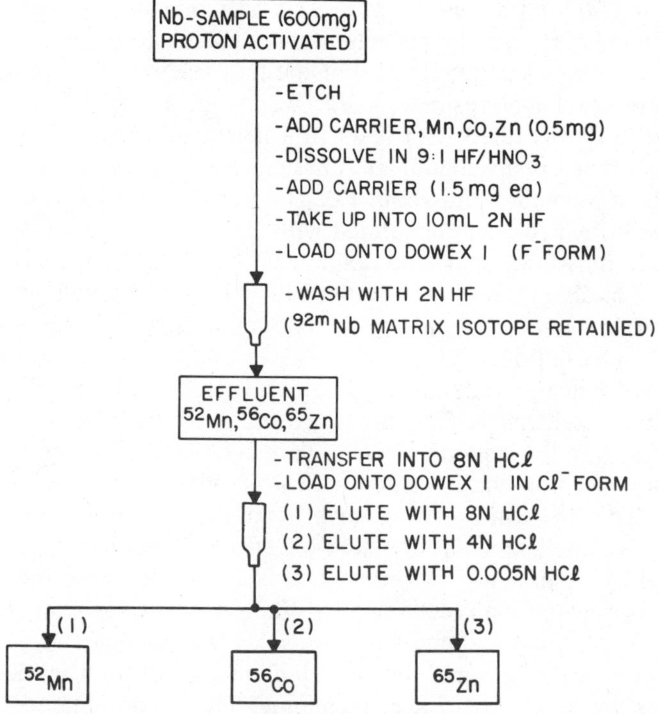

Figure 8.5. Post irradiation ion-exchange separation of trace elements from niobium.

methods have proven to be the most sensitive and valuable methods for characterizing this difficult matrix, particularly for the elements, Cr, Fe, and Cu. While instrumental proton activation analyses performed non-destructively (i.e., without separations) have provided reasonably low detection limits of 0.05 and 0.1 ppm for Cr and Fe respectively, the limits achievable are not sufficient for characterizing the most pure samples prepared by zone refining in combination with electrolytic deposition. Achieving greater sensitivity for the determination of the trace elements depended upon first separating the niobium matrix activity from ^{92m}Nb, which is generated via the reaction ^{93}Nb (p,pn) ^{92m}Nb. The nuclear transformed isotopes of each element of interest ^{52}Mn, ^{56}Co, and ^{65}Zn then in turn needed to be further separated from each other in order to obtain the highest sensitivity possible by counting in a high-efficiency well-type sodium iodide detector. The ion-exchange separation scheme formulated for the post irradiation processing of niobium is outlined in Fig. 8.5 and was based on previously reported investigations of Aubion et al. (13),

Kim et al. (14), and Kraus and Moore (15). By repeated boiling down and adding diluted HF, the sample solution was made approximately 2 *N* with HF in a volume of about 10 ml. For the first separation step, a strongly basic anion-exchange resin (F^--form) was used. A Teflon column (30 cm × 1 cm i.d.) was filled with resin to a height of 20 cm. The exchange capacity of this resin column is sufficient for a complete separation of amounts of niobium up to about 1 g. The sample solution was applied to the column, which was then eluted with 2 *N* HF at a flow rate of about 2 ml/min until 40 ml eluate was obtained. The eluate was evaporated to a volume of about 5 ml, then 10 ml of 12 *N* HCl was added before evaporation to a volume of about 2–3 ml. A 5-ml portion of 8 *N* HCl was then added. The resulting solution was used for the specific separation of the individual indicator radionuclides, which was carried out on a strongly basic anion-exchange resin in the Cl^- form. A 20-cm column of resin was filled into a Teflon column of 1-cm diameter. After applying the sample solution to the column, it was eluted with 8 *N* HCl at a speed of about 2 ml/min. The first 5 ml were removed and subsequently the manganese fraction was collected in a volume of 50 ml. The cobalt fraction was obtained by elution with 50 ml of 4 *N* HCl, and the zinc fraction by a final elution with 50 ml 0.005 *N* HCl. Each fraction was evaporated to a volume of 5 ml and counted with the NaI well-type detector.

The chemical yields and the decontamination factors were determined by using the radioactive tracer technique. The eluate obtained after the first separation step carried out using 2 *N* HF contained the elements manganese, cobalt, and zinc, and thus the indicator radionuclides of the elements chromium (^{52}Mn), iron (^{56}Co), and copper (^{65}Zn). The decontamination factors relative to the matrix nuclides were $>10^5$. The chemical yields related to the whole radiochemical separation procedure were 95.1 ± 0.5% for Mn, 98.4 ± 1.7% for Co, and 95.7 ± 1.8% for Zn. The decontamination factors for the elements specifically separated in the appropriate fractions are in the range of 10^3 to 4×10^3.

Table 8.5 shows a comparison of the limits of detection obtainable by instrumental proton activation analysis, by group separation followed by counting of the eluate after the first separation step with a high-resolution γ-ray spectrometer, and by specific separation of the individual indicator radionuclides and their counting with a well-type NaI detector. Comparable limits of detection for Cr can only be achieved presently by radiochemical neutron activation analysis after long periods of irradiation in high-neutron fluxes and, in the case of Cu, by differential pulse anodic stripping voltammetry. The sensitivity for Fe determination by radiochemical proton activation can seldom be matched by any other technique.

Table 8.5 Improved Limits of Detection for Radionuclear Methods Via Ion Exchange Separations

Performance	Element		
	Cr	Fe	Cu
Instrumental	50 ppb	0.2 ppm	—
After the first separation step. γ-ray spectrum	10 ppb	50 ppb	0.2 ppm
After specific separation NaI detector	0.2 ppb	5.0 ppb	15 ppb

Inorganic ion exchangers have also been applied very profitably in nuclear activation analyses. The extremely specific ion-exchange removal of sodium via hydrated antimony pentoxide (HAP) resins is the most important example (16). This discovery of the extraordinarily high specificity of the resin for sodium is a noteworthy contribution to practical applications of activation analyses. The ubiquity of high radiation levels from sodium-24 after neutron irradiation has made many analyses impossible, especially for isotopes with shorter halflives than that of sodium-24. In these cases, the isotope could only be detected after making a preliminary radiochemical separation of the interfering activity from sodium. High-purity sodium-calcium silicate glasses and glass-making raw materials for optical waveguide production have been characterized successfully following neutron activation, removal of sodium-24 via HAP ion exchange, and additional chemical separations (17). Elucidating the nature of the HAP specificity for sodium and scrutiny of additional possible uses of inorganic exchanges in activation analysis are investigations worthy of continuation.

8.4 SEPARATIONS IN NUCLEAR TECHNOLOGY

8.4.1 Uranium Recovery

Ion-exchange methods are fast becoming the method of choice for preconcentration or recovery of uranium from solution. The five continuous ion-exchange processes (CIX) that have found the widest acceptance in the uranium industry were recently reviewed (18). Improved U recovery by CIX processes has been achieved using specially synthesized ion-exchange resins of high density (19). For these continuous ion-exchange processes, pilot plant designs are described for treating feed solutions at rates of 5.45 m^3/hr to 24.5 m^3/hr at U concentrations of 80 ppm to 18 ppm with recoveries of 90% (20). Evaluations of ion-exchange resins for uranium recovery on commercial scales are continuing (21,22,23). Although needs for new and better resins exist (22), Ionac A-651, IRA 910, and Duolite 102D are assessed as promising (23).

8.4.2 Fission and Nuclear By-Products Control

In analyses, separations, and purifications of fission products, ion-exchange methods are also prominent. Separations of actinide and fission products of importance in the atomic energy industry were reviewed recently by Jenkins (24). Well-known conventional methods continue to be applied. For example, the recovery, separation, and purification of plutonium and neptunium were investigated comparatively, using Amberlite, Duolite, Ionac, and Dowex resins (25). Different anion exchangers have been studied for the separation of transplutonium elements (26). Rare earth fission products have been separated and determined by ion exchange (27,28). Ion-exchange methods are now substantiated as the best approach to the removal of isotopes from large volumes of nuclear reactor cooling waters. Engineering design of systems for handling radioactive ion-exchange resin beds (29) and for process waste water treatment in nuclear power plants are available (30).

The important question regarding the stability of ion exchange resins to ionizing radiation continues to be examined. Recently, a thorough review of the topic was provided by Simon (31), and degradation induced by radiation and OH^- was studied by spectroscopy (32). Resistance of resins to radiation damage becomes important in applications such as the production of kilogram quantities of ^{237}Np, where Amberlite XE-270 macroreticular, weakly basic ion-exchange resins have long been used successfully in the production plant (33).

Resins are used widely in remote radiochemical systems developed for routine multicurie radiochemical separations of radioisotopes (34). Inorganic exchangers are finding applications in fission product separations due to their high specificity and stability to intense radiation fields (35–37). In laboratory experiments, undiluted fission product waste solutions 2–3 M in HNO_3, when passed through NH_4^+ molybdophosphate/NH_4^+ tungstophosphate, undergo losses of Cs, Rb, and small amounts of Zr due to uptake by the resin. After oxidation of this effluent with NH_4 persulfate in the presence of $AgNO_3$ as a catalyst, the solution is passed through a column of MnO_2 exchanger, which retains Ce^{4+} along with the remaining Zr^{4+} and Ru^{4+}. Strontium and sodium can then be removed with hydrated antimony pentoxide (38). The nature of the inorganic ion exchangers have been compared for the separation of the fission products cesium and strontium (39). Sorption of fission product produced radionuclides on hydrated inorganic exchanges was examined recently (40).

The mechanism of $^{106}Ru(III)$, $^{106}RuNO(II)$, $^{144}Ce(III)$, $^{147}Pm(III)$, $^{85}Sr(II)$, $^{131}I^-$, $^{35}SO_4^{2-}$, and $H_3{}^{32}PO_4$ sorption on hydrated ferrous, ferric, Al, and chronic oxide was studied. In certain pH ranges, the sorption process was predominantly the ion exchange of the cationic or anionic forms of the radionuclide for a proton or hydroxyl group of the oxide. Evidence was also found for the probable sorption of colloidal forms of the radionuclide. Rapid automated and semiautomated process schemes are being developed for the separation of nuclear corrosion and fission products using combinations of inorganic and chelating resins (41). Mo(ferrocyanide), ZrP, and Chelex 100 resins were examined for the selective separation of the radionuclides, ^{46}Sc, ^{98}Zr, ^{95}Nb, ^{137}Cs, and ^{152}Eu.

8.4.3 Fuels and Waste Processing

The use of fast breeder reactor techniques to produce fissional material from nonfissional ones holds considerable promise as the basis for future nuclear energy production. There is the corresponding problem of substantially increasing the levels of nuclear wastes that must be handled in such a way as to render them as harmless as possible to the environment. Ion-exchange techniques are also useful in this area of nuclear fuels processing. After the production of ^{233}U via the reaction.

$$^{232}Th\ (n, \gamma)\ ^{233}Th \xrightarrow{-\beta^-} {}^{233}Pa \xrightarrow{-\beta^-} {}^{233}U,$$

the fissionable product, ^{233}U, must be separated from the thorium matrix and quantitated. Humphrey and Adkins (42) selectively removed traces of ^{233}U from Th and from other macroimpurities except Fe by passage of the solution through AG1-X8 anion exchange resin in the Cl^- form. Further purification of U for removal of Fe was based on a subsequent passage of the solution through another column of the same resin in the NO_3^- form. Analysis of the uranium content was then based on an alpha spectrometric counting method. ^{233}U in the range 0.1–55 ppm could be quantitated with a relative standard deviation of $\pm 1.1\%$.

The most effective treatment of highly radioactive nuclear wastes, confinement into highly inert ceramic forms, and proper storage also depend on ion exchange mechanisms. Radioactive wastes may be fixed into porous silicate glass or silica gel which contains pores associated with exchangeable alkali, Group 1B cations, and/or NH_4^+. Properly processed glasses are effective ion exchanges for radioisotope removal from wastewaters. Following sintering, the material is ready for long-term storage (43). The preparation, compounding, structure, and leaching characteristics of a crystalline ceramic TiO_2 radioactive waste form have been compared with vitrified wastes (44). The waste (25% by weight) is introduced into the titanate material by an ion-exchange process and the loaded titanates are converted into a dense ceramic by pressure sintering. Such processes have been shown to be relatively insensitive to changes in the composition of the waste feed, and material stability has exceeded that of wastes vitrified at elevated temperatures. A review of other methods that are used or underdevelopment for solidification of radioactive ion-exchange resins has been reported (45).

8.4.4 Radioisotope Production

Radiopharmaceuticals, medicinal radioisotopes for clinical diagnostics, and high specific or "carrier-free" radioisotopes for use in tracer and other analytical applications are produced via accelerator (46,47), cyclotron (48–51), and neutron activation techniques (52). A partial listing of some of the more widely used radioisotopes in medicine include ^{28}Mg, ^{47}Sc, K 42, 43, 38, ^{201}Tl, Rb 81, 82m, 86, 84; ^{99m}Tc, ^{123}I, ^{34m}Cl, Br 76, 77; and ^{97}Ru. The recent introduction of computerized axial tomography (CAT) for radioimaging has produced demands for isotopes decaying via positron annihilation. ^{18}F, ^{13}N, ^{11}C, ^{15}O, and other isotopes are being used in these improved imaging techniques for medical diagnostics (51). A host of other radioisotopes are available commercially as indicated in Fig. 8.1.

In almost every case of the generation of the desired radioisotope via electron, photon, proton, neutron, or charged particles activation of the target nuclide, the produced isotope must be chemically separated from matrix isotopes. The efficiency of the isolation of the generated radioisotope from other radioisotopic impurities is indicated by the term, radiochemical purity, while high specific activity, the maximum radiation intensity per unit weight of the isotope, is also sought. Preventing the contamination of the radioisotope by any of its stable and naturally occurring isotopes is also desirable. The term "carrier-free" has been introduced to infer that the radioisotope contains minuscule amounts of any of its naturally occurring stable isotopes. Unfortunately, the extreme precautions required to prevent the introduction of ultratrace levels of impurities, discussed previously in this chapter, are not usually exercized during the production of radioisotopes. Thus the term carrier-free literally means that no nonradioactive element (carrier) of the radioisotope was deliberately added during the process. Due to adventitious contamination, accurate determination of the carrier content still remains an extremely challenging analytical problem.

The production of reasonably radiochemically pure isotopes has been based on selective ion-exchange separations following the activation of the target nuclide. For example, ^{99}Mo is recovered from nuclear radiated MoO_3 following its dissolution in 10 *M* sodium hydroxide, reduction with 0.2 *M* I_2 in alkali solution, acidification to 0.1 − 6 *M* with acid, and formation of the $[Mo(SCN)_6]^{3-}$ complex, which is then separated by an anion exchanger with nitrilodiacetate groups (53). The anion was then eluted from the column with mineral acid. Any straightforward adaption of conventional chemistry with an appropriate ion-exchange separation can be used to isolate and recover the produced radioisotope. The separation can also be based on extraction chromatography (54,55).

The more innovative uses of ion-exchange methods makes possible the immediate generation of freshly produced radioisotopes by separating the desired daughter nuclide from its parent. These radioisotope generators are now widely used. For example, a ^{62}Cu generator based on the sequence, $^{62}Zn \xrightarrow{-\beta^+} {}^{62}Cu$, has been prepared for medical and biological uses. The parent, ^{62}Zn, was adsorbed onto Dowex 1 × 8 anion exchange resin and the daughter, ^{62}Cu, was then eluted with 0.5 *M* HCl.[56] Several hundred μCi quantities of radiochemically pure, high specific activity ^{62}Cu can be obtained within a few minutes of elution. Ion-exchange resins have been tailor synthesized to enhance specificity for separating desired radioisotopes. In the case of producing a ^{68}Ga generator, an ion-exchange resin was produced from pyrogallol and formaldehyde to obtain highly efficiency adsorption for the ^{68}Ge parent (57). The ^{68}Ge/Ga^{78} radioisotope generator operated with 10 ml 4.5 *N* HCl as eluent and showed a high yield of ^{68}Ga over a period of 200 days. During elution 80% of the ^{68}Ga activity is obtained within 4 ml of 0.5 *N* HCl eluate and contamination from ^{68}Ge was less than 1 ppm. Such unique applications of ion-exchange methods in radioisotope production will continue. Applications using alumina and chelating resins for producing radioisotope generators have also been described (58).

REFERENCES

1. J.W. Mitchell, *J. Radioanal. Chem.,* **69,** 47–105 (1982).
2. M. Zief and J.W. Mitchell, *Contamination Control in Trace Element Analysis,* Vol. 47. Wiley, New York, 1976.
3. J.W. Mitchell and V. Gibbs, *Talanta,* **24,** 741–746 (1977).

4. E.C. Kuehner, R. Alvarez, P.J. Paulsen, and T.J. Murphy, *Anal. Chem.*, **44,** 2050 (1972).
5. J.W. Mitchell, T.D. Harris, L.D. Blitzer, *Anal. Chem.*, **52,** 774–776 (1980).
6. T.D. Harris and J.W. Mitchell, *Anal. Chem.*, **52,** 1706–1708 (1980).
7. J. Ruzicka and J. Stary, *Substoichiometry in Radiochemical Analysis,* Pergamon Press, New York, 1968.
8. J.W. Mitchell, C. Herring, and E. Bylina, *Appl. Spectros.*, in press.
9. J.W. Mitchell, W.R. Northover and J.E. Riley, *J. Radioanal. Chem.*, **18,** 133 (1973).
10. J.W. Mitchell and C. McCrory, *Separations and Purification Methods,* **9,** 165–208 (1980).
11. G.H. Morrison and N.M. Potter, *Anal. Chem.*, **44,** 839–42 (1972).
12. W.G. Faix, J.W. Mitchell, and V. Krivan, *J. Radioanal. Chem.*, **43,** 97–106 (1979).
13. G. Auboin, F. Dugain, J. Laverlochere, *Bull. Soc. Chim. Fr.*, **2,** 547 (1965).
14. J.I. Kim, H. Lagally, H.J. Boin, *Anal. Chim. Acta.*, **64,** 29 (1973).
15. K.A. Kraus, G.E. Moore, *J. Am. Chem. Soc.*, **75,** 1460 (1953).
16. F. Girardi, E. Sabbioni, *J. Radioanal. Chem.*, **1,** 169 (1968).
17. J.W. Mitchell, J.E. Riley, W.R. Northover, *J. Radioanal. Chem.*, **18,** 133–143 (1973).
18. D.W. Boydell, *Proc. Natl. Meet—Inst. Chem. Eng.*, 3rd 1980, 3E/1–3E/18., Stellenbosch, S. Africa.
19. D.R. Arnold, *Proc. Natl. Meet—Inst. Chem. Eng.*, 3rd 1980, 3D/1–3D/18, Stellenbosch, S. Africa.
20. F.L.D. Cloete, *At. Energy Board, rep. PER 27,* 1980, 25 pp.
21. J.F. Bossler, R.F. Janke, *S. Texas, U. Semin.*, 1979, 78–86, AIME, New York.
22. J. Bennett, E. Byleveld, A. Himsley, *CIM Bull.*, **73,** 107–114 (1980).
23. Tsoung-Yuan Yan, U.S. 4,241,026, 23 Rec. 1980.
24. I.L. Jenkins, *Hydrometallurgy,* **5,** 1–13 (1979).
25. J.D. Navratil, and J.L. Ryan, RFP-2903, CONF-790415-18, Report (1979), 21 pp. *Energy Res. Abstr.*, **4,** Abstr. No. 37168 (1979).
26. L.I. Guseva, G.S. Tikhomirova, and V.V. Stepushkina, *Radiokhimiya,* **23,** 821–824 (1981).
27. M. Ya Kondrat'Ko, A.V. Mosesov, O.A. Teodorovich, and V.M. Shevchenko, *Issled. po Khimii, Tekhnol. i Primeueniyu Radioaktiv. Veshchestv,* L. (1980) 86–96.
28. Fa Lin, Shu Heng Yan, and Shu Lan Zhang, *Heh Hua Hsueh Yu Fang She Hua Hsueh,* **3,** 155–161 (1981).
29. S.A. Shapiro, G.L. Sotry, Symp. Manage. *Low-Level Radioact. Waste,* **1,** 563–580 (1979).

30. G.P. Simon, C. Calmon, and H. Gold, eds., *Ion Exch. for Pollut. Control,* **2,** 41–60, CRC, Boca Raton, Florida.
31. G.P. Simon, C. Calmon, and H. Gold, eds., *Ion Exch. for Pollut. Control,* **1,** 55–70, CRC, Boca Raton, Florida (1979).
32. T. Matsuura and Y. Yamamoto, *J. Chromatogr.,* **201,** 121–9 (1980).
33. W.W. Schulz, *Nucl. Technol.,* **21,** 16–25 (1974).
34. M.C. Languras-Solar and F.E. Little, *Proc. Conf. Remote Syst. Technol.,* **27,** 301–6 (1980).
35. T.S. Murthy and K.L. Narasimha, *Proc. Nucl. Chem. Radiochem. Symp.,* 309–312, India Dept. Atomic Energy, Bombay, India, 1981.
36. T.S. Murphy, and R.N. Varma, Proc. Nucl. Chem. Radiochem. Symp., India Dep. Atomic Energy, Bombay, India, 1981, 281–285.
37. A.K. Jain, Sushma Agrawal, and R.P. Singh, *Int. J. Appl. Radiat. Isot.,* **31,** 633–637 (1980).
38. H.T. Matsuda and A. Abrao, *IPEN,* **13** (1980), 12 pp. Inst. Pesqui, Energy Nucl., São Paulo, Brazil.
39. I. Sipos-Galiba and K.H. Lieser, *J. Radiochem. Radioanal. Lett.,* **42,** 329–339 (1980).
40. F. Kepak, *J. Radioanal. Chem.,* **51,** 307–314 (1979).
41. M. Szlaurova, D. Vanco, P. Galan, and M. Fojtik, *J. Radioanal. Chem.,* **51,** 281–284 (1979).
42. H.W. Humphre and E.L. Adkins, Report 1979, MLCo-1160, 42 pp. Avail. NTIS. From *Energy Res. Abstr. 1979,* **4,** Abstract NO. 55314.
43. C.J. Simmons, J.H. Simmons, P.B. Macedo, T.A. Litovitz, *Ger. Offen.,* 2,945,321, June 19, 1980, 64 pp.
44. R.G. Dosch, ACS Symp. Ser., 1979, 129–148. Sandia Labs, Albuquerque, New Mexico, 87185.
45. C. Thegerstroem, *Report 1980,* SKBF/KBS-TR-80-06, 71 pp. Available from *INIS Atomindex 1980,* **11,** Abstract NO. 555627.
46. G.A. Brinkman, *Report 1978,* Interiko-7815, 41 pp. Available from *INIS Atomindex 1979,* **10,** Abstract No. 461447.
47. A. Donnerhack and E.L. Sattler, *Int. J. Appl. Rad. Iso.,* **31,** 279–285 (1980).
48. R. Weinreich, S.M. Quaim, G. Stoecklin, *Nuklearmedizin, Suppl.* (Stuttgart), **16,** 226–231 (1978).
49. A.E. Ogard, R.E. Whipple, P.M. Wanek, P.M. Grant, J.W. Barnes, G. Bently and H.A. O'Brien, *Nuclearmedizin, Suppl.* (Stuttgart), **16,** 232–234 (1978).
50. P. Virtanen, V. Nanto, M. Viljanen, R. Kontti, B.S. Ek, Turun Yliopiston Julk, *Sard,* **13** (1981).
51. R.M. Lambrecht, B.M. Gallogher, A.P. Wolf and G.W. Bennett, *Int. J. Appl. Rad. Iso.,* **31,** 343–349 (1980).

52. F. Helus, O. Krauss, and W. Maier-Borst, *Radiochem. Radioanal. Lett.*, **15,** 225–230 (1973).
53. J. Knapp, S. Ali, and A-H Sameh, Ger. Offen. 2,758,783, July 1979. *CA*, **91:** 98892u.
54. V. Levin, USSR 668,876, June 25, 1979. *CA,* **91:** 98891t.
55. V. Levin, USSR 668,877, June 25, 1979. *CA,* **91:** 988905s.
56. M. Yagi, and K. Kondo, *Int. J. Appl. Radiat. Isot.*, **30,** 569–570 (1979).
57. J. Schuhmacher, and W. Maier-borst, *Int. J. Appl. Rad. and Iso.*, **32,** 31–36 (1981).
58. Y. Yano, T.F. Budinger, G. Chlang, H.A. O'Brien, A. Harold, and P.M. Grant. *J. Nucl. Med.*, **20,** 961–966 (1979).

CHAPTER

9

ION CHROMATOGRAPHY

ROY A. WETZEL, CHRISTOPHER A. POHL,* JOHN M. RIVIELLO

Dionex Corporation
Sunnydale, California

and JOHN C. MacDONALD

Fairfield University
Fairfield, Connecticut

9.1 INTRODUCTION

The first publication in the scientific literature on ion chromatography (IC) occurred in September 1975. That paper (1), "Novel Ion Exchange Chromatographic Method Using Conductimetric Detection," by H. Small, T.S. Stevens, and W.C. Bauman of the Dow Chemical Company, demonstrated that use of an appropriate cation-exchange resin for separation of cations followed by an appropriate anion-exchange resin to suppress eluent conductivity allows the use of a conductivity cell as a universal and very sensitive detector for ions. For anion analysis, an anion-exchange resin was used to separate anions and a cation-exchange column was used for suppression of eluent conductivity. This use of a second ion-exchange column for removal of eluent background had been suggested in 1970 by Bauman (2) who recognized this possibility of a general conductivity detector. Small and Stevens built the first ion chromatograph and their first published ion chromatogram (1) was the separation of the alkali metals. More recently, the definition of IC has been expanded to the chromatographic separation of ions, followed by automatic detection using any of the available chromatographic detectors. The second suppressor column has been replaced by the more efficient and convenient fiber suppressor.

Prior to this invention, detection of ion-exchange eluates was limited to the usual but nongeneral (specific) photometric, fluorometric and elec-

* Author to whom correspondence should be addressed.

trochemical detectors, or as in the previous chapter, radioactivity. Further, some detection methods for ions required additional chemical reactions. Patents pending at the time of the publication were granted and this new technology was licensed exclusively to Dionex Corporation where IC was immediately applied to the analysis of blood serum, tissue extracts, urine, and cerebrospinal fluid (3). Commercial application followed immediately when the Hooker Chemical Company used IC (4) to monitor Cl^- and Br^- in one of their proprietary products. Ion chromatography steadily developed and an increasing number of analytical problems were solvable by this technique.

Although IC was originally developed to solve process monitoring and process control applications, the unique features of the technique resulted in applications and technology being developed mainly in the analytical laboratory. IC capabilities enable faster sample throughput, improved sensitivity, improved accuracy, and the ability to determine ionic species previously difficult or impossible to determine when using other available methodologies. Cation analysis was a routine instrument procedure before IC was developed. Now cation and anion analysis, by IC, are routine procedures.

The number of ions determined by IC has grown to several hundred in 1984. Column and detector development, coupled with a need to improve ion analysis methodology, is responsible for this growth. In this chapter, we present the theory and practice of IC and include recent advances in columns and detectors.

We begin by discussing the originally reported method of ion-exchange separation followed by eluent suppressed conductivity detection. This has been the "standard IC method" for most laboratories during the early years of IC use. Next, we go on to describe other technologies which have expanded the capabilities of IC beyond those originally reported. For example, mobile phase ion chromatography (MPIC) is a new method for determining many ionic species difficult or impossible to analyze by standard IC. These ions include aromatic acids and amines, several hydrophobic ions, and metal complexes such as gold, iron, and cobalt cyanides. The introduction of amperometric detection in conjunction with conductimetric detection now enables sulfides and cyanides to be determined amperometrically followed by a conductivity detector for halides. Chromatography of metal ions now allows several elements to be determined at μg/L levels in a single 15-minute analysis, and this includes the great advantage of speciation of components such as Fe(II) and Fe(III). A current listing of IC capabilities is presented later in this chapter. It is because of these expanded capabilities that an ion chromatograph is now a necessary component of any analytical laboratory that endeavors to provide the most modern and effective analytical services.

9.2 ION CHROMATOGRAPHY (IC) TECHNOLOGY

9.2.1 Ion Chromatography with Eluent Suppressed Conductivity Detection

The Suppression Reaction

The fundamental aspect of this standard IC method is the suppression reaction, which is responsible for the high sensitivity as well as the selectivity of IC. The suppression reaction is also a major factor in the proper choice of separator column, suppressor, and eluent. Thus, it will be useful to examine the suppression reaction in some detail before going on to each of the components of an IC system.

Consider an anion IC system. In the simplest case, an eluent consisting of sodium hydroxide is used to displace anions from the anion exchange sites in the separator column. In such a case, as the sample anion is eluted from the column, it should be possible to detect a change in conductivity of the column effluent. However, because the concentration of the hydroxide in the eluent must be substantially greater than that of the sample anion in order to remain in the linear operating range of the separator column, any change in conductivity of the column effluent will be small compared to the conductivity of the eluent. Thus, direct conductivity detection of sample anions would be expected to suffer from very poor signal-to-noise characteristics. If, on the other hand, the column effluent is passed through a suppressor consisting of high-capacity cation-exchange resin in the hydrogen form, the sodium hydroxide will be removed from the column effluent:

$$\text{Resin}^-\text{H}^+ + \text{Na}^+\text{OH}^- \rightarrow \text{Resin}^-\text{Na}^+ + \text{H}_2\text{O} \qquad (1)$$

At the same time, sample anions passing through the suppressor column will be converted into their acid form:

$$\text{Resin}^-\text{H}^+ + \text{Na}^+\text{A}^- \rightarrow \text{Resin}^-\text{Na}^+ + \text{H}^+\text{A}^- \qquad (2)$$

The effluent of the combined separator-suppressor system then will exhibit very low conductivity and the sample anions eluting from the system will exhibit increased conductivity. The suppressed system thus exhibits improved signal-to-noise characteristics, and therefore excellent sensitivity.

Suppression reactions represent a novel class of post-column reactions. In general, post-column reactors involve chemistries which enhance detector sensitivity for the analyte without affecting the eluent. Here the suppression reaction simultaneously serves two functions. First, it maximizes the detector sensitivity to the analyte anion by converting it into the acid form because mobility of the hydrogen ion is 7 times that of the sodium. Second, the suppression reaction minimizes the detector sensitivity to the eluent. It is the combination of these two functions, which are absent in single column (unsuppressed) ion chromatography methods, which results in the superior sensitivity and signal-to-noise characteristics of standard ion chromatography.

While the neutralization reaction is one possible suppression reaction for use in anion IC, it is a special case of the general anion IC suppression reaction:

$$n\ \mathrm{Resin^-H^+} + (\mathrm{C^+})_n\mathrm{A}^{n-} \rightarrow n\ \mathrm{Resin^-C^+} + (\mathrm{H})_n\mathrm{A} \qquad (3)$$

$(\mathrm{H})_n\mathrm{A}$ is any of a number of weak acids, the salt forms of which also make useful anion-exchange eluents. Since it is important to minimize the conductivity of the eluent, the preferred anion is one which has a $\mathrm{p}K_a$ of 6 or greater.

In the cation IC mode, analogous post-column suppression chemistry is used. In this case, the eluent used is a mineral acid. The eluent, after passing through the cation-exchange separator column, is passed through a suppressor. In this case, a suppressor can consist of a high-capacity anion-exchange column in the hydroxide form. As is the case in anion IC, the suppressor removes eluent ions from the separator column effluent:

$$\mathrm{Resin^+OH^-} + \mathrm{H^+Cl^-} \rightarrow \mathrm{Resin^+Cl^-} + \mathrm{H_2O} \qquad (4)$$

At the same time, sample cations passing through the suppressor column are converted into their base form:

$$\mathrm{Resin^+OH^-} + \mathrm{C^+Cl^-} \rightarrow \mathrm{Resin^+Cl^-} + \mathrm{C^+OH^-} \qquad (5)$$

Again, the general effect of the suppression reaction is to minimize detector sensitivity towards the eluent while simultaneously increasing the detector sensitivity to the analyte cations, since the mobility of hydroxide anion is 2.6 times that of chloride ion.

Just as in the case of the anion IC suppression reaction, the cation IC suppression reaction noted above is a special case of the general cation IC suppression reaction:

$$n\ \text{Resin}^+\text{OH}^- + \text{C}^{n+}(\text{Cl}^-)_n \rightarrow n\ \text{Resin}^+\text{Cl}^- + \text{C(OH)}_n \qquad (6)$$

$C(OH)_n$ is any of a number of weak bases, the salt forms of which make useful cation-exchange eluents. It should be noted that most suppressors in use are not packed columns but ion-exchange fibers as discussed later.

The Separator Column

The suppression reaction provides substantial benefits with respect to selectivity and sensitivity. However, its use necessitates the use of a properly designed separator column which complements the requirements of the suppressor device. First, it is necessary to minimize the ion-exchange capacity of the separator column. This feature is desirable because there is an upper limit to the ion flux (approximately 50 meq/min) which can be processed within a membrane suppressor device (see below) with minimal dispersion of solute bands. Consequently, low ion-exchange capacity separator columns that allow the use of dilute eluents and low ion flux rates are used in conjunction with the suppression reaction. However, because it is necessary for the separator column to handle significant sample loads as well as sample contaminants, it is necessary to restrict the reduction of the ion-exchange capacity. In practical terms, the ion-exchange capacity of the separator column should be between 10 and 200 microequivalents per gram.

A second consideration in the choice of separator column is the pH of the eluent in suppressed IC. Generally, the pH of cation eluent will be between 2 and 5 and the pH of the anion eluent will be between 8 and 12. The proper separator column material must then be able to handle such extremes of pH. Because of this, the materials of choice in separator columns are low-capacity organic ion-exchange resins, rather than silica-based materials.

For anion IC, a binary pellicular resin is used in the separator column as in Fig. 9.1. As in the case of the cation separator resin, the core of the resin particle is polystyrene-divinylbenzene. Surrounding this core is a layer of sulfonated polystyrene-divinylbenzene. The purpose of the sulfonated layer in this resin is to provide a surface to which the outer anion exchange layer is bound via ionic bonding interactions. The outer layer consists of uniformly sized anion-exchange latex particles which are deposited on the sulfonated layer in a uniform monolayer. Due to the large

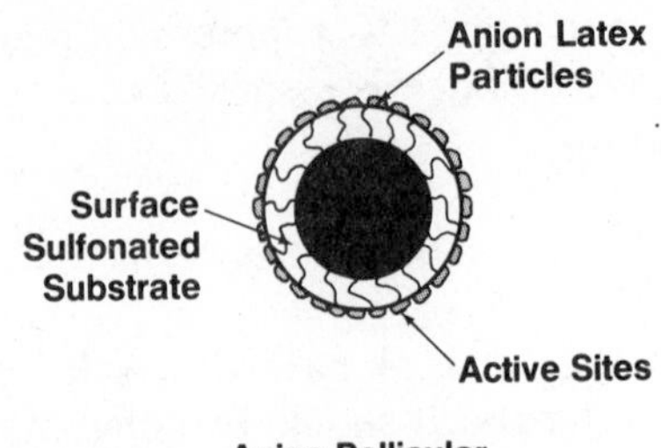

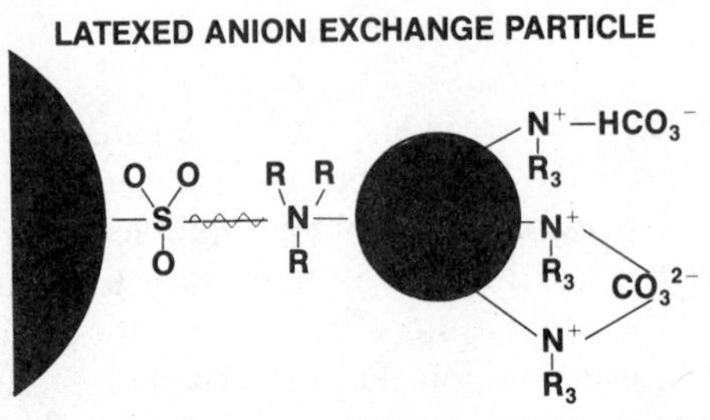

Figure 9.1. Composition of the IC anion-exchange particle. (Courtesy of Dionex Corporation)

number of ionic interaction sites between each latex particle and the resin surface, the attachment of the latex to the surface of the sulfonated layer is essentially an irreversible process. For example, treatment of the anion resin with 1 *M* NaOH removes none of the latex particles from the surface of the resin.

For cation IC, the separator resin is surface sulfonated ion-exchange resin as in Fig. 9.2 or a binary pellicular resin with cation exchangers attached to the surface of the resin spheres, analogous to the anion exchange particles in Figure 9.1. The inert polystyrene divinylbenzene core

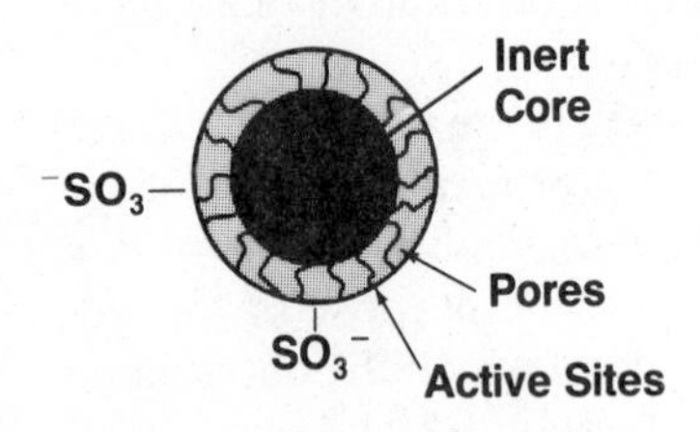

Figure 9.2. Composition of the IC cation-exchange particle. (Courtesy of Dionex Corporation)

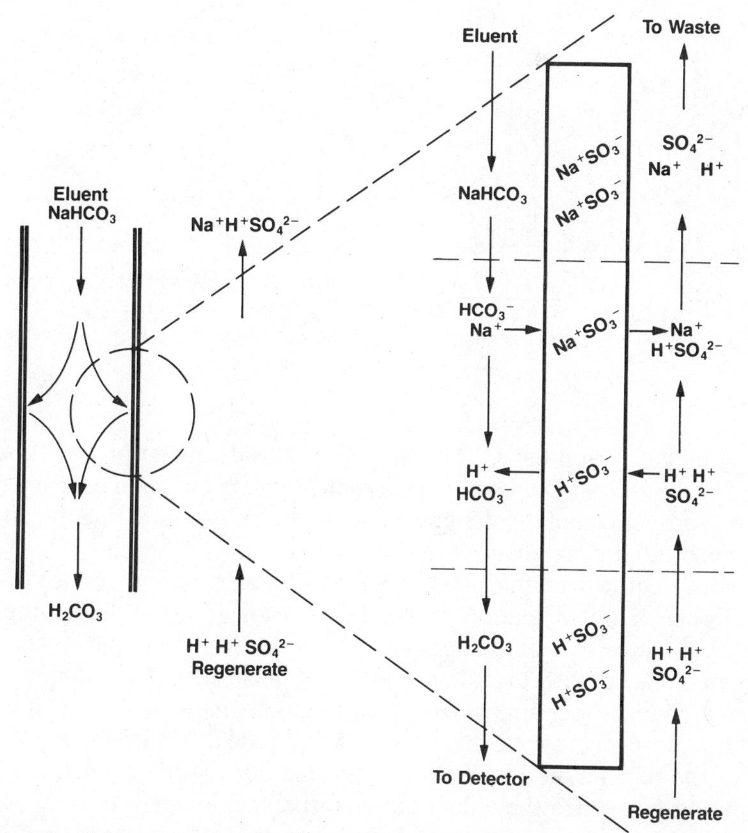

Figure 9.3. Schematic of the fiber suppressor mechanism. (Courtesy of Dionex Corporation)

of the resin particle provides a rigid support for the sulfonated outer layer. Unlike high capacity ion-exchange resins, this type of material does not change size appreciably when converted from one cation form to another.

The Fiber Suppressor

A relatively recent development in ion chromatography is the use of fiber suppressors as shown in Fig. 9.3. In these devices, the same basic function originally served by the packed bed suppressor (a detailed discussion of the packed bed suppressor follows) is accomplished with an ion-exchange fiber. However, since the exterior of the fiber is continuously bathed in a regenerating solution, there is no need to interrupt analyses for regen-

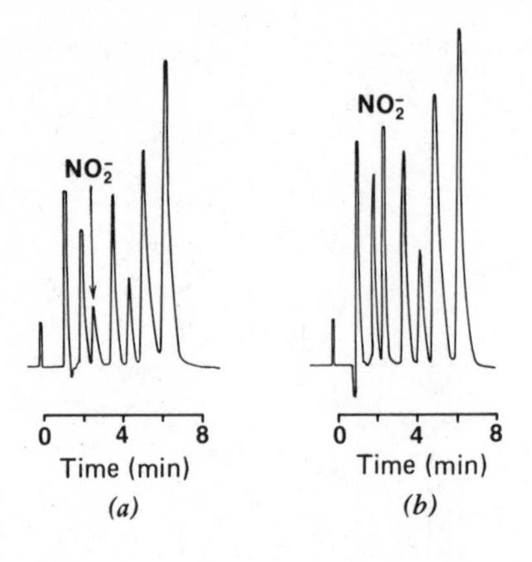

Figure 9.4. Comparison of (*a*) column suppression and (*b*) fiber suppression. The nitrite peak response increases 290 percent. (Courtesy of Dionex Corporation)

eration as with suppressor columns. Another advantage of this type of device is the lack of Donnan exclusion phenomena common to resin-type suppressors. The analytical improvement is seen by comparing the nitrite peak height and peak shape in Fig. 9.4.

Several features of the fiber suppressor should be understood when considering their application to the suppression reaction. First, the rejection characteristics of the fiber limit the concentration of the regenerant that can be used with the fiber suppressor. Consider, for example, the cation-exchange fiber suppressor used in anion analysis. The most commonly used regenerant in this case is sulfuric acid. While the hydrogen ions of the regenerant are freely transported across the cation-exchange membrane to replace the eluent cations, the sulfate anion present in the regenerant is for the most part excluded from the interior of the fiber (due to the high concentration of fixed sulfonic acid sites within the fiber). However, complete Donnan exclusion can only be expected at concentrations less than 0.01 *M*. Thus, concentrations well in excess of this will be expected to allow some regenerant anions into the fiber membrane. This counterion leakage results in slightly higher eluent background conductivity levels. As a result, the upper concentration limit of the regenerant is generally limited to the 0.10–0.01 *M* concentration range.

A second consideration for a fiber suppressor is band broadening within the suppressor. While a hollow fiber can be used in a fiber suppressor, it has been found that laminar flow within the fiber results in poor mass transport of ions from the eluent to the walls of the fiber. The result is that excessively long fibers must be used in the fiber suppressor if the fiber is hollow. The result of the long hollow fiber is that excessive band broadening is observed. The solution to this problem is to pack the fiber with beads of neutral polystyrene-divinylbenzene. This serves to improve mass transport to the walls of the fiber and thus improve performance. A packed fiber suppressor can actually provide eluent suppression while producing less band broadening than a resin packed bed suppressor.

The Suppressor Column

While the fiber suppressor described above is clearly the suppressor of choice in most suppressor applications, there are some applications where a packed bed suppressor column is preferred. The suppressor column also has a number of features which must be considered.

The primary consideration in the suppressor column is the length of time that the suppressor will function before the suppressor must be regenerated. Maximum suppressor life is achieved with high-capacity ion-exchange resins of the strong acid type for anion IC and of the strong base type for cation IC.

An additional consideration relates to the chromatographic behavior of weakly acidic anions in anion IC or weakly basic cations in cation IC. Consider, for example, the behavior of anions entering a hydrogen form suppressor column. An anion such as chloride will be converted to hydrochloric acid when the chloride reaches that part of the suppressor which is still in the hydrogen form. Because hydrochloric acid is fully dissociated under normal IC conditions, the chloride ion will pass through the remainder of the suppressor column while being confined to the interstitial volume of the suppressor column.

The chloride ion is prevented from entering the micropores of the suppressor resin because of Donnan exclusion. This is the inability of ions (at concentrations $<0.01\ M$) to enter ion-exchange materials containing a high density of ion-exchange sites of the same charge type as the ions of interest. In the current example, migration of chloride ions into the cation-exchange resin in the suppressor is prevented by Donnan exclusion due to the high density of fixed anionic sites where the concentration of sulfonic acid sites in the resin is between 2 and 3 M. The chloride ion is completely dissociated, whether it is in the sodium form or in the hydrogen form, and it will be excluded from the micropore volume of the suppressor resin regardless of where along the suppressor the chloride is converted from the sodium form to the hydrogen form.

Now, consider the effects upon a *weakly* acidic anion such as nitrite during the passage through the suppressor column. In this case, while the nitrite remains in the sodium form, it will be Donnan excluded from the suppressor resin as was chloride ion. However, as the nitrite reaches the sodium-hydrogen interface band in the suppressor, the nitrite will be converted from the sodium form to the hydrogen form. The nitrous acid thus formed is in equilibrium:

$$H^+ + NO_2^- \rightleftharpoons HNO_2 \tag{7}$$

Because the nitrous acid is a weak acid, a substantial portion of nitrous acid exists in undissociated form even in dilute solution. Neutral species such as HNO_2 are not subject to Donnan exclusion and un-ionized nitrous acid will penetrate both the interstitial volume and the micropore volume of the suppressor after that point along the suppressor where NO_2^- is converted to HNO_2. Further, because the sodium-hydrogen interface band in the suppressor is constantly moving down the suppressor column, the extent to which nitrous acid penetrates the micropores of the suppressor resin will change as a function of the relative exhaustion of the suppressor.

While retention of weak electrolytes in the suppressor column represents only a minor inconvenience, it is possible to minimize the phenomenon with proper choice of suppressor resin. In high-capacity ion-exchange resins, the micropore volume is inversely proportional to the degree of crosslinking. Thus, by maximizing the crosslinking of the suppressor resin, it is possible to minimize the effect of micropore retention of weak electrolytes in the suppressor column. Unfortunately, adsorption of weak electrolytes in the resin micropore also is directly proportional to the crosslinking of the suppressor resin. Thus, the proper choice of crosslinking for suppressor resins is a compromise between maximal crosslinking (to minimize pore volume and maximize the physical integrity of the resin) and minimal crosslinking (to minimize adsorption of weak electrolytes in the micropores). In practice, this means that suppressor resins have crosslinkings in the range of 8–12%.

IC Eluents for Suppressed Conductivity Detection

As was mentioned earlier, regarding the suppression reaction, the general requirement for an anion IC eluent is that the eluting anion must have useful affinity for anion-exchange resins and have a pK_a of >6. Table 9.1 lists the selectivities relative to chloride for common anions. Only hydroxide, borate, carbonate, and phenate meet the acidity requirement stated above. Of the four anions, phenate is generally not useful because of the ease with which it is oxidized in alkaline solution. Of the remaining three, only carbonate has a high enough selectivity coefficient to be an effective eluent at the dilute concentrations necessary for long suppressor

Table 9.1 Ion Exchange Selectivity Coefficients of Anions Relative to Chloride

	Dowex 1	Dowex 2
Hydroxide	0.09	0.65
Fluoride	0.09	0.13
Aminoacetate	0.10	0.10
Acetate	0.17	0.18
Formate	0.22	0.22
Dihydrogen phosphate	0.25	0.34
Bicarbonate	0.32	0.53
Chloride	1.00	1.00
Bisulfite	1.3	1.3
Cyanide	1.6	1.3
Bromide	2.8	2.3
Nitrate	3.8	3.3
Benzene sulfonate	—	4.0
Bisulfate	4.1	6.1
Phenoxide	5.2	8.7
Iodide	8.7	7.3
p-Toluene sulfonate	—	13.7
Thiocyanate	—	18.5
Perchlorate	—	32
Salicylate	32.2	28

Adapted from Peterson (1954).

life. Figure 9.5 shows an IC chromatogram of the common anions with a carbonate eluent system.

For cation IC there are basically two eluent systems used. In the case of monovalent cations, the normal eluent is 0.005 M HCl. As can be seen from Table 9.2 of cation-exchange selectivities, the selectivity coefficient of hydrogen ion is close enough to that of the other monovalent cations to be a useful eluting cation. Figure 9.6 shows a separation of monovalent cations by IC.

However, the concentrations of hydrogen ion required for divalent cations are such that hydrogen ion is an impractical eluent for divalent cations. Instead, the preferred cation for the alkaline earth cations is *m*-phenylenediamine (mixed with an equal molar amount of HCl). The divalent nature of *m*-phenylenediamine makes it an efficient eluent for other divalent cations, while its weakly basic character results in very little conductivity when it is converted to the free base form in the suppressor. Figure 9.7 shows a separation of divalent cations using a *m*-phenylenediamine eluent.

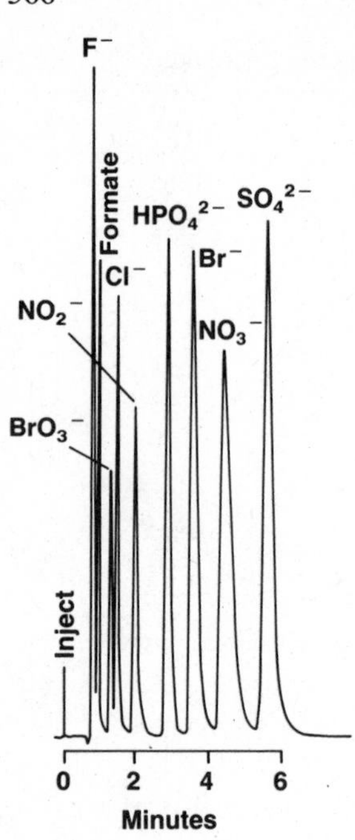

Figure 9.5. Rapid separation of nine anions. Concentrations (in ppm) are fluoride (3), formate (8), bromate (10), chloride (4), nitrite (10), monohydrogen phosphate (30), bromide (30), nitrate (30), sulfate (25). (Courtesy of Dionex Corporation)

Recently, the zinc cation has been proposed (5) as an alternative eluent for divalent cations. In this case, eluent suppression of zinc is accomplished via the precipitation of zinc in the form of zinc hydroxide. The author claims that no pressure increase in the suppressor column is observed due to this precipitate and that the suppressor is easily regenerated.

9.2.2 Ion Chromatography Exclusion (ICE)

The ICE Mechanism

The effects of incomplete Donnan exclusion from the packed bed suppressor are described in detail above. While this phenomenon may be the source of problems in standard IC, this same phenomenon is put to use as a separation method in ICE, using deionized water as the eluent.

In anion ICE, the separator column is a comparatively large column (9 × 250 mm) containing high-capacity cation-exchange resin in the hydrogen form. Such large volume columns are typical of separation methods based on exclusion phenomena. In ICE, the retention of anions must fall between the total exclusion volume (corresponding to the interstitial volume of the packed column) and the total permeation volume (corre-

Table 9.2 Ion Exchange Selectivity Coefficients of Cations Relative to H^+ On Differently Cross-Linked Dowex 50

	4% DVB	8% DVB	16% DVB
Univalent ions			
Li	0.76	0.79	0.68
H	1.00	1.00	1.00
Na	1.20	1.56	1.61
NH_4	1.44	2.01	2.27
K	1.72	2.28	3.06
Cs	2.02	2.56	3.17
Ag	3.58	6.70	15.6
Bivalent ions			
UO_2	0.79	0.85	1.05
Mg	0.99	1.15	1.10
Zn	1.05	1.21	1.18
Co	1.08	1.31	1.19
Cu	1.10	1.35	1.40
Cd	1.13	1.36	1.55
Ni	1.16	1.37	1.27
Mn	1.15	1.43	1.54
Ca	1.39	1.80	2.28
Sr	1.57	2.27	3.16
Pb	2.20	3.46	5.65
Ba	2.50	4.02	6.52
Trivalent ions			
Cr	1.6	2.0	2.5
Ce	1.9	2.8	4.1
La	1.9	2.8	4.1

Adapted from Bonner et al., (1957) and (1958).

sponding to the sum of the interstitial volume and the micropore volume of the packed column). Only if other retention processes (such as adsorption) are a factor can retention of anions exceed the total permeation volume.

Retention of anions in anion ICE is controlled by the extent to which the anion is ionized in solution:

$$HA \rightleftharpoons H^+ + A^-$$

$$K_a = \frac{[H^+][A^-]}{[HA]} \tag{8}$$

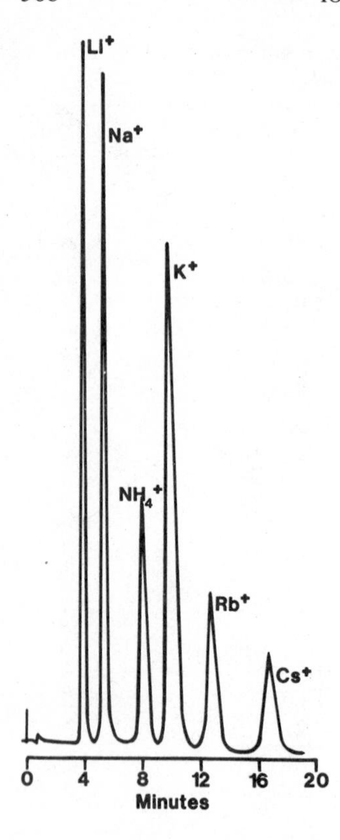

Figure 9.6. Separation of monovalent cations with cation fiber suppression. (Courtesy of Dionex Corporation)

As indicated by the equations, the concentration of the neutral species is determined primarily by the dissociation constant of the acid. On the basis of this equation, one would expect that anions would elute in order of increasing pK_a. Anions with very low pK_a values, such as chloride or sulfate, will be totally excluded and elute in the total exclusion volume of the column. Anions with high pK_a values, such as carbonate, will elute near the total permeation volume.

Additional separation power can be added to the ICE separation by using dilute acid eluents instead of deionized water. In this case, the addition of hydrogen ion to the eluent shifts the dissociation equilibrium to the left. The result is that some anions which may have been totally excluded when water was chosen as the ICE eluent will now experience considerable retention.

The ICE Suppressor

In ICE, the suppression method must, of necessity, be different from that of standard IC. When water is used as the eluent, no suppression is nec-

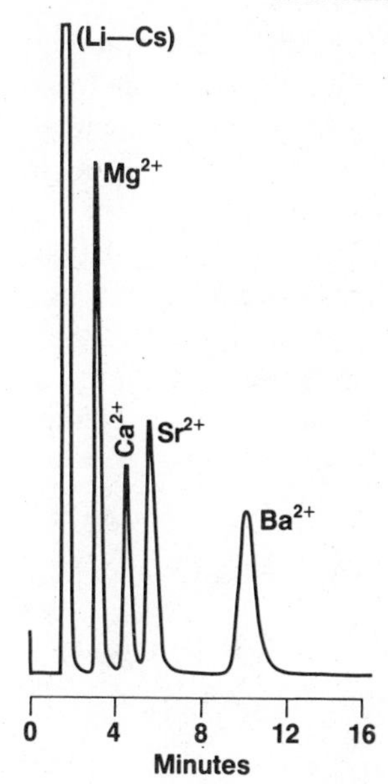

Figure 9.7. Separation of alkaline earths. Concentrations (in ppm) are Mg^{2+}(3), Ca^{2+}(3), Sr^{2+}(10), Ba^{2+}(25). (Courtesy of Dionex Corporation)

essary. When, however, an acid is added to the eluent in order to enhance the retention of moderate pK_a electrolytes, a suppressor is needed. Unlike anion IC, where the suppression reaction exchanges the counterion of the eluent ion, the object of the suppression reaction in anion ICE is to remove the anion, which is the hydronium ion source in the separation step when a suppression column is used, or to remove the hydronium ion itself when a fiber suppressor is used.

The original method for anion ICE eluent suppression was the precipitation reaction between hydrochloric acid and resin in the silver form:

$$\text{Resin}^-\text{Ag}^+ + \text{H}^+\text{Cl}^- \rightarrow \text{Resin}^-\text{H}^+ + \text{AgCl (s)} \tag{9}$$

While the reaction described in Equation 9 removes the hydrochloric acid from the eluent, the sample anions are converted to the silver form via the following reaction:

$$n\ \text{Resin}^-\text{Ag}^+ + (\text{H}^+)_n\text{A}^{n-} \rightarrow n\ \text{Resin}^-\text{H}^+ + (\text{Ag}^+)_n\text{A}^{n-} \tag{10}$$

As was mentioned earlier, detection sensitivity is enhanced by maintaining anions in their hydrogen form. Thus, an H-form postsuppressor can be added downstream of the silver suppressor in order to convert the sample anions back to their acid forms:

$$n\ \text{Resin}^-\text{H}^+ + (\text{Ag}^+)_n\text{A}^{n-} \rightarrow n\ \text{Resin}^-\text{Ag}^+ + (\text{H}^+)_n\text{A}^{n-} \tag{11}$$

Since only the sample anions reach the postsuppressor, it, in practice, lasts for months before regeneration is necesssary. Because this suppression reaction is a precipitation reaction, it is generally not practical to regenerate the ICE suppressor column. Thus the suppressors used in anion ICE are disposed of once they are exhausted.

More recently, a fiber suppressor compatible with anion ICE analysis has been developed. The fiber is a sulfonic acid form ion exchange membrane saturated with tetrabutylammonium ions. An alkyl sulfonic acid eluent is then suppressed by the following reaction as tetrabutylammonium ions exchange into the eluent stream and hydronium ions exchange into the regenerant stream:

$$\underset{\text{Eluent}}{RSO_3^- H^+} + \underset{\text{Regenerant}}{TBA^+OH^-} \xrightarrow[\text{Fiber}]{} \underset{\text{Eluent, Suppressed}}{RSO_3^- - TBA^+} + \underset{\text{Regenerant, waste}}{H_2O} \tag{12}$$

9.2.3 Mobile Phase Ion Chromatography (MPIC)

The Separation Method

A general class of ions to which standard IC is not well suited is hydrophobic ions. Such ions fall into two basic categories. The first type of hydrophobic ion is that which has low hydration energy. This type of ion includes inorganic anions such as iodine, fluoroborate, or perchlorate. While such ions are generally not referred to as hydrophobic, it is this minimal hydration sphere that results in their tendency to be retained via both coulombic and Van der Waals adsorption interactions within the ion exchange resin used for their separation.

A second type of hydrophobic ion that is not well suited to low-capacity IC separator resins is the large organic ion. Regardless of whether the ionic portion of such molecules are characterized by high hydration energy, the hydrophobic portion of the molecule tends to dominate the kinetics of the molecule's interaction with the separator resin environment. The result is that such molecules are difficult if not impossible to elute from a standard IC separator resin.

MPIC represents the method of choice for hydrophobic ions with pK_a values less than 7. The resin used in the MPIC column is a highly crosslinked, high-surface-area macroporous resin. Retention is accomplished

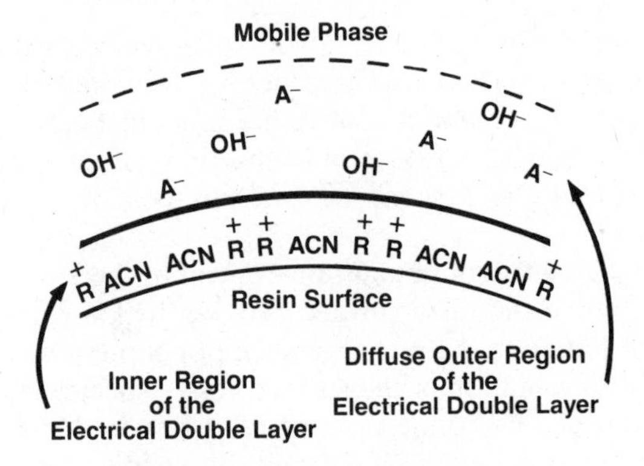

Figure 9.8. Schematic representation of the electrical double layer in mobile phase ion chromatography. (Courtesy of Dionex Corporation)

by the addition of a variety of eluent additives that affect the resin surface in such a way as to determine how and where anions or cations will be retained.

In reversed phase systems such as MPIC, retention of ions is accomplished usually via one of two intimately related retention processes. Under certain circumstances, both processes occur simultaneously.

The first retention process involves the direct adsorption of surface active ionic species at the resin-mobile phase interface. Included in the category of surface active ionic species are a number of species that the surface chemist would not generally consider to be surfactants but which do indeed lower the surface tension in the reverse phase system. In such a process, the hydrophobic portion of the surface active molecule will be adsorbed on the resin surface with the ionic portion of the molecule oriented toward the polar mobile phase (6). The adsorbed ionic surface active species, by virtue of its ordered orientation, sets up an electrical double layer with oppositely charged counterions forming a diffuse outer layer surrounding the primary layer of adsorbed surface active ions as shown in Fig. 9.8.

The two major factors which control the concentration of adsorbed ionic surface active species, and thus the retention of such species, are (a) the presence of nonionic molecules such as methanol or acetonitrile, which compete for adsorption sites on the resin surface and (b) the type of counterion present in the outer portion of the electrical double layer. Although the exact cause of the latter factor is still uncertain, it has been proposed that the phenomenon is due to the effect that various counterions have on the density of the diffuse outer portions of the electrical double layer, and thus upon the concentration of surface active ions at the resin surface (6).

While the exact cause of counterion control of the surface active species concentration at the resin surface may not be known, certain trends are known which dictate the proper choice of counterion in this mode of MPIC. In general, retention of the surface active species is increased and peak shape improved by counterions of relatively low hydration energy. Thus, perchlorate is a useful eluent additive in the analysis of surface active cations.

The second retention process in MPIC is nearly always the one used for inorganic ions. These ions generally have no surface active properties. Such ions are retained by the inverse of the retention process described above. In this second process, ionic surface active materials are added to the eluent. These ionic surface active agents are adsorbed at the resin surface as in the previously described retention process. In this case, however, the analyte ions (opposite in charge to the surface active ions) are retained via interaction with the adsorbed surface active layer. The retention of analyte ions takes place only in the diffuse outer portion of the electrical double layer due to the absence of any surface active character in the analyte ion. Selectivity in this mode of MPIC is the result of the relative ability of ions in the outer portion of the electrical double layer to affect the surface tension (and thus the surface concentration) of adsorbed surface active ions.

Experimentally, selectivity in MPIC is dominated by two factors:

1. The hydrophobic nature of the resin surface.
2. The high surface tension of the liquid–solid interface.

Thus, while ion-exchange provides excellent selectivity for ions with large hydration energies, the above factors dictate the MPIC will exhibit poor selectivity for ions with high hydration energies, except in cases where such ions also contain significant hydrophobic components. For example, as shown in Fig. 9.9, HPIC exhibits superior selectivity for fluoride, chloride, and nitrate when compared to MPIC because of the high hydration energies associated with such anions. However, MPIC exhibits superior peak shape and selectivity for ions with lower hydration energy such as iodide and thiocyanate as shown in Fig. 9.10.

The proper choice of eluents in MPIC is dictated by three factors:

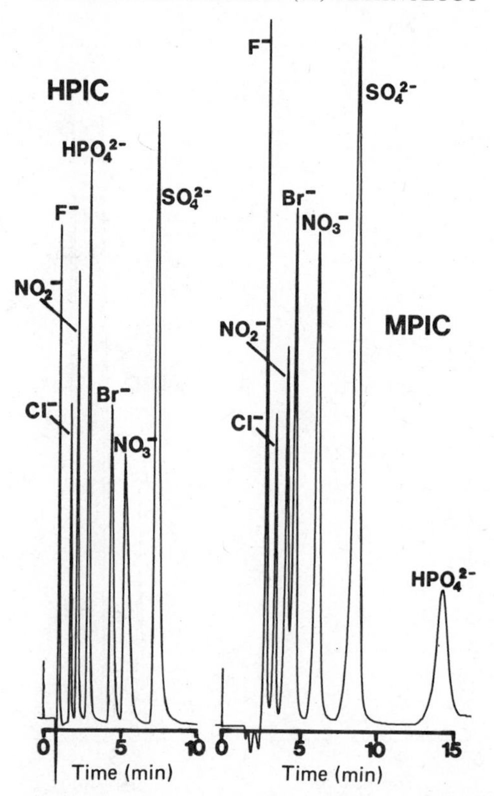

Figure 9.9. Comparison of high-performance ion chromatography (HPIC) and mobile phase ion chromatography (MPIC). Note the positions of HPO_4^{2-}. (Courtesy of Dionex Corporation)

1. The ion separation mode (i.e., whether or not the ion is surface active).
2. The supressor compatibility of various potential eluent additives.
3. The ease with which various ions can be removed from the suppressor column once the suppressor has been exhausted.

Table 9.3 lists a number of eluent choices which have been found to be useful in MPIC. Note that while only a relatively small number of eluent choices have been shown in Table 9.3, a larger number of eluents are possible. In addition to the known eluent systems shown, a large number of other possible MPIC reagents, counterion additives, and solvents are under investigation.

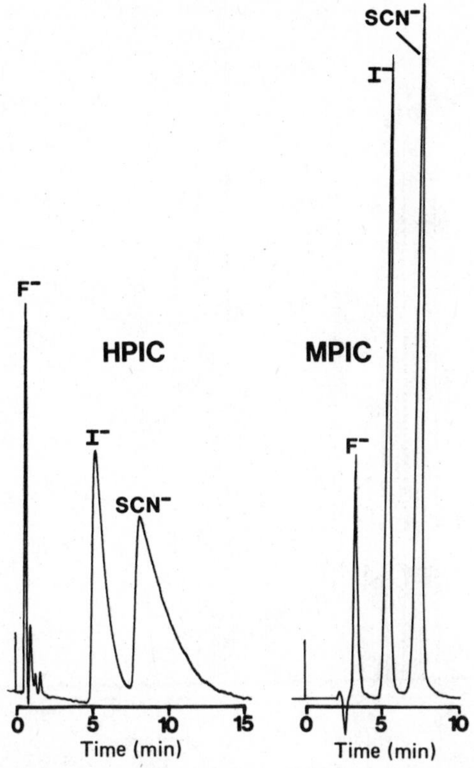

Figure 9.10. Effect of ion type upon peak shape in high-performance ion chromatography and mobile phase ion chromatography. (Courtesy of Dionex Corporation)

The suppressors used in MPIC are similar to those used for standard IC. Fiber suppressors are used with most MPIC eluent systems; however, suppressor columns are used with eluents containing ≥2 m*M* of large quanternary ammonium ions such as tetrabutylammonium ion or large alkylsulfonates such as octane sulfonates.

9.2.4 IC with Post-Column Derivatization Before Detection

Introduction

Post-column derivatization is an established technique which chemically modifies the separated components so that they are detectable by one of the common chromatographic detectors (e.g., absorbance, fluorescence). In post-column derivatization, a reagent is continually added to the sep-

Table 9.3 Useful Eluents in MPIC

	Nonsurface Active Anions	Surface Active Anions	Nonsurface Active Cations	Surface Active Cations
Typical examples	I^-, SCN^-, BF_4^-, $S_2O_5^{2-}$,	Alkylsulfates,	Alkylamines,	Quaternary amines,
	$Fe(CN)_6^{4-}$, ClO_4^-	Alkylsulfonates	Alkanolamines	Phosphonium ions
MPIC reagent	Tetrabutylammonium hydroxide[a]	Ammonium hydroxide[a]	Hexane sulfonic acid[a]	Hexane sulfonic acid
	Tetrapropylammonium hydroxide	Potassium hydroxide	Octane sulfonic acid	Perchloric acid
MPIC reagent counterion additives	Na_2CO_3, $Na_2B_4O_7$, Boric acid, Glutamic acid	—	HCl	—
Solvents	Methanol, acetonitrile,[a] isopropanol	Methanol, acetonitrile,[a] isopropanol	Methanol, acetonitrile,[a] isopropanol	Methanol, acetonitrile, isopropanol

[a] Generally preferred in MPIC eluents.

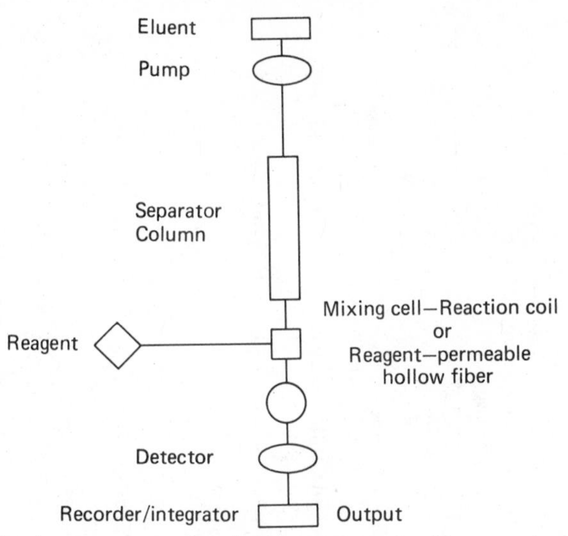

Figure 9.11. Basic components of an ion chromatograph with pressurized single reagent postcolumn reactor. (Courtesy of Dionex Corporation)

arator column effluent. The reagent reacts selectively with the sample species to form a detectable product. In conventional post-column derivatization techniques, the reagent is delivered to a mixing cell which combines the post-column reagent and column effluent. The reagent and effluent mixture continues into a reaction coil to complete mixing and the associated derivatization reaction. The reaction coil containing the reagent-effluent mixture is directly connected to a detector for the monitoring of the separated derivatized analyte. Figure 9.11 is a schematic of a column post-column reactor system.

Post-column derivatization incorporates the chemistry of "wet" analysis into an on-line chromatographic detection scheme. The success of post-column derivatization methods depends upon optimizing the derivatization reaction and upon careful design of the post-column reactor. The derivatization reaction should be fast (<1 min.) and the background signal of the reagent should be as low as possible. Reducing the dead volume of the post-column reactor to minimize band broadening is an important design consideration. The mixing cell and reaction coil must include a minimal dead volume but still allow efficient mixing of the derivatizing reagent and separated analyte ions. Any nonuniformity of the reagent-effluent mixture will result in background noise because the reagent produces a small background signal. Thus, a constant effluent and reagent flow and an efficient mixing system are required in a post-column reactor. The requirements for a successful, conventional post-column reaction system have been reviewed (7).

Recently, hollow fiber technology has been extended to post-column derivatization techniques (8). By using special hollow fibers which are permeable to the reagent(s), the reagent can be continually added to the column effluent stream that passes through the interior of the hollow fiber. The exterior of the fiber is bathed in the reagent solution and is prevented from "leaking" out of the fiber by the continual flow of reagent into the fiber interior. The fiber allows for more homogeneous addition of the reagent than does a mixing cell. The use of reagent-permeable hollow fibers has greatly improved the simplicity and sensitivity of post-column derivatization techniques.

The detection of amino acids via reaction with ninhydrin or o-phthalaldehyde is an example of a widely used post-column derivatization method. The determination of sequestering agents such as polyphosphates, NTA, and EDTA has also been performed using anion exchange separation followed by addition of $Fe(NO_3)_3$ to permit visible absorbance detection in the ultraviolet.

Recently, a post-column reaction has been applied for the detection of transition metals after separation on an ion exchange column (9–11). The post-column reagent is a metallochromic indicator (4-(2-pyridylazo)-resorcinol). Other post-column chemistries have been developed and the choice depends on the metal ions of interest.

Post-column derivatization offers additional selectivity and sensitivity in the chromatographic scheme and has greatly broadened the scope of ion chromatography.

Ion Chromatography of Transition Metals

In 1953, Kraus and Moore (12) demonstrated the usefulness of ion exchange chromatography for the separation of transition metals. High-performance liquid chromatography (HPLC), thin-layer chromatography (TLC) and gas chromatography (GC) also have been used. Many earlier applications of HPLC (13), TLC (14), and GC (15) for transition metal determination were limited to the separation of metal complexes or chelates rather than the "free" transition metal ions. These methods typically require solvent-extraction. Ion chromatography is now a very useful alternative analytical tool for the separation and detection of both free transition metal ions and metal complexes.

Due to the strong affinity of transition metal ions for the cation exchange site of the stationary phase, eluents of high ionic or acid strength are normally required for elution of metals. For sensitive conductimetric detection of the analyte, the eluent must be chemically modified to reduce the background conductivity of the eluent. In suppressed cation IC, this is achieved by using a cation suppressor which, from above, can be a packed bed post-column reactor containing a high-capacity, strongly basic anion-exchange resin in the hydroxide (OH^-) form (16). More recently, fiber suppressor devices using ion exchange membranes have been used to perform the eluent suppression reactions. Suppressed cation IC, however, is not compatible with transition metal determination because the insoluble hydroxides of the metals are formed in the suppressor.

The development of UV/VIS detection of transition metal complexes coupled with the use of low-capacity, pellicular ion exchange resins now enables the routine determination of transition metal ions by IC. This method has sensitivity comparable to graphite furnace atomic absorption spectroscopy. Figure 9.12a shows a separation of transition metals using post-column derivatization. Note that different oxidation states of the same metal (Fe), which are stable in solution, can be individually determined.

To understand the mechanism of separation, consider a strong cation-exchange resin to be used for the separation of two transition metal ions, A^{n+} and B^{n+}. When an eluent containing monovalent cations such as Na^+ or H_3O^+ is used, the metal ions often show similar chromatographic behavior and thus coelute. For metal ions with similar charge-to-size ratios, their chromatographic separation by pure cation exchange under equilibrium conditions is very difficult. By employing an eluent that contains complexing agents (ligands or chelating agents), the ion-exchange properties of the metal ions can be radically altered. Choosing complexing agents for which the formation constants of different metal ion complexes are significantly different allows for rapid and complete separation of most metal ions. The competing equilibria of complex ion formation in the ion-exchange process allows for a wide variety of separations of metal ions using anion- and cation-exchange resins.

The ion-exchange behavior of a metal ion complex is primarily dependent on the charge of the complex. In general, the greater the difference in the charge of the metal ion and metal ion complex, the more effective the chromatographic separation. If the eluent contains neutral complexing agents (e.g., ammonia or ethylenediamine), the charge of the metal ion will not change when complexed. This type of complexation does not significantly affect the exchange selectivity.

If complexation with a charged ligand or chelating agent results in a neutral complex, a dramatic change occurs in the exchange equilibrium. The metal ion is now bound in the neutral complex, which cannot be ion exchanged or retained by the chromatographic packing. If complexation

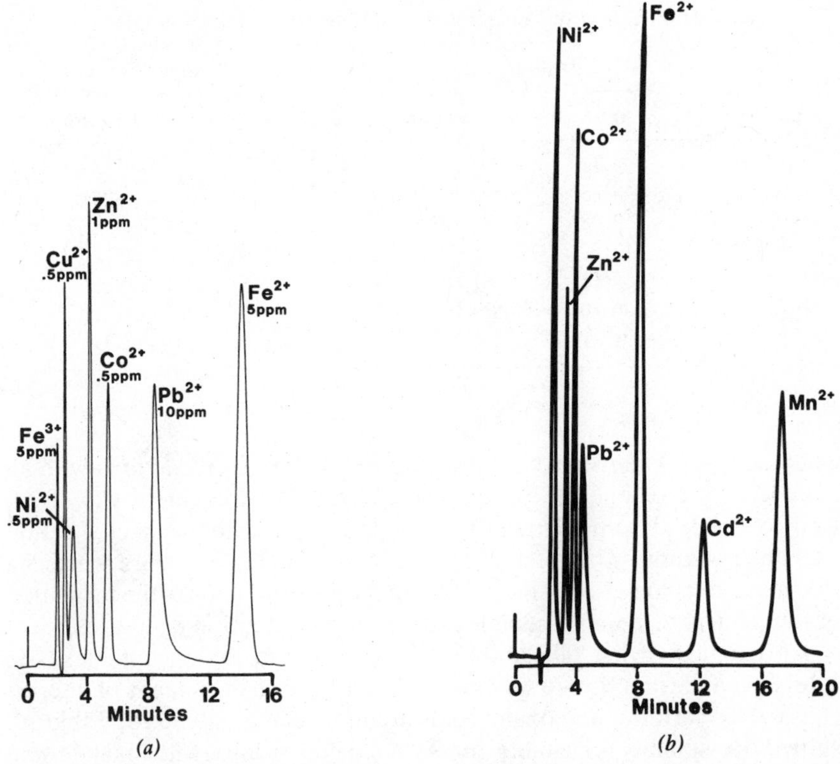

Figure 9.12. (*a*) Cation-exchange separation of transition metals using a citrate-oxalate eluent. Concentrations (in ppm) are Fe^{3+}(5), Cu^{2+}(0.5), Ni^{2+}(0.5), Zn^{2+}(1), Co^{2+}(0.5), Pb^{2+}(10), Fe^{2+}(5). (*b*) Cation-exchange separation of transition metals using a citrate-tartrate eluent. Concentrations (in ppm) are Ni^{2+}(1), Zn^{2+}(1), Co^{2+}(1), Fe^{2+}(3), Cd^{2+}(3), Mn^{2+}(2). (Courtesy of Dionex Corporation)

continues another step with the complexing anion, the positively charged metal ion is now bound in an anionic complex and the ion-exchange equilibrium is again radically altered. Because stoichiometry (hence charge) of the metal ion complex will depend upon the metal, oxidation state, complexing agent, and pH, the resulting complex ion-formation and ion-exchange equilibria can only be approximated in theoretical calculations.

With low-capacity pellicular ion-exchange resins, the complexing agents which offer the maximum separations of metal ions are weak organic acids. Acids such as tartaric, citric, oxalic, and phthalic have been used as moderate strength chelating agents in ion-exchange separations. Optimum selectivity is obtained when two weak acids are present in the

Table 9.4 Stability Constants for Oxalate and Citrate Complexes

	Citrate ($HCit^{2-}$)		Oxalate (Ox^{2-})	
Metal Ion	Log K_f	Formula	Log K_f	Formula
Co^{2+}	4.8	Co HCit	7.0	$Co(Ox)_2^{2-}$
Cu^{2+}	4.35	Cu HCit	10.3	$Cu(Ox)_2^{2-}$
Fe^{2+}	3.08	Fe HCit	5.15	$Fe(Ox)_2^{2-}$
Fe^{3+}	12.5	Fe $HCit^+$	18.5	$Fe(Ox)_3^{3-}$
Ni^{2+}	5.11	Ni HCit	5.16	Ni(Ox)
Pb^{2+}	6.50	Pb HCit	5.82	$Pb(Ox)_2^{2-}$
Zn^{2+}	4.71	Zn HCit	6.40	$Zn(Ox)_2^{2-}$

eluent. Figure 9.12*a* showed the separation of various metal ions with an oxalate-citrate eluent. In Fig. 9.12*b*, a citrate-tartrate eluent was used. Both of these chromatograms were obtained on a low-capacity cation exchange column. The retention of the cationic metals occurs when the free metal exchanges onto the chromatographic packing. As the formation of the neutral or anionic complexes increases, retention decreases.

An exhaustive list of stability constants for metal-ion complexes is available, and from this compilation one can choose the ligands of interest (17). Citric, tartaric, and oxalic acids are most commonly used. Table 9.4 shows the stability constants for selected metal ions with oxalate and citrate along with the stoichiometric formulas of the complexes.

The ability of transition metals to form anionic complexes is not only useful in separations by cation exchange, but can also be used for anion exchange. Much of the early and later work (18,19) in the separation of transition metals used anion exchange resins with hydrochloric acid eluents. The separations are accomplished by the formation of anionic chloro complexes with step gradient elution. Recently, excellent separations have been obtained on a pellicular anion-exchange column using oxalic acid eluents. The chromatogram in Fig. 9.13 was obtained with a pellicular anion-exchanger as described in Section 9.2.1. Both cation- and anion-exchange mechanisms occur in this separation (20). The cation-exchange capacity of the column results from the pellicular cation exchange substrate resin.

Because weak organic acids are effective chelating agents only when ionized, eluent pH has a dramatic effect on the separation and retention of the metal ions. In cation exchange, an increase in the eluent pH to the pK_a of the acid will result in a decrease in the retention time of the metal ion. This occurs because the increase in pH results in an increase in the

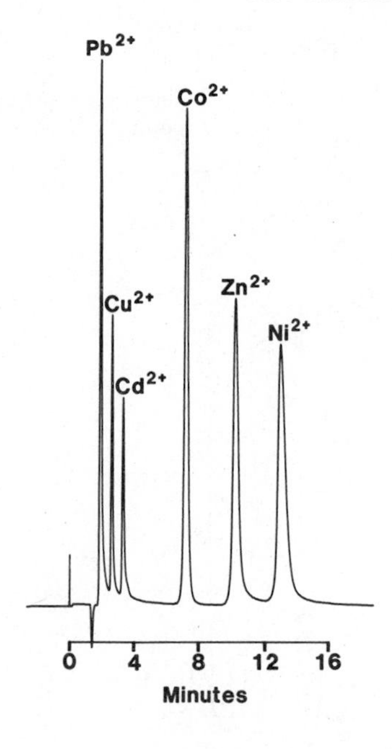

Figure 9.13. Anion-exchange separation of transition metals using an oxalic acid eluent. Concentrations (in ppm) are Pb^{2+}(3), Cu^{2+}(0.25), Cd^{2+}(3), Co^{2+}(0.5), Zn^{2+}(1), Ni^{2+}(0.5). (Courtesy of Dionex Corporation)

concentration of the chelating form of the acid. Thus, the residence time of the metal ion in the eluent increases due to enhanced complexation. This shifts the equilibrium of the metal ion from the stationary phase (resin) to the mobile phase (complexing eluent), thus decreasing retention. The exchange equilibrium can be altered by a change in the eluent concentration or by a change in the eluent pH. The interested reader is referred to an excellent review for additional information concerning the use of complexing agents in ion exchange chromatography (21).

After separation, the metals enter a post-column reactor which adds a reagent that selectively complexes with the metal ions to yield absorptions that commonly lie in the visible region. Finally, the post-column reactor effluent is monitored by a UV/VIS chromatographic detector. Ion chromatography offers the analytical chemist a rapid, sensitive, and selective method for the simultaneous multielemental analysis of transition metals.

9.2.5 IC with Multiple Detectors

In order to increase the information obtained from a sample, two detectors can be simultaneously used for two different groups of ions. Figure 9.14 shows two chromatographic traces obtained from a single injection. The upper trace was obtained by amperometric detection with the cell located

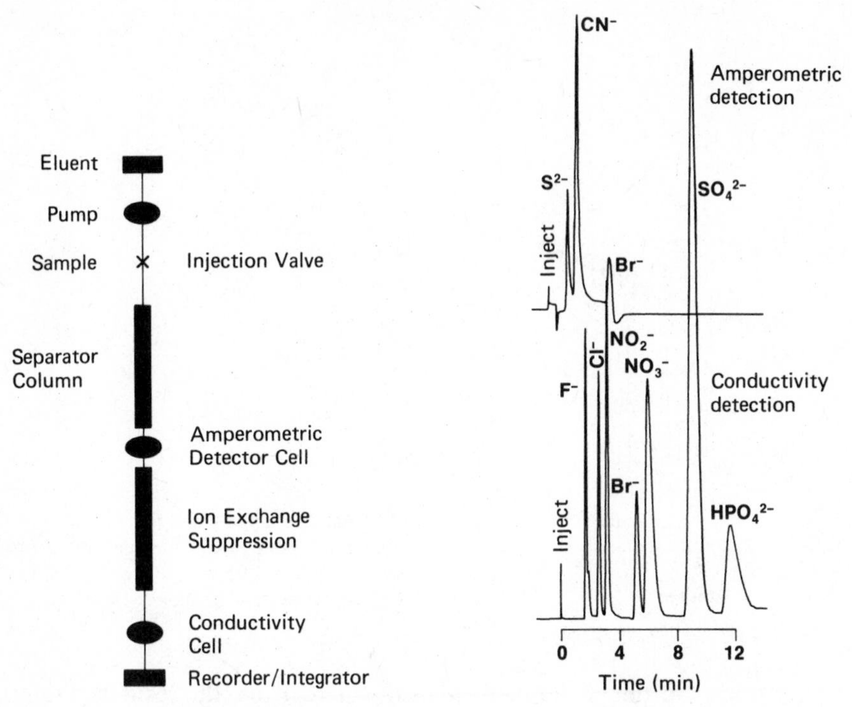

Figure 9.14. Dual detection ion chromatographic analyses. (Courtesy of Dionex Corporation)

between the separator and suppressor. This position is preferred because the high salt concentration of the unsuppressed eluent is required for efficient amperometric detection. Compounds that facilitate the oxidation of the silver electrode at the set potential are detected at high sensitivity before the separator effluent reaches the suppressor. Nonelectrochemically active species pass through the suppressor and are detected by their conductivity response in the usual manner. Table 9.5 lists electroactive species commonly determined electrochemically by IC.

Conductivity and UV/VIS absorbance can also be used simultaneously. As with the amperometric detector, the UV cell is placed before the suppressor for most applications. When using wavelengths <220 nm, the cell is placed after the suppressor to eliminate absorbance interference from the unsuppressed eluent. This analytical scheme greatly increases the ions determinable by IC. Table 9.6 is a list of ions commonly determined by IC using UV/VIS detection.

Table 9.5 Electroactive Ions

Ions	Electrode
Oxidations	
CN^-, HS^-, I^-, Br^-, Cl^-, $S_2O_3^-$, SCN^-	Ag
Sugars, alcohols	Au, Pt
Ar—NH_2, Ar—OH, N_2H_4, ascorbic acid	GC,[a] Pt
Reductions	
Transition metals	Hg
OCl^-, N_2H_4	Pt
Polyaromatics, —S—S—, R—NO_2	Hg, Au

[a] Glassy carbon.

9.2.6 IC Coupled Chromatography

IC coupled chromatography is a hyphenated technique combining any two of the three ion chromatographic modes which can increase the information obtained from a single chromatographic system with little increase in labor. This occurs by passing the sample ions through two different separator columns.

Major advantages of IC coupled chromatography include:

1. Different classes of ions can be separated from one another; one class then can have individual constituents separated.

Table 9.6 Anions Determinable by Ultraviolet Absorption

S^{2-}	NO_2^-
SO_3^{2-}	NO_3^-
SCN^-	N_3^-
$S_2O_3^{2-}$	Cl^- (very weak)
SeO_3^{2-}	Br^-
SeO_4^{2-}	I^-
$SeCN^-$	IO_3^-
AsO_3^{3-}	BrO_3^-
AsO_4^{3-}	ClO_3^- (very weak)
	ClO_2^-

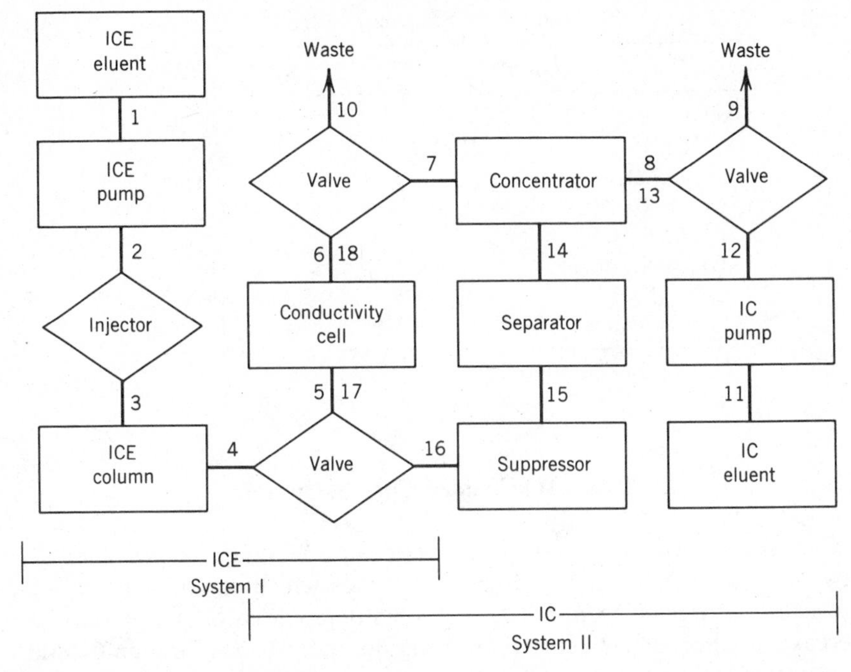

Figure 9.15. Ion chromatography exclusion coupled with suppressed ion chromatography. Ion chromatography exclusion, 1–9; ion chromatography, 10–18. (Courtesy of Dionex Corporation)

2. "On-column" sample pretreatment can be performed with subsequent automatic injection onto the second separator column.
3. Recycle chromatography can be performed.
4. Analyte ions can be concentrated through repetitive injection before separation.

Instrumental Configuration

Figure 9.15 shows the flow diagram employed for coupled chromatography. The two flow systems of the chromatography module are coupled by one of the auxiliary valves. The conductivity detector effluent from ion exclusion (System I) enters the auxiliary valve where it is directed to either waste (normal operation) or to an ion-exchange "concentration" column which replaces the ion chromatography injection loop. As ions are detected in ICE, they can be selectively collected for reinjection for IC (System 2). Collection may be performed manually or automatically.

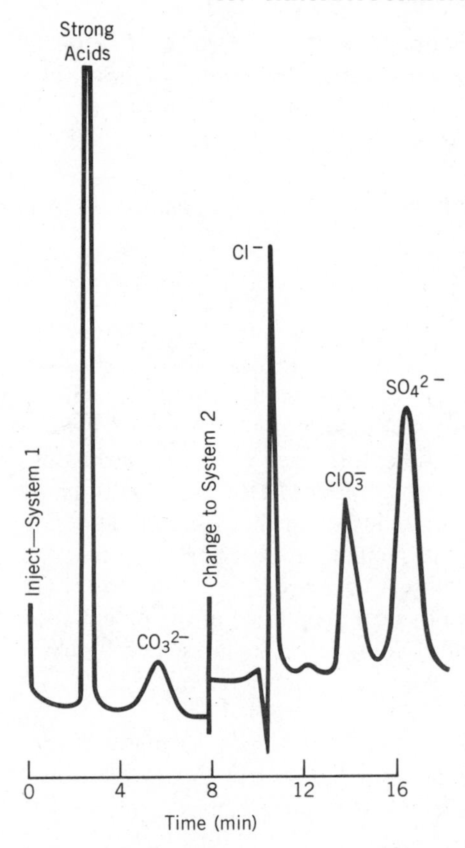

Figure 9.16. Application of coupled chromatography. The separation of anions in 4% caustic. (Courtesy of Dionex Corporation)

Examples of IC Coupled Chromatography

Different classes of ions can be separated using ICE coupled to IC. Figure 9.16 shows a 4% NaOH sample analyzed for CO_3^{2-}, Cl^-, ClO_3^-, and SO_4^{2-}. Using only standard IC, the large OH^- concentration would cause a large negative baseline dip. This dip, along with the CO_3^{2-} peak, would render Cl^- quantitation impossible.

As the sample passes down the ICE separator column (System 1) in an eluent of deionized (DI) H_2O, the H^+ form resin neutralizes the sample by exchange of H^+ for Na^+ converting the NaOH to H_2O. Also, the strong acid anions are separated as a group from the weak acid CO_3^{2-}. After ICE, the strong acid anions are collected on a small anion-exchange "concentration" column. After collection, the auxiliary valve directs the effluent to waste, and CO_3^{2-} is detected after normal ICE elution.

Simultaneously with CO_3^{2-} elution, the strong acid anions are injected into the System 2 eluent flow and separated by standard IC. The result is sample cleanup, separation of weak acid from strong acid anions, and separation of the strong acid anions from one another with one sample injection. If so desired, this process can easily be automated.

9.2.7 Single Column Methods in Ion Chromatography

Introduction

A large number of single column methods (no suppression of the eluent before detection) have been developed as an alternative to suppressor-based IC methods. Before describing each method in detail, it will be useful to review some general aspects of detection.

There are basically two types of detectors used in liquid chromatography (22). The first type is the solute property detector. As its name implies, the solute property detector relies upon the use of a detection property of the solute. When eluent systems are properly chosen, the detector is sensitive only to the solute. As a result, such detectors are generally insensitive to changes in physical properties of the eluent such as temperature or eluent composition. Such detectors are generally characterized by high sensitivity and a linear dynamic range of three to four orders of magnitude.

While a solute property detector would appear to have many advantages in the detection of ions, such detectors suffer from a lack of general applicability, which lessens their usefulness in ion analysis. Thus, as will be seen in subsequent discussions, direct detection of ions via a UV/VIS absorption detector, a fluorescence detector, or an amperometric detector provides a sensitive detection method for only a relatively few ions.

The second type of detector is the bulk property detector. Such a detector is designed to continuously monitor some physical property of the column effluent. While such a detection mode necessarily makes the bulk property detector more general, it also results in significantly reduced sensitivity and selectivity when compared to solute property detectors. In addition, bulk property detectors are characterized by high sensitivity to changes in eluent temperature, eluent composition, and flow rate, and tend to be characterized by a rather limited sensitivity range, usually less than two orders of magnitude (22).

While there are really only two basic types of liquid chromatography detectors, two additional detector modes exist. First, under certain conditions a bulk property detector can be made to function as a solute property detector. An example of this is IC using eluent suppression with conductivity detection. Using a suppressor to remove the eluent ions to which the conductivity detector is sensitive, the conductivity detector becomes sensitive to only the solute ions and thus behaves as a solute property detector. In so doing, suppressed conductivity detection gains 2 to 3 orders of magnitude in linear range and 1 to 2 orders of magnitude in sensitivity when compared to single column nonsuppressed conductivity detection using the same ion-exchange column.

The second additional detector mode is the use of a solute property detector under conditions whereby it functions as a bulk property detector. As will be discussed later, inverse photometric chromatography (IPC) represents such a case. However, a UV/VIS detector used to perform IPC has all of the limitations of the bulk property detectors noted above.

Single-Column IC Using Solute Property Detectors

The most common solute property detector used in single column (nonsuppressed) IC is the UV/VIS absorption detector. While many authors have used this detector for aromatic ions (23), its use for inorganic ions is limited, since few absorb in the visible or in the near UV regions; however, several inorganic ions do absorb strongly in the low wavelength region of the UV. Use has been made of UV detection in all three of the major modes of IC: the ion-exchange mode, the ion-exclusion mode (24), and the ion-pair mode (25,26). In some cases, such as for nitrate, iodide, or aromatic ions, direct UV detection exhibits superior sensitivity to standard (suppressed) IC, but, of course, a significant number of important ions do not absorb in the UV. Such ions include sulfate, phosphate, carbonate, alkyl sulfates, alkyl sulfonates, alkyl amines, and phosphonium ions.

Another important solute property detector is the amperometric detector. As with other solute property detectors, the sensitivity of this detector exceeds that of IC with suppressed conductivity detection. For example, when using an amperometric detector with a silver electrode, increased sensitivities are observed for many inorganic anions such as iodide, cyanide, sulfide, and ferrous cyanide (27). However, as is the case with most solute property detectors, electrochemical detectors offer high sensitivity for only a small portion of the total number of ions that can be detected with standard IC.

Single-Column IC Using Bulk Property Detectors

By far the most commonly used bulk property detector in single-column methods is the conductivity detector. Its popularity stems from the inherent generality of such a detector for ions. Use of the conductivity detector has been applied to all three major modes of ion chromatography: ion-exchange (28–30), ion-exclusion, and ion-pair chromatography (31).

As mentioned earlier, the use of a conductivity detector in single column ion chromatography can be accomplished only with considerable loss of linearity and sensitivity when compared to standard IC. The use of a conductivity detector in single column IC evolved primarily because of the desire by chromatographers to eliminate the effects of increased column band broadening and adsorption effects due to the suppressor column. However, recent improvements in suppressor design, such as the fiber suppressor described earlier, minimize, or eliminate this band broadening and adsorption.

An important characteristic of the conductivity detector when used in single-column IC is that the magnitude of the detector noise is directly related to the eluent conductivity. Thus, in an effort to maximize sensitivity, most work in this mode has made use of ion-exchange columns with capacities as low as or lower than those used in standard IC. This requires the use of rather dilute eluents which, of course, exhibit lower conductivity and thus allow for higher sensitivity. Such an approach, however, merely shifts the linear range of the method to lower concentrations without actually increasing the linear range. In general, the linear range of conductivity detection in the single column mode is one to two orders of magnitude, regardless of the actual eluent concentration used.

Another bulk property detector used in single column IC is the refractive index detector, which, like the conductivity detector, can be used in all three modes of ion chromatography: ion-exchange (32), ion-exclusion, and ion-pair chromatography. Significant use of the refractive index detector has been thus far only in the ion-exchange mode.

Single-Column IC Using a Solute Property Detector in a Bulk Property Mode

Several authors have reported new single-column methods making use of optical absorption detectors in a bulk property detector mode (33–35). In the technique of inverse photometric chromatography (IPC) the eluents for use in an ion-exchange separation are those which absorb strongly at UV or visible wavelengths. Because the detector is being utilized under conditions in which it is sensitive to the eluent, it confers upon the optical absorption detector all of the properties of a bulk property detector. In the general case, solute ions are detected by a stoichiometric *decrease* in concentration of eluent ions during the time that the solute ions are eluting. Thus, solute "dips" are observed in IPC rather than the conventional peaks observed in most modes of chromatography.

Another single column IC method where an absorbance detector is used as a bulk property detector was pioneered by Shill (36). In this technique, a reversed-phase column is used along with an eluent containing a UV absorbing surface active ion. In such a system, ions with charges opposite to those of the eluent ions are detected as an increase in the background absorbance as the solute ions elute. Solute molecules of the same charge as the UV absorbing eluent ions are detected as "dips" in the background absorbance.

9.3 METHODS DEVELOPMENT

9.3.1 Separation and Detection Mode

Ion chromatography methods development begins with the selection of the separation and detection mode. While there may seem to be a confusing number of separation modes possible in IC, the proper separation mode can be chosen in a straightforward manner by considering the basic characteristics of the analyte ions. The second step in IC methods development is sample preparation. Although IC generally requires less preparation than most other types of liquid chromatography, the very nature of ion chromatography often requires special measures for certain sample types. The final step in IC methods development is the optimization of instrumental conditions. Of course, the choices made in the first two steps in methods development largely determine the appropriate steps in the optimization process.

Figure 9.17 lists the separation and detection modes generally used for most groups of ions. The detailed mechanisms for each separation mode were discussed in Section 9.2. While there is certainly some overlap in the areas of applicability of various modes, general guidelines can be used. For the most part, the optimum separation mode can be determined by considering the hydrophobic character as well as the hydration energy of the sample ion. Ions characterized by significant hydration energy and low hydrophobicity, such as chloride or potassium, are best done by standard IC. Ions characterized by low hydration energy or high hydrophobicity, such as perchlorate or tetrabutylammonium ion, are best done by MPIC. Ions that have some hydrophobic character but also a significant hydration energy, such as acetate or propionate, are best done by ICE. Of course, for ICE to be useful, the pK_a of the parent acid of the analyte anion must be between 1 and 7.

In addition to the guidelines above, two other areas of ion analysis must be mentioned. First, amphoteric ions such as amino acids can be

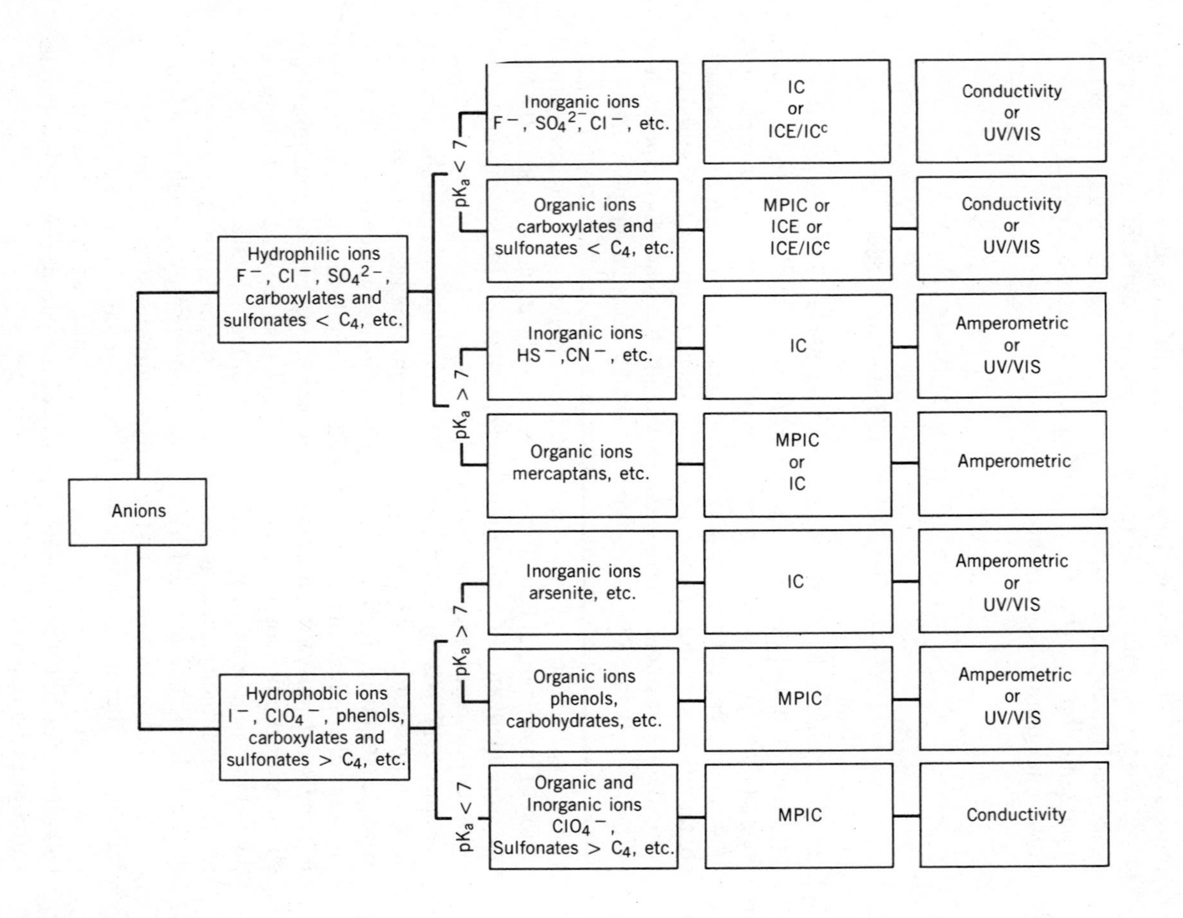

Anions
Hydrophilic ions F −, Cl −, SO4 2−, carboxylates and sulfonates < C4, etc.
Hydrophobic ions I −, ClO4 −, phenols, carboxylates and sulfonates > C4, etc.
pKa < 7
pKa > 7
pKa > 7
pKa < 7
Inorganic ions F −, SO4 2−, Cl −, etc.
Organic ions carboxylates and sulfonates < C4, etc.
Inorganic ions HS −, CN −, etc.
Organic ions mercaptans, etc.
Inorganic ions arsenite, etc.
Organic ions phenols, carbohydrates, etc.
Organic and Inorganic ions ClO4 −, Sulfonates > C4, etc.
IC or ICE/IC^c
MPIC or ICE or ICE/IC^c
IC
MPIC or IC
IC
MPIC
MPIC
Conductivity or UV/VIS
Conductivity or UV/VIS
Amperometric or UV/VIS
Amperometric
Amperometric or UV/VIS
Amperometric or UV/VIS
Conductivity

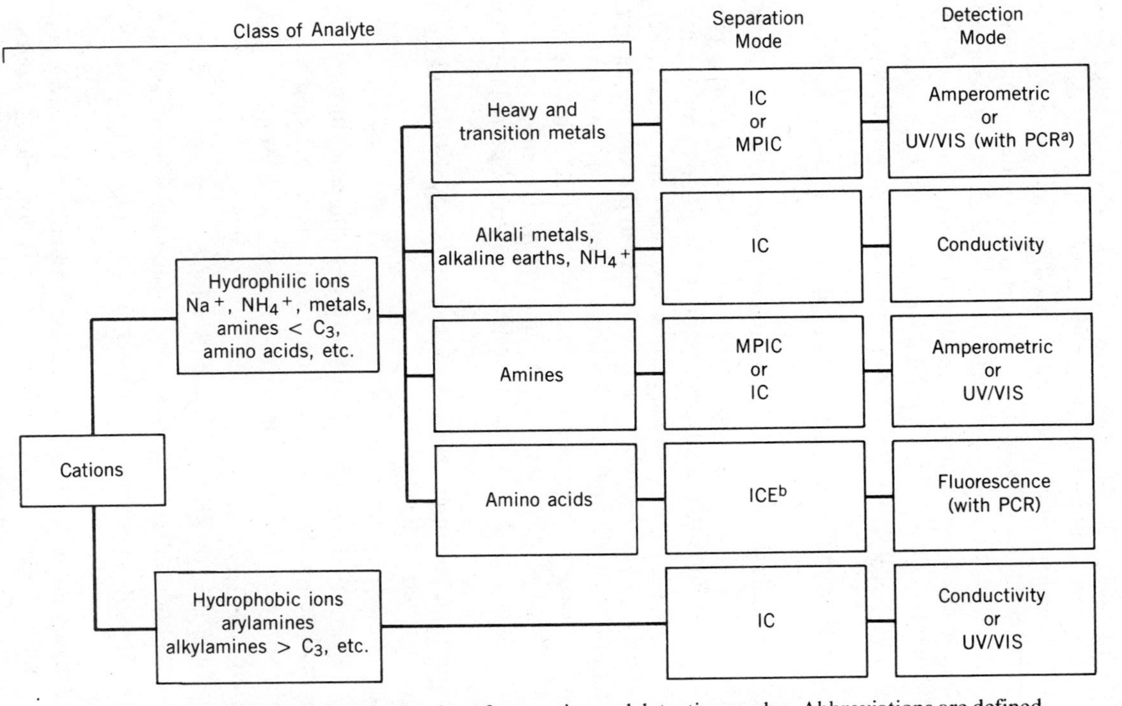

Figure 9.17. Flow diagram for selection of separation and detection modes. Abbreviations are defined in the text. [a] Post-column reaction, [b] separation is primarily ion-exchange, not ion-exclusion, [c] coupled chromatography. (Courtesy of Dionex Corporation)

separated on either anion or cation exchange resins. Second, the separation of transition metals is possible using the separation modes discussed in Section 9.2.4.

Figure 9.17 also lists the detection modes used for each group of ions. Again, while there are a number of cases where more than one detection mode is possible, general guidelines can be used to choose the most appropriate detection mode. Conductivity is the preferred detection mode for most nonabsorbing, strongly dissociated acids and bases. Amperometry is used for most electroactive, weakly dissociated ions, and UV/VIS or fluorescence is used for absorbing ions or complexes which do not readily respond to the electrochemical detectors just mentioned.

9.3.2 Sample Information

To obtain an IC analysis in the shortest possible time, begin each IC analysis by collecting information about the sample to be analyzed. Frequently, the analyst finds that analytical standard solutions produce reasonable results, but that "real" samples produce various unwanted effects in the chromatogram. These effects are usually due to the sample matrix. In order to minimize or eliminate these sample effects, it is important to know, whenever possible, the sample matrix and the effects that the matrix can be expected to have on the chromatogram. The most efficient approach to the analysis is to begin by collecting information about the sample.

Several important questions should be asked and answered about each species to be analyzed:

1. Is the sample soluble in water or in water-alcohol mixtures?
2. Is the analyte ion an anion, a cation, or an amphoteric ion?
3. Can the analyte species be converted into ionic form or into a more desirable ion?
4. What is the pK_a or pK_b for the analyte ion?
5. What ions and nonionic species exist in the sample matrix?
6. Has the species or a similar one been analyzed before?

Because samples must be in solution for an IC analysis, the answer to the first question may determine whether analysis is possible. Of course, a number of techniques are available to get sample ions into solution when the sample itself is insoluble in water or water-alcohol mixtures (37–39). Although aqueous samples are preferred because they have less effect on the resins used in IC, samples dissolved in organic solvents are commonly analyzed. It should be remembered, however, that the magnitude of ionic hydration is an integral part of any ion retention process. As such, the presence of organic solvents in the sample can be expected to disturb the hydration equilibrium of the chromatographic system. Therefore, when performing an analysis where the sample is dissolved in an organic solvent, it is best to use a standard dissolved in the same organic solvent.

The selection of the IC detection mode is largely dependent upon the pK of the parent compound of the sample ion. Strongly ionic species usually are detected best by conductivity. If the pK is greater than 7, amperometric or spectrophotometric detection may be preferred because of the low conductivity inherent in such experimental conditions. The pK of many compounds can be found in reference tables (40,41).

The answer to the fifth question gives knowledge of the sample matrix. This may save time and prevent incorrect interpretation of analytical results. If interfering species are present, steps may be taken to minimize interferences, either through pretreatment or through specific choice of operational parameters.

If the answer to the sixth question is positive, conditions from previous analyses may be used as a guideline for the present analysis. While such an approach will not always be useful, eluent conditions often are similar for many related ions. As will be seen later, the valence of the ion of interest is a determining factor in its retention characteristics, and thus an ion cannot be considered "similar" to another ion unless the two ions are of the same valence.

9.3.3 Sample Pretreatment

Pretreatment of a sample may be necessary for many possible reasons:

1. To dissolve the sample so that it may be injected.
2. To form an ionic species from the sample species.
3. To decrease the species concentration so that the separator column capacity is not exceeded and so that the calibration curve for the concentrations to be determined is linear.
4. To remove particulates which cause blockage in the system and a resultant rise in pressure.

Some samples can be easily prepared for analysis by dissolving in DI H_2O or in eluent prior to analysis. Also, many ionic species can be extracted into an organic phase, for example toluene, and then back extracted into DI H_2O or eluent. Extraction of acidic organic compounds may be accomplished with bicarbonate (HCO_3^-), which puts the sample into an ideal matrix for IC analysis. If a solid sample is not completely water soluble, species may be extracted from the insoluble matrix with water and ultrasonic agitation. Dissolving of species by this method depends on sample characteristics such as particle size and wettability. These characteristics must be investigated for each different sample type. At times, more drastic conditions are necessary; for example, the Schöniger combustion is used to convert heteroatoms of organic compounds into inorganic species.

The high sensitivity of IC requires a dilution of most samples. For unknown samples, it is good general practice to dilute the sample by a factor of 100 in order to minimize the chances of column overloading. In many cases, filtration will be necessary to remove particulates from the sample. It is extremely important to note, however, that many filtration media are contaminated with significant quantities of inorganic ions. Prewashing of filtration media may be necessary in order to eliminate these interferences.

9.3.4 Selection of Chromatographic Conditions

Standard Ion Chromatography

Eluent selection in standard IC is determined by two factors:

1. The relative affinity of the eluent and analyte ions for the ion-exchange column.
2. The compatibility of the eluent ion with the suppression reaction.

These two factors effectively limit the useful eluting ions in this mode of IC to $H_2BO_3^-$, OH^-, HCO_3^-, and CO_3^{2-}, which are listed in order of increasing eluent strength in anion exchange; H^+, Zn^{2+}, and *m*-phenylenediamine ion are used in cation exchange. In both anion and cation IC, the use of monovalent eluent ions is generally limited to the elution of monovalent ions. In anion IC, the use of carbonate ion or bicarbonate ion/carbonate ion mixtures is usually preferred because of its general utility in the elution of both monovalent and polyvalent ions. In cation IC, *m*-phenylenediamine ion and sometimes Zn^{2+} (5) are used for the elution of divalent ions. In cation IC, however, the use of such divalent eluting ions results in essentially no retention of all monovalent ions.

In IC, some control of selectivity can be obtained with the addition to the eluent of various additives that can effect interactions, other than ion exchange, between the resin phase and the analyte ions. To a limited extent, addition of polar organic solvents such as methanol or acetonitrile to the mobile phase will reduce retention and improve peak symmetry for ions that tend to interact strongly with the resin matrix. Ions in this category include polarizable ions, such as I^- and ClO_4^-, as well as hydrophobic ions, such as benzoic acid or triethylamine. A generally more effective method of reducing such undesirable non–ion-exchange interactions with the resin matrix is the addition of suppressor compatible ions that are also subject to adsorption with the resin matrix. An example of this type of eluent additive is *p*-cyanophenol, which is commonly used in anion IC for selectivity control.

A useful feature of ion-exchange chromatography is the effect that eluent concentration has upon selectivity. While this effect is of limited use in cation IC, due to the nonexistence of a single eluent system for the analysis of both mono and divalent cations, the phenomenon is extremely useful in anion IC in tailoring the eluent system to various mixtures of anions. In general, the selectivity coefficient K_c in ion-exchange phenomena is expressed as

$$K_c = \frac{[A^{a+}]_r \, [B^{b+}]}{[A^{a+}] \, [B^{b+}]_r} \tag{13}$$

where $[A]$ = the concentration of analyte ion in the mobile phase
$[A]_r$ = the concentration of analyte ion in the resin phase
a = the valency of analyte ion
$[B]$ = the concentration of eluent ion in the mobile phase
$[B]_r$ = the concentration of eluent ion in the resin phase
b = the valency of eluent ion

The distribution coefficient D for an ion exchange separation is given by

$$D = K_c^{a/b} \frac{[B]_r^a}{[B]^a} \tag{14}$$

The distribution coefficient is related to a more readily determined chromatographic parameter, the capacity of factor k':

$$k' = \frac{DV_m}{V_s} = \frac{(K_c)^{a/b}[B]_r^a V_m}{[B]^a V_s} = \frac{t_1 - t_0}{t_0} \tag{15}$$

where V_m = the volume of the mobile phase in the column
V_s = the volume of the stationary phase

t_1 = the retention time of the analyte ion
t_0 = the retention time of an unretained peak

Examination of the capacity factor equation reveals two fundamental facts about ion-exchange selectivity. First, when all analyte ions have the same valency and the eluent concentration is varied, one should expect no changes in selectivity because all factors in the capacity factor equation are held constant except $[B]$. Second, when analyte ions have different valencies and the eluent concentration is varied, one should expect significant changes in selectivity because both a and $[B]$ are variables. For example, if one analyte ion is monovalent and another analyte ion is trivalent, then doubling the concentration of eluent ion should reduce the retention of the monovalent ion to one half the previous k' value, whereas the retention of the trivalent ion should be reduced to one eighth the previous k' value.

In view of the dramatic effect that the eluent concentration has on selectivities of ions of different valencies, another related method of selectivity control is the adjustment of the valence of the analyte ion by pH adjustment. While the available range over which the eluent pH may be adjusted is somewhat limited, due to the restriction that for suppressor compatibility the eluent anion must have a pK_a greater than 7, adjustment of the eluent pH is frequently an effective method of selectivity control.

The choice of ion-exchange columns used for the analysis also controls selectivity. Several different cation and anion exchange columns are available and differ significantly in functional group and cross-linking. The proper choice of column is sample dependent. Efficient separations are usually possible by choosing a column that exhibits good selectivity for the analyte ions in a given sample matrix. The literature contains extensive references to columns used for specific analyses.

The analyst must also consider the selection of the suppressor. In general, the use of a suppressor demands the use of eluents that are 0.01 M or less. Such eluents generally result in acceptable performance with a fiber suppressor at eluent flow rates of three milliliters per minute or less. For special analytical problems where there is a need for a suppressor column (e.g., where samples intended for anion analysis contain large amounts of metal ions that may strongly bind to the ion-exchange fiber), it is usually recommended that a 6 mm i.d. × 60 mm suppressor column be used. For more concentrated eluents (greater than 0.01 M), a 9 mm i.d. × 100 mm suppressor column may be required in order to achieve reasonable suppressor use before regeneration is required.

When using the fiber suppressor, one rule must be observed: the total number of equivalents of regenerant ion per unit time must exceed that of the eluent by a factor of two or more. This can be accomplished by altering either the regenerant flow rate or its concentration. It must be remembered, however, that incomplete Donnan exclusion of ions from the fiber must be considered when choosing the regenerant ion concentration. Maximal Donnan exclusion of the regenerant counterion is observed when the size and valency of the regenerant counterion is maximized. Thus, sulfuric acid is preferred over hydrochloric acid as a regenerant in anion suppression due to its higher valence. For minimal leakage when regenerant concentrations must be above 0.01 *M*, large ions usually are preferred. Thus, 0.05 *M* toluene sulfonic acid or 0.1 *M* alkylbenzenesulfonic acid can be used in an anion suppressor, and 0.04 *M* tetramethylammonium hydroxide in the cation fiber suppressor.

In cases where weak bases are being analyzed, 0.02 *M* potassium carbonate is used. The suppression reaction in this case produces a more highly ionized salt than the hydroxide form of the amine produced when a hydroxide regenerant is used. For example, ammonium hydroxide, produced with a tetramethylammonium hydroxide regenerant system, is in equilibrium with the nonconducting ammonia molecule, resulting in lower sensitivity to the ammonium cation. However, the ammonium bicarbonate formed when a potassium carbonate regenerant is used exhibits higher conductivity because virtually all of the ammonia is ionized.

Single-Column Ion Chromatography

In choosing eluents for single-column IC in the ion-exchange mode, two factors must be considered:

1. The effect of eluent concentration on the noise level.
2. The effect of eluent ion choice on sensitivity.

The choice of eluent depends on the detector chosen for single column IC. For example, the baseline noise level is a function of eluent concentration in both conductivity detection and indirect UV detection (IPC). Eluent ion concentration does not affect noise levels in refractive index or direct UV detection. Obviously, this means that eluent choices are different for each detection mode.

In single-column IC with conductivity detection, noise is minimized by minimizing the conductivity of the eluent. This is accomplished via minimization of the eluent ion concentration. Two main approaches are used to facilitate this:

1. The use of eluent ions that have a significantly higher affinity for the resin than the analyte ions. These ions tend to limit the analyst's

ability to separate a large group of ions during a single chromatographic run.

2. The use of very low capacity columns, which tend to overload when used for analyses involving complicated matrices.

While each of these approaches has serious disadvantages, they are both necessary in order to minimize the noise problems inherent to single-column IC.

The sensitivity of single column IC is determined by the difference between the conductivity of the eluent ions and the conductivity of the sample ions (18). Because of this, it is important to choose an eluent ion that satisfies the requirements for minimal noise and at the same time exhibits relatively low ionic mobility. In practice, most analysts have chosen one of several different aromatic acids, such as benzoate or phthalate in the anion-exchange mode, and mineral acids or ethylenediamine in the cation-exchange mode.

In single-column IC with UV detection of a UV absorbing eluent (IPC), eluents are again chosen so as to minimize noise and maximize sensitivity. Noise is minimized by reducing eluent UV absorbance, but since this adversely affects sensitivity UV absorbance usually is adjusted to the 1.2–0.8 absorbance units range by adjusting the wavelength of the detector. Sensitivity is maximized by choosing an eluent ion with the minimum concentration that still provides the necessary absorbance and elution characteristics. While a large number of eluents are possible in IPC, in practice the most commonly used eluents are phthalate and trimesate in the anion-exchange mode, and benzyltrimethylammonium ion and Cu^{2+} in the cation-exchange mode.

In single column IC with refractive index detection, eluent choice is somewhat simpler. Since eluent ion concentration has very little effect on noise, the primary consideration is the refractive index difference between the eluent ion and the analyte ion. For the most part, eluent ions that exhibit large refractive index differences from inorganic anions are aromatic acids such as phthalic acid, hydroxybenzoic acid, or sulfosalicylic acid. No work has been reported for the cation version of this technique.

Single-column IC with UV/VIS detection represents the simplest of the various single-column methods. Eluent selection is limited only to the extent that the eluent ion should not absorb at the observation wavelength. Sensitivity is maximized by setting the detection wavelength at the absorbance maximum of the analyte ion.

Ion Chromatography Exclusion (ICE)

As with other modes of suppressed ion chromatography, the suppressor plays a major role in determining the eluents in ICE. As mentioned in Section 9.2.2, a fiber suppressor or suppressor column is used. The most commonly used eluents are alkyl sulfonic acids with a fiber suppressor, and HCl with suppressor columns.

Retention in ICE is controlled through adjustment of the eluent pH. Ions with pK_a values above or near the eluent pH will be retained. Ions with pK_a values below the eluent pH will be excluded from the resin pores. Selectivity in ICE is primarily a result of the pK_a values of the analyte ions, although adsorption phenomena frequently play a role in retention. As such, changes in eluent conditions (other than pH) rarely affect the selectivity of an ICE system. When ICE is performed with a suppressor column, the upper limit of eluent concentration is generally considered to be 0.01 *M* HCl. This limits the applicability of ICE to ions with pK_a values between about 1.5 and 7.

While the resolving power of an ICE system is determined for the most part by the pK_a values of the analyte ions and the pH of the eluent, one aspect of the separator column, the crosslinking, can also significantly affect resolution in ICE. This is because the charge density of a cation-exchange column drops as the cross-linking of the cation-exchange resin decreases. The charge density of a cation-exchange resin determines how effectively anions are excluded from the resin. Thus, cation-exchange resins of low cross-linking (i.e., low-charge density) exhibit improved resolving power for ions that are almost totally excluded on a cation-exchange resin of higher cross-linking. On the other hand, high cross-linked cation-exchange resins generally exhibit superior resolution for anions with high pK_a values because adsorption, which increases as cross-linking increases, contributes to the separation process.

Ion Exclusion Without Eluent Suppression

Refractive index, conductivity, and direct UV/VIS have been used as detection modes in nonsuppressed ion exclusion chromatography. The use of the conductivity mode is rather limited, due to the high conductivity of most eluents used in ion-exclusion work. Refractive index and UV/VIS detectors, however, are widely used because of their ability to handle the higher eluent concentrations necessary for ion-exclusion chromatography. Indeed, concentrations as high as 0.1 *M* H_2SO_4 can be used with either of these detectors. Further, since ionization of the analyte is unnecessary in either UV/VIS detection or RI detection, the useful range of ion-exclusion in either mode includes ions with pK values of 0.5 and greater.

Mobile-Phase Ion Chromatography (MPIC)

Due to the fact that eluent additives substantially determine the nature of the stationary phase where retention occurs, MPIC offers the chromatographer a wider choice of eluent conditions than either standard IC or ICE. While the number of possible eluent systems in MPIC may at first seem confusing, in fact the most useful eluent systems in MPIC are relatively few in number.

MPIC reagents that have been used for the retention of nonsurface active anions include: tetramethylammonium hydroxide, tetraethylammonium hydroxide, tetrapropylammonium hydroxide, ammonium hydroxide, tetrabutylammonium hydroxide, and hexadecyltrimethylammonium hydroxide. The general trend observed with these reagents is that retention increases as the number of methylene groups per molecule increases. Reagent ions with too few methylene groups retain only anions that are relatively hydrophobic, thus demonstrating relatively narrow scope. Reagent ions with too many methylene groups retain all anions but suffer from loss in selectivity and excessively long equilibration times. For the majority of applications, tetrabutylammonium hydroxide provides optimum retention and selectivity characteristics.

MPIC reagents that have been used for the retention of nonsurface active cations include butane sulfonic acid, pentane sulfonic acid, hexane sulfonic acid, heptane sulfonic acid, octane sulfonic acid, dodecane sulfonic acid, toluene sulfonic acid, perfluorodecanoic acid, camphor sulfonic acid, and bis-2-ethylhexylphosphonic acid. For a homologous series, such as the alkyl sulfonic acids, retention increases as the number of methylene groups increases. When comparing different reagents for selectivity, it becomes difficult to predict which will provide the optimum selectivity. However, the optimum reagent is usually a relatively poor surfactant. This is because the retention of nonsurface active ions in MPIC takes place via the enhanced effectiveness (i.e. the ability to lower-surface tension) that these nonsurface active ions provide for the surface active MPIC reagent. An effective surfactant will exhibit only slight increases in effectiveness as its counterion is changed, and thus it will exhibit poor selectivity in MPIC. Hexane sulfonic acid provides optimum retention and selectivity characteristics for most applications.

MPIC reagents of high purity are essential in order to perform MPIC with acceptable background conductivity levels. While a number of suppliers claim to sell reagents such as hexane sulfonic acid, it has been the authors' experience that these reagents invariably consist of sodium hexane sulfonate acidified with acetic acid. Obviously, the sodium content of these reagents is unacceptable for MPIC. MPIC reagents of the highest purity are required for efficient separations, and are commercially available.

The choice of solvent in MPIC is generally the same as in any other reversed-phase system. While methanol, acetonitrile, and isopropanol are the most frequently used in MPIC, any of the other solvents used in reversed-phase chromatography can be used. Acetonitrile is generally preferred because it forms lower viscosity mixtures with water and because it has an endothermic heat of mixing with water. The latter factor generally minimizes eluent degassing problems.

An important method of retention control in MPIC is the use of reagent counterion additives. Table 9.3 lists those most commonly used. While the amount of solvent in the eluent is the most common method of retention control, addition of counterions to the eluent serves to speed the elution of analyte ions in much the same way that increasing eluent strength affects retention in ion-exchange chromatography.

MPIC counterion additives are also useful for controlling selectivity because changes in counterion concentrations affect elution order of analyte ions just as in ion-exchange. Furthermore, addition of a divalent counterion additive can be useful in improving the chromatographic performance of polyvalent ions, which otherwise tend to exhibit low efficiency and poor peak symmetry.

MPIC analyses use fiber suppressors or suppressor columns. Generally, fiber suppressors are used with eluent reagent concentrations ≤ 2 m*M*. Because MPIC reagent concentrations rarely exceeded 5 m*M* and because flow rates are usually 1 ml/min, suppressor column lifetimes are much greater than with standard IC. Consequently, the most commonly used suppressor column is the smaller 6 mm i.d. $\times$ 60 mm column.

In the analysis of amines, an alternative to the conventional hydroxide form suppressor is frequently used. The problem with the hydroxide form suppressor in amine analysis is that the following equilibrium exists for amines other than quaternary amines:

$$R_3NH^+ + OH^- \rightleftharpoons R_3N + H_2O \qquad (16)$$

The free base form of the amine contributes nothing to the conductivity of the sample ion, and thus the extent to which the free base form of the amine is present will increase the detection limit of the amine. While the above equilibrium does not seriously limit detection sensitivity for strongly basic amines such as trimethylamine, the detection of weakly basic amines, such as triethanolamine, are severely hampered by the hydroxide form suppressor. The solution to the general detection problem of weakly basic amines is to use a borate form suppressor. In this case, the pertinent equilibrium is

$$R_3NH^+ + H_2BO_3^- \rightleftharpoons R_3N + H_3BO_3 \tag{17}$$

while this equilibrium also results in the formation of neutral amine, this equilibrium generally lies far to the left because boric acid is more acidic than water. The use of the borate form suppressor also helps minimize exclusion phenomena that occur when amines are present in the free base form. It should be remembered that the mobility of hydroxide ion is three times that of the borate ion, resulting in greater conductivity detection sensitivity for amines where equilibrium 16 lies far to the left. Therefore, the borate form suppressor is recommended only for samples that contain a cation with a pK_a of 5 or greater.

Ion-Pair Chromatography

As in other nonsuppressed modes, a major consideration in eluent choice is minimizing noise and maximizing sensitivity. When conductivity detection is used, two techniques are available to minimize the eluent conductivity. First, the concentration of ion-pair reagent is kept to an absolute minimum. The problem with this approach is that column equilibrium times tend to be extremely long if concentrations much below 2 millimolar are used. Second, weakly basic cations and weakly acidic anions are used in place of more conventional ion-pair reagents. These reagents also tend to be of limited usefulness, due to the long equilibrium times they require.

Refractive index detection has also been used as a detection method with nonsuppressed ion-pair chromatography. The general method of sensitivity enhancement in this mode is to maximize the refractive index difference between the analyte ion and the ion-pair reagent. Because of sensitivity limitations, this approach has seen little development.

When indirect UV detection (IPC) is used as a detection mode for non-UV absorbing ions, the general approach has been to select a UV absorbing ion-pair reagent. Sensitivity is maximized by minimizing the concentration of the ion-pair reagent, although low concentration of ion-pair reagents do result in long column equilibrium times. An interesting feature of this technique is that both anions and cations can be detected simultaneously. Ions with a charge identical to that of the ion-pair reagent will produce a negative vacancy peak. Ions with a charge opposite that of the ion-pair reagent will produce positive peaks. Sensitivity for the negative peaks is maximized when the absorbance of the eluent is at a practical maximum (0.3–1.2 A.U.), while sensitivity for positive peaks is maximized when the absorbance is minimized. As a compromise when detecting both anions and cations, a value of 0.3–0.6 A.U. is used.

When direct UV is used, only one major consideration is important in eluent selection. In this mode, the eluent should not absorb at the observation wavelength. While some eluent absorbance generally exists, sensitivity is generally maximized if the absorbance of the eluent is kept to a minimum.

9.4 ION CHROMATOGRAPHY INSTRUMENTATION

Ion chromatographs are specialized high-performance liquid chromatographs optimized to perform ion analysis. Because IC and LC instrumentation have many similarities and LC instrumentation was discussed in Chapter 4, here we will discuss only the features unique to IC.

9.4.1 Noncorroding Hydraulics

Because IC uses many corrosive solvents (including strong acids and bases) for both elution and suppressor regeneration, IC employs a chemically inert liquid flow system. All high-pressure components of an IC are made of chemically inert, nonmetallic materials. The electrodes in the detector cell are the only exception. A nonmetallic flow system has the following features:

1. Components cannot corrode and produce metal ions that will interfere with analysis and/or contaminate cation exchange column packing.
2. Undesired metal ions will not contribute to unwanted baseline noise.

9.4.2 Styrene/Divinylbenzene Column Packings

Another major difference from LC is that for IC columns to be efficient and durable column packings must be styrene/divinylbenzene (S/DVB) copolymers, whereas LC columns typically use silica-based packings.

These carbon based packings are stable over the entire pH range. The results of using S/DVB column packings are:

1. There is no limit to sample or eluent pH.

2. The columns can be cleaned with strong acids and bases, thus maximizing their useful lifetime.

9.4.3 Eluent Suppression

Eluent suppression is the major difference between IC and LC and is discussed in detail in Section 9.2.1. Basically, eluent suppression removes unwanted eluent ions from the column effluent before they enter the detector cell, thus decreasing baseline noise and maximizing sensitivity and selectivity.

9.4.4 Detector System

Most IC determinations use eluent suppression followed by conductivity detection. However, because weakly disassociated ions do not exhibit a strong conductivity response, other techniques such as amperometric, UV/VIS absorption and post column derivatization are used for various groups of ions.

Ion chromatographs and IC detectors have been designed to permit the simultaneous use of two or more detectors. Special low-volume, high-pressure cells permit placement of amperometric and UV/VIS detector cells in the flow system before or after the suppressor. The special cell design also minimizes band broadening.

Because conductivity detection is so frequently used, special high-performance conductivity detectors have been designed that include full compensation for ambient temperature changes in the laboratory. Ideally, the IC conductivity detector has an adjustable input, the "Eluent Temperature Coefficient," which enables the detector microprocessor to apply individual temperature compensation programs, and circuitry based on the eluent being used.

9.4.5 Dual Channel Operation

Because a complete ion profile requires the determination of both anions and cations, most analysts prefer a dual-channel IC. Consequently, ion chromatographs have evolved as dual-channel chromatographs such as in Fig. 9.18. Although IC pumps and detectors are single channel, the chromatography module, which is a temperature isolated module housing the injection valve and other hydraulic components, is dual channel. This dual-channel module permits simultaneous independent or coupled operation of two chromatographic systems. When two or more detectors are used on each system, several chromatograms are obtained that pro-

Figure 9.18. The Dionex System 2120i. A fully automated dual channel ion chromatograph. (Courtesy of Dionex Corporation)

vide a detailed scan of most sample anions and cations. The chromatography module also offers temperature control of the separator column from a single system. The column heater is used:

1. To improve baseline stability.
2. For reversed-phase separations.

Controllers and computers are available to automate ion chromatographs and permit independent or coupled operation of the two chromatographic systems.

9.5 APPLICATIONS

9.5.1 Literature of Ion Chromatography

Because ion chromatography is relatively new, analysts who are just beginning to use the technique can review the literature rapidly. Further,

Table 9.7 Dionex Application Notes

A. Ion Chromatography in Energy and Power Production
B. Ion Chromatography in the Semiconductor Industry
C. Ion Chromatography in the Pulp and Paper Industry
D. Ion Chromatography in the Food Industry
E. Ion Chromatography in the Electroplating and Metal Finishing Industry
1. Sulfate Analysis in a Non-Aqueous Media
2. Analysis of Nitrate and Sulfate Collected on Air Filters
3. Trace Sulfate and Phosphate in Brine
4. Analysis of Engine Coolants Using Ion Chromatography
5. Analysis of Oxalate in Industrial Liquors
6. Analysis of Ammonia and Amines in Dimethylformamide
7. Separation of Large Anions
8. Sulfur and Nitrogen in Soil Extracts
9. Analysis of Organic and Inorganic Ions in Milk Products
10. Analysis of Small Aliphatic Carboxylic Acids
11. Analysis of Industrial Wastestreams
12. Analysis of Ions in Flue Gas Scrubber Solution
13. Determination of Thermal Decomposition Products of Polymeric Materials
14. Azide Determination in Complex Matrices
15. Anion Analysis of Fuel Combustion Products
16. Determination of Dibutylphosphoric Acid (DBP) in Nuclear Reprocessing Streams
17. New Improved Separation of Divalent Cations
18. Determination of Perchlorate
19. Analysis of Strong and Weak Acids in Coffee Extracts
20. Determination of Sulfate and Chromate in Electroplating Baths
21. Analysis of Organic Acids and Cationic Species in Wine
22. Determination of As^{+III} and As^{+V}
23. Determination of Monochloracetyl Chloride
24. Determination of Formaldehyde as Formate Ion
25. Analysis of Inorganic Anions and Organic Acids in Carbonated Beverages
26. Routine Determination of Chromate
27. Analysis of Sulfur Dioxide in Malt
28. Determination of Inorganic Acids in Pickling Baths and Contaminants in Fresh Mineral Acids
29. Analysis of Hypochlorite and Chlorate in Bleach Using Combined Conductivity and Electrochemical Detection
30. Determination of Sulfur Species and Oxalate in Kraft Liquors Using Ion Chromatography Combined with Electrochemical Detection
31. Determination of Anions in Acid Rain
32. Determination of μg/L Levels of Chloride in High Purity Methanol
33. Determination of Ionic Contaminants in Encapsulation Plastics

Table 9.7 (*continued*)

34.	Determination of Ionic Contaminants in DI Water, Process Water, Aqueous Rinses, and Extracts of Semiconductor Devices
35.	Determination of Trace Ionic Contaminants in Electronic Chemicals and Toxic Vapors in Industrial Atmospheres
36.	Determination of Oxalate in Urine
37.	Determination of Iodide by Ion Chromatography
38.	Analysis of Infant Formula by Ion Chromatography
39.	Determination of Ethanolamines in Refinery Water
40.	Separation of Gold Complexes
41.	Determination of Major Constituents in Aluminum Treating Baths
42.	Determination of Aluminum by Ion Chromatography
43.	Determination of Organic Acids in Vinegar
44.	Determination of Sequestering Agents
45.	Fatty Acid Analysis
46.	Ion Chromatography: A Versatile Technique for the Analysis of Beer
47.	Determination of Glucose and Galactose in Serum
48.	Determination of Uranium by Ion Chromatography
49.	Ion Chromatography Applied to Ni and Ni Alloy Plating Baths
50.	Copper Plating Solution Analysis by Ion Chromatography

since IC is a post-1970 development, literature references have been placed in computer memory and are available over telephone lines as discussed in Chapter 10. Several excellent monograms (42–44) and review articles (2,45–48) are available. Vendor companies in ion chromatography maintain literature files and application laboratories; a telephone call to a vendor may rapidly solve diverse problems in IC. Table 9.7 illustrates the kind of analytical information that is available. Lastly, one literature source that is frequently overlooked is the doctoral dissertation. These are current and available at nominal cost as microfilm or photostats from Dissertation Abstracts, University of Michigan.

9.5.2 Range of Applications

Several hundred ions have been determined by ion chromatography. Many of these ions are listed in Tables 9.8 and 9.9. Analytical protocols are available for ions in a wide variety of matrices and some examples are in Table 9.7. Since the invention of ion chromatography, the perception of the analyst towards ion analysis by column chromatography has changed. The analyst no longer associates fraction collection with chromatography of ions. The analyst now thinks of continuous instrumental

Table 9.8 Inorganics Determined by Ion Chromatography

Inorganic Ions		Inorganic Complexes
Aluminum	Mercury	Chrome EDTA
Ammonium	Molybdate	Cobalt EDTA
Arsenate	Monofluorophosphate	Copper EDTA
Arsenite	Nickel	Lead EDTA
Azide	Nitrate	Nickel EDTA
Barium	Nitrite	Cobalt Cyanide
Borate	Percholorate	Gold (I, II) Cyanide
Bromide	Periodate	Iron (II, III) Cyanide
Cadium	Phosphite	Palladium Cyanide
Calcium	Platinum	Platinum Cyanide
Carbonate	Potassium	Silver Cyanide
Cesium	Pyrophosphate	
Chlorate	Rhenate	
Chlorite	Rubidium	
Chromate	Selenate	
Cobalt	Selenite	
Copper	Silicate	
Cyanide	Sodium	
Cyanate	Strontium	
Dithionate	Sulfate	
Fluoride	Sulfide	
Gold	Sulfite	
Hydrazine	Tetrafluoroborate	
Hypochlorite	Thiocyanate	
Hypophosphite	Thiosulfate	
Iodate	Tripolyphosphate	
Iodide	Tungstate	
Iridium	Uranium	
Iron (II, III)	Vanadate	
Lead	Zinc	
Lithium		
Magnesium		

detection (conductivity, amperometry, atomic absorption, spectrophotometry, and fluorometry) of the effluent ions after ion exchange, ion exclusion or reversed-phase separation with or without eluent suppression. Ion analysis can also be extended to nonionic compounds, which can be converted to an ion or which can behave as ions in solution.

The general application areas of IC are shown in Fig. 9.17. Anions and cations are hydrophobic or hydrophilic. The anions are from weak (pK_a

Table 9.9 Organics Ions Determined by Ion Chromatography

Class	Examples
Amines	(Methyl amine, diethanolamine)
Amino acids	(Alanine, threonine, tyrosine)
Carboxylic acids	(Acetate, oxalate, citrate, benzoate, trichloroacetate)
Carbohydrates	(Lactose, sucrose, xylitol, cellobiose, maltononose)
Chelating agents	(EDTA, NTA, DTPA)
Quaternary ammonium compounds	(Tetrabutylammonium ion, cetylpyridinium ion)
Nucleosides	(Adenosine monophosphate, guanidine monophosphate)
Phenols	(Phenol, chlorophenol)
Phosphates	(Dimethylphosphate)
Phosphonates	(Dequest 2000®, Dequest 2010®)
Phosphonium compounds	(Tetrabutylphosphonium ion)
Sulfates	(Lauryl sulfate, lauryl sulfate)
Sulfonates	(Linear alkyl benzene sulfonate, hexane sulfonate)
Sulfononium compounds	(Trimethylsulfonium ion)
Vitamins	(Ascorbic acid)

> 7) or stronger acids ($pK_a < 7$). Based on these properties, the analyst selects the separation mode: IC, ICE, or MPIC. The subsequent selection of detector depends upon the analyte ions and the separation mode. Detection modes also are summarized in Fig. 9.17.

Because sample preparation is minimal, usually consisting only of dilution and filtration for liquid samples, and because several ions are determined at high sensitivity at rates of approximately one ion per minute, the method has been applied to hundreds of ions in a variety of fields: chemical, energy, environmental, electronics, plating, food, geology, industrial hygiene, medicine, paper, petrochemical, textile, and others.

The ability to closely monitor several ions can be of great benefit in an industrial setting; for example, plating baths containing expensive metals can be monitored to increase efficiency, thus minimizing the cost associated with plating and manufacturing with defective parts.

Any ion in any sample can now be thought of as a target for ion chromatography. Many of these applications are summarized in the primary and secondary literature (42–48). We here select a few of the lesser-known published applications to illustrate the versatility of the technique. Many of the following include sample preparation that can be used to increase the usefulness of IC.

Determination of Borate

Boric acid is too weak to be determined at high sensitivity by standard IC. Chemical reaction of borate with 10% HF to form BF_4^- is done after concentrating borate on Amberlite XE-243 (49). The lower limit for the resulting ion chromatographic method is 0.05 mg/L, and analysis was possible in the range 0.05–500 mg/L with a relative standard deviation of 0.5–2.6%. The inherent sensitivity of standard IC was thus achieved by chemical reaction. Now borate is determined directly using an ICE column with octane sulfonic acid eluent, anion fiber suppressor, and tetramethylammonium hydroxide regenerant. The minimum detectable quantity of borate is 0.03 mg/L. This method eliminates the sample preparation step that forms BF_4^-.

Determination of Organic Compounds

Standard IC with conductivity detection can determine organic compounds if they are ionized or can form ionized compounds. Organic acids in body fluids have been determined (50). Aldehydes have been oxidized to acids and then determined by standard IC (51). Chloroacetyl chloride in industrial air is collected in silica gel sampling tubes, desorbed by $NaHCO_3$, and the resulting acid then determined by ion chromatography (52). Previously, the acetylchloride was converted to an ester and then determined by electron capture gas chromatography. The new useful limit of detection is about 0.4 μg/ml or about 0.01 ppm for a 20-L air sample. Sensitivity can be improved when necessary by increased air sampling volume or increased injection volume.

More recently, MPIC has been applied to the determination of sulfonates, phosphonates, and amines, and pulsed amperometric detection combined with ion-exchange enables the determination of carbohydrates. Amino acid analysis has also been greatly improved with the use of pellicular cation-exchange resins.

Determination of Silica

The determination of silica at low concentrations, less than 1 ppm, in commercial borate products by the conventional colorimetric measurement of the silicomolybdate complex is time-consuming and subject to chemical interferences. Improvement by direct application of ion chromatography is not feasible. Ion chromatography following chemical reaction of SiO_2 with HF to form H_2SiF_6 allowed detection down to 0.010 ppm SiO_2, with linearity extending at least to 20 ppm SiO_2 (53). More recently, SiO_2 is determined by applying the wet chemical procedure on an ion chromatograph fitted with a post-column reactor to form the absorbing complex after ion-exchange separation. Detection limits are approximately 10 ppb.

Determination of Sulfide and Cyanide

In the past, methods for S^{2-} and CN^- were subject to interferences. Using ion-exchange separation with amperometric detection rather than conductimetric detection, analyses in basic solution eliminate formation of H_2S and HCN. Detection limits are in the ppb range (54).

Determination of EDTA and NTA

Many products (e.g., iodine supplement solutions) have EDTA present as a preservative, and analysis is necessary for quality control. EDTA and NTA are strongly retained on ion exchangers. Two methods have been developed for determining EDTA (55,56), and one for NTA (56). In the former, using an eluent of 0.004 *N* HCl and 0.025 *M* Zn^{2+}, EDTA chromatograms with acceptable peak shape are obtained. EDTA enters the conductivity detector as the free base and is determinable at the 1 ppm level.

The second method uses a specially formulated ion-exchange resin that resolves EDTA, NTA, and a number of polyphosphates before postcolumn addition of $Fe(NO_3)_3$ to permit absorbance detection at 330 nm.

Determination of HCN in Air

Environmental air analysis is done by trapping the HCN in an alkaline impinger and converting cyanide to the more acidic formate. Analysis by standard IC gives 100% recovery at the ppm level with a relative standard deviation of 2.5% (57).

Determination of Chloride in Cadmium Sulfide

A pure substance doped with a small amount of inorganic additive is an ideal sample for ion chromatography. Here 100 mg of CdS is dissolved in 100 ml of 1:1 mixture of carbonate buffer solution (0.003 *M* $NaHCO_3$ and 0.0018 *M* Na_2CO_3) and 30% H_2O_2. An aliquot is then injected directly into the ion chromatograph for analysis (58).

9.5.3 Analytical Techniques

For difficult determinations, analysts improve protocols for IC analyses by utilizing some of the following techniques.

Pre-Column Reactions

SO_2 can be determined by conversion to SO_3^{2-} followed by ion chromatography. The direct method is subject to error because of oxidation to SO_4^{2-}. A different reaction has been used wherein SO_2 reacts directly with formaldehyde to yield hydroxymethane sulfonic acid (59), which is then determined by ion chromatography. The determination of aldehydes by oxidation to organic acids is another example of pre-column reactions that lead to ion chromatography. These applications indicate the potential value of pre-column reactions in IC method development or method improvement.

Post-Column Reactions

A limitation of any analytical protocol is the sensitivity of the detector. If a signal is inherently low, or low because of small concentrations, the value of the procedure is compromised. Chemical conversion of the analyte, however, can lead to new protocols. Examples already accomplished include:

1. Reaction of primary amines with o-phthaldehyde in the presence of 2-mercaptoethanol to form fluorescent derivatives.
2. Reaction of eluting separated cations with spectrophotometric reagents such as PAR.
3. Reaction of anionic sequestering agents with Fe^{III} to form detectable colored complexes (56).

Microchemical Analysis

Analysis for the heteroatoms Cl, Br, P, S in milligram samples is time-consuming using previously available techniques, particularly so when chemical separations are necessary. Ion chromatography incorporates separation and universal detection in a simple procedure: 5–10 mg of the sample is oxidized in a 500-ml Schöniger flask containing 10 ml of distilled water plus 3 drops of 30% H_2O_2. After reaction, the solution is diluted to 100 ml and 0.100 ml injected into the ion chromatograph with appropriate operating parameters (60). Ion chromatography for microanalysis gives results comparable to those of a laboratory specializing in organic analysis using classical precipitation methods (61). The great advantage of ion chromatography for determining heteroatoms is the sensitivity of the method. If homogeneity is not a sampling problem, then less than one mg is required for Schöniger combustion. Indeed, the limitation is the presence of trace heteroatoms in the filter paper flag in the Schöniger combustion (61) and not in ion chromatography per se.

Ion Chromatography with Detection by Atomic Absorption

Several arsenic compounds have been determined using a standard IC separation followed by detection by atomic absorption. These include arsenites (AsO_3^{3-}), arsenates (AsO_4^{3-}), monomethyl arsonates ($CH_3AsO_3^{2-}$), dimethylarsinates ($(CH_3)_2AsO_2^-$), and *p*-aminophenylarsonates (p-$NH_2(C_6H_4)AsO_3^{2-}$). These compounds are important as herbicides, pesticides, and wood preservatives. Millions of pounds are used annually and methods to determine each arsenic compound, rather than total arsenic, are necessary. For this determination, compound separation occurs in an ion-exchange column, and detection is by atomic absorption after reduction to arsine, AsH_3 (62). The method has a sensitivity of at least 3 ng/ml for methanearsonate and 20 ng/ml for arsenate, and is faster and more sensitive than previous ion chromatography/atomic absorption methods (63,64).

9.6 ANALYTICAL PERSPECTIVE

Since its introduction in 1975, ion chromatography has evolved into a major analytical technique. The original reason for its success was the ability to rapidly and simultaneously determine several inorganic anions at high sensitivity with minimal sample pretreatment, which gave it a decisive advantage over previous techniques such as manual or automatic wet chemistry.

Once the value of IC became recognized, the continuous development of additional columns and detectors greatly expanded the usefulness of the technique. ICE and MPIC added the capability to determine organic ions. Additional detectors have enabled the determination of weakly ionized species, transition metals, carbohydrates, and amino acids.

With the development of ion chromatography, advantages inherent in modern chromatographic analysis were introduced into inorganic analysis. These advantages include chromatographic resolution (selectivity) for multicomponent analysis, ease and speed of multiple analysis, sensitivity to at least the ppb level, and on-line compatibility with automation. For example, the determination of nine ions (F^-, $HCOO^-$, BrO_3^-, Cl^-, NO_2^-, HPO_4^-, Br^-, NO_3^- and SO_4^{2-}) in Fig. 9.5 at μg/ml concentrations only 6 minutes after sample injection, can be a routine procedure in ion chromatographic analysis. These advantages will increase in degree as the improvements detailed above and in the literature are incorporated into analytical procedures.

REFERENCES

1. H. Small, T.S. Stevens, and W.C. Bauman, *Anal. Chem.*, **47**, 1801 (1975).
2. J.C. MacDonald, *Amer. Lab.*, **11**, (1), 45 (1978).
3. C. Anderson, *Clin. Chem.*, *22*, 1424 (1976).
4. J.F. Colaruotolo and R. Eddy, *Anal. Chem.*, **49**, 824 (1977).
5. J.W. Wimberly, *Anal. Chem.*, **53**, 2137 (1981).
6. F.F. Cantwell and S. Pugh, *Anal. Chem.*, **51**, 623 (1979).
7. R.S. Deelder et al., *J. Chromatogr.*, **149**, 669 (1978).
8. J. Riviello and C. Pohl, "Ion Chromatography at Transition 'Metals, Paper No. 239, 25th Rocky Mtn. Conf. of Analytical Chemistry, Denver, Colorado, Aug. 1983.
9. H. Freiser and J.R. Jezorek, *Anal. Chem.*, **51**, 366 (1979).
10. R.M. Cassidy and S. Elchuk, *Anal. Chem.*, **51**, 1434 (1979).
11. J.S. Fritz and J.M. Story, *Anal. Chem.*, **46**, 825 (1974).
12. K.A. Kraus and G.E. Moore, *J. Am. Chem. Soc.*, **75**, 1460 (1953).
13. J.F. Huber and J.C. Krauk, *Anal. Chem.*, **44**, 1554 (1972).
14. R.W. Frei and V. Miletukova, *J. Chromatog.*, **47**, 427 (1970).
15. R.W. Moshier and R.E. Sievers, *Gas Chromatography of Metal Chelates*, Pergamon Press, Oxford, 1965.
16. E.L. Johnson and C.A. Pohl, *J. Chromatogr. Sci.*, **18**, 442 (1980).
17. A.E. Martell and R.M. Smith, *Critical Stability Constants*, Vol. 1–4, Plenum Press, New York, 1976.
18. W.H. Delphin and E.P. Horwitz, *Anal. Chem.*, **50**, 843 (1978).
19. M.D. Seymour and J.S. Fritz, *Anal. Chem.*, **45**, 1394 (1973).
20. J. Riviello and C. Pohl, *Busenki Kagaku*, in press.
21. E. Wanninen, *Essays on Analytical Chemistry*, p. 351, Pergamon Press, Oxford, 1977.
22. R.P.W. Scott, *Liquid Chromatography Detectors*, Elsevier Scientific Publishing Co., New York, 1977.
23. R. Woods, L. Cummings, and T. Jupille, *J. Chromatgr. Sci.*, **18**, 551 (1980).

24. V.T. Turkelson and M. Richards, *Anal. Chem.*, **50**, 1420 (1978).
25. P. Jandera and J. Churacek, *J. Chromatogr.*, **197**, 181 (1980).
26. R.C. Meyer, *J. Pharm. Sci.*, **69**, 1148 (1980).
27. R. Rocklin and E.L. Johnson, *Anal. Chem.*, **55**, 4 (1983).
28. D.T. Gjerde, J.S. Fritz, and G. Schmuckler, *J. Chromatogr.*, **186**, 509 (1979).
29. D.T. Gjerde, G. Schmuckler, and J.S. Fritz, *J. Chromatogr.*, **187**, 35 (1980).
30. K.M. Roberts, D.T. Gjerde, and J.S. Fritz, *Anal. Chem.*, **53**, 1691 (1981).
31. R.M. Cassidy and S. Elchuk, *Anal. Chem.*, **54**, 1558 (1982).
32. F.A. Buytenhuys, *J. Chromatogr.*, **218**, 57 (1981).
33. H. Small and R.E. Miller Jr., *Anal. Chem.*, **54**, 462 (1982).
34. A. Laurent and R. Bourdon, *Am. Phar. Fr*, **36**, (9–10), 453 (1978).
35. R.A. Cochrane and D.E. Hillman, *J. Chromatogr.*, **241**, 392 (1982).
36. M. Denkert, L. Hackzell, G. Shill, and E. Sbgren, *J. Chromatogr.*, **218**, 31 (1981).
37. I.M. Kolthoff, E.B. Sandell, E.J. Meehan, and S. Bruckenstein, *Quantitative Chemical Analysis*, 4th ed., Macmillan, Toronto, 1969.
38. R.L. Pecsok, L.D. Shields, T. Cairns, and I.G. McWilliam, *Modern Methods of Chemical Analysis*, 2nd ed., Wiley, New York, 1976.
39. D.A. Skoog and D.M. West, *Fundamentals of Analytical Chemistry*, 2nd ed., Holt, Rinehart & Winston, New York, 1969.
40. J.A. Dean (Ed)., *Lange's Handbook of Chemistry*, 11th ed., McGraw-Hill, New York, 1973.
41. R.C. West, (Ed)., *Handbook of Chemistry and Physics*, 53rd ed., Chemical Rubber Co., Cleveland, 1972.
42. E. Sawicki, J.D. Mulik, and E. Sittgenstein, *Ion Chromatographic Analysis of Environmental Pollutants*, Ann Arbor Science Publishers, Ann Arbor, Michigan, 1979.
43. F.C. Smith Jr. and R.C. Chang, *The Practice of Ion Chromatography*, Wiley, New York, 1982.
44. J.S. Fritz, D.T. Gjerde, and C. Pohlandt, *Ion Chromatography*, Hüthig, New York, 1982.
45. R.A. Wetzel, *Env. Sci. Tech.*, **13**, 1214 (1979).
46. R.A. Wetzel, *Ind. Res. Devel.* **24**(4), 92 (1982).
47. F.C. Smith Jr. and R.C. Chang, *CRC Crit. Rev. Anal. Chem.*, August 1980.
48. C.A. Pohl and E.L. Johnson, *J. Chromatogr. Sci.*, **18**, 442 (1980).
49. C.J. Hill and R.P. Lash, *Anal. Chem.*, **52**, 24 (1980).
50. W. Rich, E. Johnson, L. Lois, P. Kabra, B. Stafford, and L. Morton, *Clin. Chem.*, **26**, 1492 (1980).
51. J. Lorrain, C.R. Fortune, and B. Dellinger, *Anal. Chem.*, **53**, 1302 (1981).
52. P.R. McCullough and J.W. Worley, *Anal. Chem.*, **51**, 1120 (1979).

53. J.P. Wilshire, Abstract No. 244, Pittsburgh Conference on Analytical Chemistry and Applied Spectroscopy, 1982.
54. R. Rocklin and E. Johnson, *Anal. Chem.,* **55,** 4 (1983).
55. R.C. Buechele and D.J. Reutter, *Anal. Chem.,* **54,** 2113 (1982).
56. Dionex Application Note No. 44.
57. T.W. Dolzine, G.G. Esposito and D.S. Rinehart, *Anal. Chem.,* **54,** 470 (1982).
58. W.F. Koch and J.W. Stolz, *Anal. Chem.,* **54,** 340 (1982).
59. P.K. Dasgupta, K. DeCesare, and J.C. Ulrey, *Anal. Chem.,* **52,** 1912 (1980).
60. J.F. Colaruotolo and R.S. Eddy, *Anal. Chem.,* **49,** 824 (1977).
61. F.C. Smith Jr., A. McMurtrie, and H. Galbraith, *Microchem. J.,* **22,** 45 (1977).
62. R. Dagani, *Chem. Eng. News,* Feb. 16, 1981, p. 29.
63. F.E. Brinckman, *J. Chromatogr.,* **191,** 31 (1980).
64. E.A. Woolson, *J. Assoc. Off. Anal. Chem.,* **63,** 523 (1980).

CHAPTER

10

COMPUTER ONLINE DATABASE LITERATURE SEARCHING AND EXAMPLES

EVELYN R. SAVITZKY*

Perkin-Elmer Corporation
Norwalk, Connecticut

JOAN T. OVERFIELD

Fairfield University
Fairfield, Connecticut

10.1 INTRODUCTION

The chemical literature explosion is clearly illustrated by Barker's (1) statistics, which announce that in 1980 *Chemical Abstracts* published 475,739 abstracts of papers and patents plus 72,937 patent equivalents for a total of 548,676 cited documents. While a scientist's specific area of research naturally restricts the universe of relevant documents, simply coping with the sheer bulk of literature output is a formidable task when one attempts to distill the pertinent information. Online database literature searching offers a viable solution to the researcher's quest for both current and retrospective documents, and is a vital enhancement to traditional information-gathering methods.

10.2 HOW LITERATURE SEARCHING IS DONE BY THE SCIENTIST

The active research scientist maintains a collection of information sources pertinent to current research projects as well as those planned for the future, according to Dillon (2). Sources may include journal articles, extracts from books, technical reports, reprints, correspondence, personal evaluations, and annotations. The scientist may also gain information from the "invisible college," in which ideas are communicated by experts through face-to-face discussions at seminars, conferences, meetings, and

* Present address: Silvermine Associates, Wilton, Connecticut.

classroom discussions. A recognized scientist may be called upon to referee articles submitted to journals and thus gain information that others will not learn of until several months later. That person might also referee proposals for research grants or contracts submitted to federal agencies such as the National Science Foundation. The scientist may be active in a new area of interest in which a very select group share their interests by exchanging preprints or publishing newsletter reports, according to Arnett (3). The traditional use of the chemical literature (4) is no longer sufficient.

Maintaining a record-keeping system for these various information sources is both time consuming and difficult. The information gathered may be either current or retrospective. Online literature searching cannot replace some components, such as the "invisible college" or refereeing, but it can fill other needs for current or retrospective awareness. The three major vendors of online services Lockheed (Dialog), Systems Development Corporation (SDC Orbit) and Bibliographic Retrieval Services, Inc. (BRS) offer SDI (Selective Dissemination of Information) services that consist of regularly scanning new bibliographic data as they are entered in the data files, comparing this data with the interest profiles entered by the researcher, and printing the results on a regular basis. The researcher entering a profile is automatically alerted to new documents meeting interest requirements. SDI vastly reduces the manual searching time needed in order to keep abreast of current developments.

A scientist may wish to do a retrospective search to be sure an important document is not missing in a personal research collection. A retrospective search also functions as a thorough literature search for a scientist embarking upon a new area of interest.

10.3 TYPES OF INFORMATION ACCESSIBLE ONLINE

It is an often-repeated fact that new information on any technical subject appears in the journal literature. However, many other types of literature play an important part in a scientist's search of the literature. Though books cannot be as current as journal articles, they do play an important role. They are essential when looking for overall coverage of a subject up to a particular point in time or when looking for background information on a subject. The journal literature can then be used to update the material in a book or to treat a specific aspect of a subject in greater detail. Edited books with chapters by different authors often bring together many specialists on a subject, each dealing with the one aspect of the subject that he knows best. Some databases are beginning to index edited books.

Another important form of review of a subject is the review article found in many journals. These are long articles, often with extensive bibliographies, which cover a subject for a particular length of time—whether for a year, five years, or an even longer time. Many online databases make it possible, by special codes or the word "review," to list the review articles available. Such items, coupled with the specific subject, will retrieve just the articles needed.

The proceedings of symposia are also a very important source of current information on a subject. The papers at a symposia are often an advanced look at work in progress; the authors of these papers are important researchers in their field and often worth following, as well as checking retrospectively. Such papers have always been difficult to access manually, but now several of the online databases index proceedings in depth. There is one particular file in Dialog, File 77, Conference Papers Index, that has listings from 1973 to the present.

The report literature, which consists of work reported on as the result of contracts, usually with government agencies, is also an important part of the body of current literature on a subject. Agencies that index such reports are DTIC (Defense Technical Information Center), NASA (National Aeronautic and Space Administration), NTIS (National Technical Information Services), and SSIE (Smithsonian Scientific Information Exchange). The latter does not reference reports, but indicates work in progress under various agency grants. The information on the Dialog SSIE file includes the size of the grant, the chief program scientists and managers, the location where the work is being done, the length of the contract, and a description of the project. This can be a very important source of technical information as well as an important marketing tool. NTIS indexes all unclassified and unlimited reports done under government contract. NTIS will cover some NASA and some DTIC materials. NASA and DTIC can be accessed directly by representatives of government contractors who have received training on these special databases, and cover not only the reports in NTIS but material that is limited or classified. The NASA database also includes IAA (International Aerospace Abstracts) and the book and report holdings of the various NASA centers.

10.4 HOW TO ACCESS INFORMATION ONLINE

Online database searching theory involves a few basic concepts of operation and uses Boolean logic, which is simply the logic of sets. Hoover

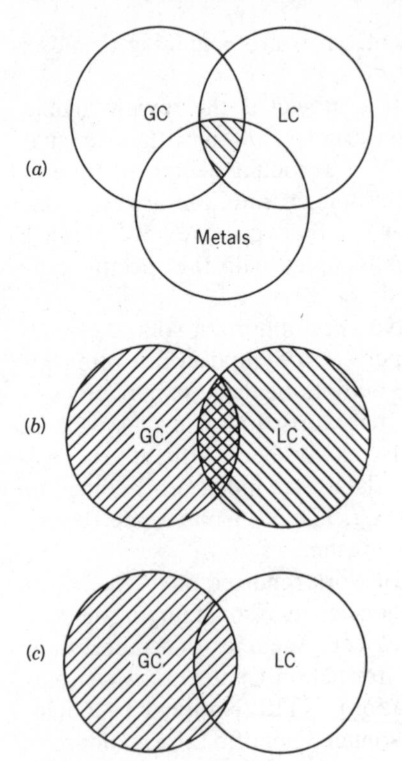

Figure 10.1. Venn diagrams for LC, GC and metals. (a) AND; (b) OR, (c) NOT.

(5) states "set logic combines sets to include or exclude the concept of one set from another." The logical operators are "and," "or," and "not." A researcher composes a list of terms, phrases, compounds, registry numbers, or procedures, and a search is structured combining the selected terms and the Boolean operators to form search statements. Examples of Boolean statements are:

1. AND Find a set that includes items from sets, GC, LC, and metals.
2. OR Find a set that includes either GC, LC, or both.
3. NOT Find a set that includes GC, but specifically excludes LC.

Visually, these sets may be shown by overlapping circles called Venn diagrams, as in Fig. 10.1 for GC, LC, and metals.

These are simple logical processes most people employ in daily thought. Some systems also employ "relational operators" requiring certain word or term proximity, such as "adjacent to," "in the same field," "equal to," or "greater than."

Online database searching is an interactive process. As the terminal operator enters a search statement, the computer responds with the number of "hits" or "postings" located. Further statements may then be entered to expand upon or modify the original statement, or a totally new concept may be typed in. The operator may choose to display or type out a few of the citations to check the strategy, or print them all and end the search. The operator may also store the search in the computer for later use, request the search be run as an SDI, as previously explained, or have the citations printed remotely and mailed. The operator may also wish to order copies of some of the cited documents or articles immediately, and this too may be done while still online.

10.4.1 Setting Up a Search

Due to the vast array of databases available and the various peculiarities of each one, it may be most expeditious for the scientist to use an information specialist as an intermediary, rather than to become personally proficient at online searching. Communication becomes the vital element in this association. The scientist must indicate clearly stated needs so that the information specialist can then select the best data files and adapt the question to the specific procedures indicated in the database file guidelines.

The information specialist must not only know how to search the large number of on-line files now available but must also know how to conduct what is known as the "presearch reference interview." With the advent of computer searching, the reference interview has become even more important than before, because omissions of terms or other parameters or an unclear understanding of the subject cannot be corrected at the terminal without increasing cost. The user must understand the limitations of a computer search, and the searcher must have a clear understanding of the problem.

The user should be briefed on the use of Boolean logic as a means of searching online, as well as how the particular databases to be used are accessed. For example, some databases include abstracts, and all words in the abstract can be searched. In other databases this is not possible. The user should know the time span covered by the particular databases to be used, indexing policies, subject coverage, languages covered, and types of documents that can be recovered. It is also important to discuss with the user the difference between free text searching and searching using the controlled vocabulary terms listed in the various thesauri available. Many databases also rely heavily on many types of classification codes and concept codes.

The searcher, on the other hand, should have some knowledge of the subject before setting up the search strategy with the user. This will be important as one explores all possible synonyms, concepts, and search strategies worked out in advance, in case the search yields too few or too many references. The searcher also must know enough about the subject to suggest the various databases to be searched and how they should be restricted, whether by date, language, type of document, and so forth.

When performing a complicated search, it may be necessary for user and searcher to clarify the search strategy before going online, or to run a test search before refining the strategy. An experienced searcher might suggest that the user be at the terminal at the time of the search. This enables the user to understand the search process, to be aware of decisions that need to be made online, and to help make these decisions at the time needed. The user can also check the relevancy before off-line prints are ordered (6).

Under a successful collaborative approach, the scientist and the librarian or information specialist combine their individual areas of expertise to produce the citations that satisfy the criteria originally set forth in the search statement.

There are many advantages to online database services for the scientist. Most obvious is the speed at which vast amounts of literature may be searched. Another is defining the specificity and/or flexibility of the data by the combination or exclusion of terms. Currentness is another advantage, for most material is online before the printed counterparts are published. Hoover (5) also cites comprehensiveness, as a library may have access to more online databases than its book budget can support for the printed counterparts.

All that is necessary to begin a search is a terminal and a regular telephone, so that searches may be conducted anywhere that is convenient. While database connect charges range from $16–$300 an hour, the system will still be cost-effective in comparison to work-hours of manual searching.

Until systems become more "user friendly," the necessity of a trained intermediary will remain a disadvantage for some scientists. One must also remember that the computer can only match the terms one specifies. It cannot make value judgments about the relevancy of the retrieved documents to the user's needs. Also, some researchers do not think of their own time in money terms and consider the service too expensive. Naturally, scientists must weigh the advantages and disadvantages of on-line

```
File411:DIALINDEX(tm)

? b411; s files 308,320,311,309,5,10,12,13,19,23,24,25,28,34,35,40,41,6,65,50,
51,55,60,72,73,74,76,77,79,94,110,115,125,132,136,223,224,225

File5:BIOSIS Previews 81-84/Feb BA7703;RRM2603
File6:NTIS - 64-84/Iss02
File10:AGRICOLA - 1979-83/Oct & 1979 Supplemental
File12:INSPEC - 1969 thru 1976
File13:INSPEC - 77-84/Iss02
File19:Chem. Industry Notes 1974-83Iss52
File23:CLAIMS/US Patent Abstracts 1950-1970
File24:CLAIMS/US Patent Abstracts 1971-1981
File25:CLAIMS/US Patent Abstracts Jan 1982 - Oct 1983
File28:OCEANIC ABSTRACTS - 64-84/Jan
File34:SCISEARCH - 81-83/Wk46
File35:DISSERATION ABSTRACTS ONLINE 1861 TO Dec 83
File40:ENVIROLINE - 71-83/Dec
File41:Pollution Abstracts - 70-83/Nov
File50:CAB Abstracts - 72-83/Nov
File51:FSTA - 69-84/Jan
File55:BIOSIS Previews - 1977 thru 1980
```

Figure 10.2. File descriptions using DIALINDEX.

retrieval systems against their own needs, but the currency, comprehensiveness, and speed offered by online database systems cannot be ignored.

10.4.2 Examples Involving Inorganic Chromatographic Analysis

In exploring the various databases applicable to this volume, we chose to work with the vendor known as Dialog Information Services, Inc. Of over 180 databases, we chose 38 that experience has shown cover chemical information. These consisted of the following Dialog files in Fig. 10.2.

In order to see how many postings each database contains and, therefore, determine inexpensively if it is worth exploring, the vendor provides a file called DIALINDEX. By entering the list of files and a one-statement strategy it is possible to get the number of postings in each file in advance.

A simple statement that combined (inorganic or metallic or organometallic) and (chromatography or chromatographic or ion exchange)

```
File60:CRIS/USDA - 75-83/Nov
File65:SSIE Current Research - 78-82/Feb
File72:EMBASE (Excerpta Medica) 80-83/Iss39
File73:EMBASE (Excerpta Medica) In Process 83/Wk42
File74:IPA - 70-83/Oct!(Copr. ASHP)
File76:Life Sciences Collection - 78-83/Sep(Copr. Cambridge Scientific Abs.)
File77:Conference Papers Index - 73-83/Oct
File79:FOODS ADLIBRA - 74-83/Sep
File94:SCISEARCH - 78-80
File110:AGRICOLA - 70-78/Dec
File115:Surface Coatings Abstracts - 76-83/Dec
File125:CLAIMS/US PATENT ABSTRACTS WEEKLY - Nov 1 1983 TO Dec 13 1983
File132:Standard & Poors Daily News, 79-84
File136:Federal Register - Mar 1977 to Dec 21 1983
File223:CLAIMS/UNITERM - 1950-1970
File224:CLAIMS/UNITERM - 1971-1981
File225:CLAIMS/UNITERM
File308:CA Search - 1967-1971
File309:CA Search - 1972-1976
File311:CA Search 1982-83 UD=09926
File320:CA Search - 1977-1979
```

Figure 10.2. ***(continued)***

showed immediately whether a file was worth further exploration. In File 308, for example, the postings appeared as follows:

Set Number	Number of Postings	Terms
1	51112	Inorganic
2	8239	Metallic
3	13822	Organometallic
4	13365	Chromatography
5	8386	Chromatographic
6	5904	Ion exchange
7	2970	(1 or 2 or 3) and (4 or 5 or 6)

The postings of set number 7 in the 38 files chosen ranged from a high in File 308 (Chem Abstracts 1967 through 1971) of 2970 postings to a low

```
? ss (inorganic or metallic? or organometallic?) and (chromatography or chromato
graphic or ion(w)exchange)
        (5)  BIOSIS Previews 81-84/Feb BA7703;RRM2603
              3097 INORGANIC
               506 METALLIC?
                34 ORGANOMETALLIC?
             25669 CHROMATOGRAPHY
              5938 CHROMATOGRAPHIC
              2306 ION(W)EXCHANGE
               150  (1 OR 2 OR 3) AND (4 OR 5 OR 6)
        (6)  NTIS - 64-84/Iss02
             11303 INORGANIC
              5142 METALLIC?
              1168 ORGANOMETALLIC?
              5646 CHROMATOGRAPHY
              2477 CHROMATOGRAPHIC
              2564 ION(W)EXCHANGE
               626  (1 OR 2 OR 3) AND (4 OR 5 OR 6)
       (10)  AGRICOLA - 1979-83/Oct & 1979 Supplemental
              1367 INORGANIC
                77 METALLIC?
                32 ORGANOMETALLIC?
              2325 CHROMATOGRAPHY
              1150 CHROMATOGRAPHIC
               176 ION(W)EXCHANGE
                 7  (1 OR 2 OR 3) AND (4 OR 5 OR 6)
       (12) INSPEC - 1969 thru 1976
             28118 INORGANIC
              8980 METALLIC?
               419 ORGANOMETALLIC?
              1025 CHROMATOGRAPHY
               364 CHROMATOGRAPHIC
               499 ION(W)EXCHANGE
                54  (1 OR 2 OR 3) AND (4 OR 5 OR 6)
       (13) INSPEC - 77-84/Iss02
             32209 INORGANIC
             15546 METALLIC?
               673 ORGANOMETALLIC?
              1494 CHROMATOGRAPHY
               440 CHROMATOGRAPHIC
               910 ION(W)EXCHANGE
               147  (1 OR 2 OR 3) AND (4 OR 5 OR 6)
```

```
(19)  Chem. Industry Notes 1974-83Iss52

      1565 INORGANIC
       715 METALLIC?
        33 ORGANOMETALLIC?
       101 CHROMATOGRAPHY
        18 CHROMATOGRAPHIC
       392 ION(W)EXCHANGE
        10  (1 OR 2 OR 3) AND (4 OR 5 OR 6)

(23)  CLAIMS/US Patent Abstracts 1950-1970

      4929 INORGANIC
      7809 METALLIC?
       495 ORGANOMETALLIC?
       295 CHROMATOGRAPHY
       351 CHROMATOGRAPHIC
      1044 ION(W)EXCHANGE
        92  (1 OR 2 OR 3) AND (4 OR 5 OR 6)

(51)  FSTA - 69-84/Jan

      1182 INORGANIC
       502 METALLIC?
        10 ORGANOMETALLIC?
     12730 CHROMATOGRAPHY
      4473 CHROMATOGRAPHIC
      1936 ION(W)EXCHANGE
       138  (1 OR 2 OR 3) AND (4 OR 5 OR 6)

(55)  BIOSIS Previews - 1977 thru 1980

      3760 INORGANIC
       570 METALLIC?
        40 ORGANOMETALLIC?
     29541 CHROMATOGRAPHY
      6606 CHROMATOGRAPHIC
      2918 ION(W)EXCHANGE
       214  (1 OR 2 OR 3) AND (4 OR 5 OR 6)

(60)  CRIS/USDA - 75-83/Nov

       291 INORGANIC
        21 METALLIC?
         3 ORGANOMETALLIC?
      1122 CHROMATOGRAPHY
       385 CHROMATOGRAPHIC
       147 ION(W)EXCHANGE
        28  (1 OR 2 OR 3) AND (4 OR 5 OR 6)
```

```
(65) SSIE Current Research - 78-82/Feb
      33773 INORGANIC
       2949 METALLIC?
        859 ORGANOMETALLIC?
       5183 CHROMATOGRAPHY
       1009 CHROMATOGRAPHIC
        633 ION(W)EXCHANGE
       1730  (1 OR 2 OR 3) AND (4 OR 5 OR 6)
(72) EMBASE (Excerpta Medica) 80-83/Iss39
       8907 INORGANIC
       3996 METALLIC?
         23 ORGANOMETALLIC?
      20971 CHROMATOGRAPHY  DC=0009269
       4727 CHROMATOGRAPHIC
       1453 ION(W)EXCHANGE
        335  (1 OR 2 OR 3) AND (4 OR 5 OR 6)
(73) EMBASE (Excerpta Medica) In Process 83/Wk42
        378 INORGANIC
        282 METALLIC?
          3 ORGANOMETALLIC?
       2646 CHROMATOGRAPHY  DC=0009269
        692 CHROMATOGRAPHIC
        169 ION(W)EXCHANGE
         20  (1 OR 2 OR 3) AND (4 OR 5 OR 6)
(74) IPA - 70-83/Oct!(Copr. ASHP)
        166 INORGANIC
         63 METALLIC?
          3 ORGANOMETALLIC?
       4831 CHROMATOGRAPHY
       1969 CHROMATOGRAPHIC
        176 ION(W)EXCHANGE
         26  (1 OR 2 OR 3) AND (4 OR 5 OR 6)
```

Figure 10.3. Inorganic literature search of selected files.

```
ss (chromatography or chromatographic) and metal? and fossil(w)fuel?

          1 12452 CHROMATOGRAPHY  (SEE ?GENERAL)
          2  3734 CHROMATOGRAPHIC
          3 99319 METAL?
          4 26088 FOSSIL(W)FUEL?
          5    11  (1 OR 2) AND 3 AND 4
? t5/3/1-5
5/3/1
  95222521    CA: 95(26)222521w    JOURNAL
  High-speed  ion-exchange  chromatographic  separation  of  alkaline earth
metals and rapid determination  of  calcium,  zinc,  and  barium  in  oil
additives
  AUTHOR(S): Fan, Bi Wei; Liu, Man Cang; Hu, Zhi De
  LOCATION: Dep. Chem., Lanchow Univ., Lanchow, Peop. Rep. China
  JOURNAL:  Hua Hsueh Tung Pao  DATE:  1981  NUMBER:  7  PAGES:  25-8,  39
CODEN: HHTPAU  ISSN: 0441-3776  LANGUAGE: Chinese

5/3/2
  95206464    CA: 95(24)206464z    DISSERTATION
  The  development  of  on-line  liquid  chromatograph-inductively  coupled
plasma-atomic emission spectrometer for the analysis of  trace  organically
bound metals and metalloids in coal derived products
  AUTHOR(S): Hausler, Douglas William
  LOCATION: Virginia Polytech. Inst. and State Univ., Blacksburg, VA, USA
  DATE: 1981  PAGES: 299 pp.  CODEN:  DABBBA  LANGUAGE:  English  CITATION:
Diss. Abstr. Int. B 1981, 42(3), 1006  AVAIL: Univ. Microfilms Int.,  Order
No. 8118696
```

Figure 10.4. The five most recent of eleven postings from Chemical Abstracts, 1980–81 for a chromatography *and* metal *and* fossil fuel search.

of 1 in File 79 (Foods ADLIBRA-1974-1981). Figure 10.3 is part of that search in File 411, DIALINDEX.

As can be seen, many files that at first might seem unrelated to the subject yielded a sizable number of postings. In the files with very high postings the searcher must refine the search to reduce the number of

5/3/3

95135554 CA: 95(16)135554r JOURNAL

Organically bound metals in a solvent-refined coal: metallograms for a Wyoming subbituminous coal

AUTHOR(S): Taylor, Larry T.; Hausler, Douglas W.; Squires, Arthur M.

LOCATION: Dep. Chem., Virginia Polytech. Inst. and State Univ., Blacksburg, VA, 24061, USA

JOURNAL: Science (Washington, D. C., 1883-) DATE: 1981 VOLUME: 213 NUMBER: 4508 PAGES: 644-6 CODEN: SCIEAS ISSN: 0036-8075 LANGUAGE: English

5/3/4

95027564 CA: 95(4)27564k JOURNAL

Size exclusion chromatography of organically bound metals and coal-derived materials with inductively coupled plasma atomic emission spectrometric detection

AUTHOR(S): Hausler, D. W.; Taylor, L. T.

LOCATION: Dep. Chem., Virginia Polytech. Inst. State Univ., Blacksburg, VA, 24061, USA

JOURNAL: Anal. Chem. DATE: 1981 VOLUME: 53 NUMBER: 8 PAGES: 1227-31 CODEN: ANCHAM ISSN: 0003-2700 LANGUAGE: English

5/3/5

94142437 CA: 94(18)142437n JOURNAL

Trace metal distribution in fractions of solvent-refined coal by ICP-OES and implications regarding metal speciation

AUTHOR(S): Hausler, Doug W.; Hellgeth, John W.; Taylor, Larry T.; Borst, Jane; Cooley, W. Brad

LOCATION: Dep. Chem., Virginia Polytech. Inst. and State Univ., Blacksburg, VA, 24061, USA

JOURNAL: Fuel DATE: 1981 VOLUME: 60 NUMBER: 1 PAGES: 40-6 CODEN: FUELAC ISSN: 0016-2361 LANGUAGE: English

Figure 10.4. ***(continued)***

```
? ss (gas(w)chromatogr? or chromatography(w)gas or gc or glc or gas(1w)liquid(w)
chromatogr?) and organometallic?
          1  2987 GAS(W)CHROMATOGR?
          2  3525 CHROMATOGRAPHY(W)GAS
          3   471 GC
          4   102 GLC
          5   557 GAS(1W)LIQUID(W)CHROMATOGR?
          6 12008 ORGANOMETALLIC?
          7    29  (1 OR 2 OR 3 OR 4 OR 5) AND 6
? t7/3/1-5
7/3/1
  99205330    CA: 99(24)205330y    JOURNAL
  Elemental microanalysis of organic and organometallic compounds using
pyrolysis-gas chromatography
  AUTHOR(S): Sullivan, J. F.; Grob, R. L.
  LOCATION: Dep. Chem., Villanova Univ., Villanova, PA, USA
  JOURNAL: J. Chromatogr. DATE: 1983 VOLUME: 268 NUMBER: 2 PAGES:
219-27 CODEN: JOCRAM ISSN: 0021-9673 LANGUAGE: English

7/3/2
  99195134    CA: 99(23)195134w    JOURNAL
  Application of gas chromatography for the determination of the structure
of alkyl groups in zinc bis(O,O'-dialkyldithiophosphate)s
  AUTHOR(S): Zimmermann, Volker; Kempe, Jochen; Jaeger, Guenter
  LOCATION: Sekt. Chem., Friedrich-Schiller-Univ., Jena, Ger. Dem. Rep.
  JOURNAL: Chem. Tech. (Leipzig) DATE: 1983 VOLUME: 35 NUMBER: 7 PAGES:
359-62 CODEN: CHTEAA ISSN: 0045-6519 LANGUAGE: German
```

Figure 10.5. A search demonstrating the influence of terminology upon results.

postings. Since there is now about 10–15 years worth of literature in most files, ways must be found to deal with such large databases. Certain files, such as Chemical Abstracts, now consist of five separate files broken down to follow the five-year Collective Indexes, beginning with 1967–1971. Even this makes for too large a file, and the tenth Collective Index has been broken down into two files, one for 1977–1979 and another for 1980–1981. The current eleventh Collective Index begins with 1982. It is necessary, therefore, to seek other limits to reduce the number of citations. Such delimiters have been described before; they could be by type

```
7/3/3
  99195052    CA: 99(23)195052t    JOURNAL
  Use  of  gas  chromatography  to  study  the  thermal  decomposition  of
organogallium peroxide compounds
  AUTHOR(S): Stepanova, I. G.; Chikinova, N. V.; Makin, G. I.; Aleksandrov,
Yu. A.
  LOCATION: Nauchno-Issled. Inst. Khim., Gorkiy, USSR
  JOURNAL: Fiz.-Khim. Metody Anal.  DATE: 1982  PAGES: 49-51  CODEN: FKMADK
  LANGUAGE: Russian

7/3/4
  99195051    CA: 99(23)195051s    JOURNAL
  Use  of  gas chromatography for the study of the thermal decomposition of
organogallium peroxide compounds
  AUTHOR(S): Stepanova, I. G.; Chikinova, I. V.; Makin, G. I.; Aleksandrov,
Yu. A.
  LOCATION: USSR
  JOURNAL: Fiz.-khim. Metody Analiza, M.  DATE: 1982  PAGES:  49-51  CODEN:
D3RAPP  LANGUAGE: Russian  CITATION:  Ref.  Zh.,  Khim.  1983,  Abstr.  No.
14G240

7/3/5
  99164433    CA: 99(20)164433p    JOURNAL
  Studies on the separation of metal acetylacetonates by gas chromatography
  AUTHOR(S): Maslowska, J.; Starzynski, S.
  LOCATION: Inst. Gen. Food Chem., Tech. Univ. Lodz, 90-924, Lodz, Pol.
  JOURNAL:   Chromatographia   DATE:  1983  VOLUME:  17  NUMBER:  8  PAGES:
418-20  CODEN: CHRGB7  ISSN: 0009-5893  LANGUAGE: English
```

Figure 10.5. ***(continued)***

of article, such as review articles, theoretical articles, preparation of material, or patents, Other delimiters could be by year or language. Each database will explain ways of limiting a search.

An example of refining a search follows wherein "metal" and "fossil(w)fuel" are used as delimiters. The search was in File 310 (CA Search—1980–1981). There were eleven postings and the five most recent are reproduced in Fig. 10.4.

Another example, Fig. 10.5, of a search in File 311 shows the importance of using all possible synonyms for a term, in this case the term gas chromatography as it is used in Chemical Abstracts. The five most recent postings are listed.

10.5 CREATIVE USES FOR ONLINE SYSTEMS

Not only are the online databases invaluable for retrospective searching, but they can be used in several other ways. Among these creative uses are current awareness searches or SDI (Selective Dissemination of Information). A search can be entered into the system and saved so that it is automatically run against the latest update to selected files, or a search strategy can be saved by the searcher and run at regularly chosen intervals determined by the searcher and researcher.

Another very efficient use of online files is to follow the work of competitors and colleagues. It is possible to search not only by author but also by the corporate source, the place where the author works. It is also possible to search a company name that is mentioned in titles or abstracts of papers in certain databases. This could provide very useful applications information to engineers and scientists developing new instruments for the chemical industry, or to people in the pharmaceutical industry searching for trade names in the literature.

In certain databases, it is further possible to retrieve all papers that cite key papers in one's field of research. Locating key people in a field can be achieved similarly by those desiring to identify possible consultants or speakers. Other researchers check online before they embark on a new project to see if the same idea, or one similar, has been explored previously.

A new group of databases are now taking their place alongside bibliographic files. These are files using graphics, such as CAS Online, produced by Chemical Abstracts. This file allows one to do substructure searching graphically, based on the CAS Chemical Registry file, which has over one million substances in it and 6000 new substances added weekly. This database is available directly from Chemical Abstracts Service, and is best used by the end user.

To the advantage of today's scientist, the online database industry has mushroomed in the past ten years, organizing and making information more readily accessible. As one who constantly explores new ideas, it behooves the scientist to be educated in this powerful information resource.

Table 10.1 Outline for Introductory Course in Chemical Literature Searching

1. Some basic concepts
2. Information flow and communication patterns in chemistry
3. Search strategy
4. Keeping up-to-date—current awareness programs
5. How to get access to articles, books, patents, and other documents quickly and efficiently
6. The Chemical Abstracts service
7. Other abstracting and indexing services
8. Computer-based online and off-line information retrieval systems and services
9. Reviews
10. Encyclopedias and other major reference books
11. Patents
12. Safety and related topics
13. Locating and using physical property and related data
14. Chemical marketing and business information sources
15. Process information

10.6 INSTRUCTION IN COMPUTER SEARCHING

Chemists have available many sources on the chemical literature (7–10) and some chemistry departments have formal courses (11,12). The chapter outline, Table 10.1, of Reference 7 is the course outline for such an introductory course at Indiana University and is a typical model of such

Table 10.2 Outline for Advanced Course in Chemical Literature Searching

1. General introduction
   - A. Overview
   - B. Information science and computers
   - C. Chemical nomenclature
   - D. Chemical Abstracts and computers
2. Computer-based searches for current awareness
   - A. Current awareness searches
   - B. Standard interest profiles
3. Online systems
   - A. Online bibliographic search systems
   - B. Structure codes: the basis of online nonbibliographic chemical information retrieval systems
   - C. Nonbibliographic databases

courses. The American Chemical Society also provides an audiovisual course on the use of the chemical literature (13).

In addition, and from the examples above, interaction of the chemist with a computer information specialist is necessary for efficient use of computer databases. The advanced course C401, "Chemical Information Storage and Retrieval Methods and Techniques," at Indiana University, is exemplary in doing this. Established in 1979, this course is tied to that Chemistry Department's Chemical Information Center. Such a connection contributes to the success of the course, which is described in the literature (12) wherein the content and assigned readings are listed. These are summarized in Table 10.2 and show the resources available to the chromatographer and all chemists for computer literature searching.

REFERENCES

1. D.B. Barker, *C & E News,* **59**(22), 29 (1981).
2. M. Dillon, *Special Libraries,* **12,** 218 (1981).
3. E. McC. Arnett, *Computer-Based Chemical Information,* Marcel Dekker, New York, 1973, p. 10.
4. M.G. Mellon, *Chemical Publications: Their Nature and Use,* 4th ed. McGraw-Hill, New York, 1965, p. 83.
5. R.E. Hoover in R.E. Hoover, ed., *The Library and Information Manager's Guide to Online Services,* Knowledge Industry Publications, White Plains, New York, 1973, p. 199, p. 19.
6. Arleen N. Sommersville, "The Place of the Reference Interview in Computer Searching: The Academic Setting," *Online,* **1**(4), 14 (1977).
7. R.E. Maizell, *How to Find Chemical Information,* Wiley, New York, 1979.
8. A. Antony, *Guide to Basic Information Sources in Chemistry,* Wiley, New York, 1979.
9. R.T. Bottle, ed., *Use of Chemical Literature,* 3rd ed., Butterworths, London, 1979.
10. H. Skolnik, *The Literature Matrix of Chemistry,* Wiley, New York, 1982.
11. G. Gorin, *J. Chem. Ed.,* **59,** 991 (1982).
12. O. Wiggins, *J. Chem. Ed.,* **59,** 994 (1982).
13. J.T. Dickman, M.P. O'Hara, and O.B. Ramsey, *Chemical Abstracts: An Introduction to Its Effective Use,* American Chemical Society, Washington, D.C., 1979.

INDEX

Vol. 67. **An Introduction to Photoelectron Spectroscopy.** By Pradip K. Ghosh

Vol. 68. **Room Temperature Phosphorimetry for Chemical Analysis.** By Tuan Vo-Dinh

Vol. 69. **Potentiometry and Potentiometric Titrations.** By E. P. Serjeant

Vol. 70. **Design and Application of Process Analyzer Systems.** By Paul E. Mix

Vol. 71. **Analysis of Organic and Biological Surfaces.** Edited by Patrick Echlin

Vol. 72. **Small Bore Liquid Chromatography Columns: Their Properties and Uses.** Edited by Raymond P. W. Scott

Vol. 73. **Modern Methods of Particle Size Analysis.** Edited by Howard G. Barth

Vol. 74. **Auger Electron Spectroscopy.** By Michael Thompson, M. D. Baker, Alec Christie, and J. F. Tyson.

Vol. 75. **Spot Test Analysis: Clinical, Environmental, Forensic, and Geochemical Applications.** By Ervin Jungreis.

Vol. 76. **Receptor Modeling in Environmental Chemistry.** By Philip K. Hopke

Vol. 77. **Molecular Luminescense Spectroscopy: Methods and Applications—Part 1.** Edited by Stephen G. Schulman

Vol. 78. **Inorganic Chromatographic Analysis.** Edited by John C. MacDonald